# Impact of Light on Horticultural Crops

# Impact of Light on Horticultural Crops

Editors

**Athanasios Koukounaras**
**Filippos Bantis**

MDPI • Basel • Beijing • Wuhan • Barcelona • Belgrade • Manchester • Tokyo • Cluj • Tianjin

*Editors*
Athanasios Koukounaras
Aristotle University
Greece

Filippos Bantis
Aristotle University
Greece

*Editorial Office*
MDPI
St. Alban-Anlage 66
4052 Basel, Switzerland

This is a reprint of articles from the Special Issue published online in the open access journal *Agriculture* (ISSN 2077-0472) (available at: https://www.mdpi.com/journal/agriculture/special_issues/light_horticulture_crops).

For citation purposes, cite each article independently as indicated on the article page online and as indicated below:

LastName, A.A.; LastName, B.B.; LastName, C.C. Article Title. *Journal Name* **Year**, *Volume Number*, Page Range.

**ISBN 978-3-0365-7398-4 (Hbk)**
**ISBN 978-3-0365-7399-1 (PDF)**

# Contents

# About the Editors

**Athanasios Koukounaras**

Athanasios Koukounaras is an Associate Professor of the Lab of Vegetable Crops, Department of Horticulture and Viticulture, Aristotle University, in Thessaloniki, Greece. He specializes in issues related to plant physiology, vegetable production, nutritional quality and postharvest handling. He has published several articles in peer-reviewed journals in the fields of agriculture, environment, horticulture and biotechnology, and he has participated in several national and EU research projects.

**Filippos Bantis**

Filippos Bantis is an Assistant Professor of Vegetable Crops' Science at the Department of Agriculture, University of Patras, Greece. His research interests lie in plant photomorphogenesis, plant physiology, vegetable production and nutritional quality. He has participated in several scientific publications, as well as in a plethora of national and EU research projects.

agriculture

Editorial

# Impact of Light on Horticultural Crops

**Filippos Bantis [1,*] and Athanasios Koukounaras [2]**

1 Department of Agriculture, University of Patras, 30200 Messolonghi, Greece
2 Department of Horticulture, Aristotle University, 54124 Thessaloniki, Greece
* Correspondence: fbantis@upatras.gr

Light is an essential factor for the growth and quality of horticultural plants and its effects depend upon parameters such as duration, intensity and quality. It is an energy source for photosynthesis as well as a signal triggering plant photomorphogenesis and physiological, biochemical and molecular responses. However, solar light strongly differs between winter and summer conditions, with excess light in open field cultivations imposing severe stress on plants, especially during summer months, while supplementary light sources are implemented in greenhouse crop production to complement natural light when it is insufficient. On the other hand, artificial lighting is used as the sole lighting source in plant factories (PFALs) and nurseries (e.g., healing chambers for grafted seedlings). Technological innovation such as the emergence and growth of light-emitting diodes (LEDs), heating and cooling systems, disinfection materials, and renewable energy sources enabled and facilitated horticultural crop production in closed production systems. In order to enhance sustainability and profitability, light must be studied and efficiently applied within horticultural crop production. The abovementioned novel technologies showcase the critical role of light interacting with plants from the level of seed germination to growth rate, product quality and post-harvest storage. Moreover, from an investor's and producer's point of view, these technologies offer the possibility to reduce electrical consumption, to balance the carbon footprint of fresh products, and ultimately to reduce production costs.

This Special Issue collects recent research findings dealing with a wide range of topics related to light effects on horticultural crops. All contributions are original research articles dealing with a wide range of subjects. These subjects include seed treatment with light and laser to enhance caper germination [1], light quantity and quality effects on the production of grafted and ungrafted vegetable seedlings [2–5], application of light in a greenhouse or a PFAL system to enhance the yield and nutritional value of lettuce varieties [6–8] and watercress [9], light colour impact on Chrysanthemum cuttings' rooting and growth [10], as well as a study of the potential of LEDs in combination with adaptive lighting control protocols and greenhouse-integrated photovoltaics to enable year-round crop cultivation in the Nordics [11].

Caper (*Capparis spinosa* L.) seeds show considerable difficulties in their germination, with light offering the potential to withdraw the dormancy in certain seeds. Foschi et al. [1] analysed the germination response of caper seeds after exposure to light and He-Ne laser. According to their results, the authors concluded that light (they tested white, red, blue, and red + blue wavelengths, and darkness) during the germination process did not provoke a response to caper seeds. Therefore, to save energy, darkness is the logical way for caper germination. On the other hand, irradiation of pre-soaked caper seeds with a He-Ne laser increased their germination percentage but only after application of a gibberellic acid solution.

In the northern hemisphere, solar light reaching inside a greenhouse can be insufficient during the winter months and may result in inferior vegetable seedling quality. Supplemental lighting is a means to encounter this constrain, and light quality can have a significant effect. Yan et al. [2] tested LEDs with different spectral emissions (including white, blue

**Citation:** Bantis, F.; Koukounaras, A. Impact of Light on Horticultural Crops. *Agriculture* **2023**, *13*, 828. https://doi.org/10.3390/agriculture13040828

Received: 30 March 2023
Accepted: 3 April 2023
Published: 4 April 2023

and UV-A) as supplemental lighting options for the production of cucumber (*Cucumis sativus* L. cv. Tianjiao No. 5) seedlings. According to their findings, solar light enhanced the plant height and the hypocotyl length but decreased the leaf area and thickness compared to the supplemental LED combinations. Moreover, the shoot and root fresh weight, and the seedling quality index were enhanced by the white, blue, and UV-A lights, while the white/blue/UV-A combinations led to increased cellulose content and stem firmness. Overall, the white/blue/UV-A combinations led to the development of compact cucumber seedlings with superior mechanical properties, and preferable growth performance. Therefore, this light recipe could be applied to achieve the desired morphological and quality characteristics of cucumber seedlings.

Throughout the world, watermelon (*Citrullus lanatus* L.) is propagated in high percentages through grafting, and seedlings are then subjected to a healing stage where conditions such as light are controlled. Bantis et al. [3] examined the impact of different light qualities (applied at the healing stage) for the field cultivation of grafted watermelon (cv. Celine F1 scion grafted onto *Cucurbita maxima* × *C. moschata* cv. TZ-148 rootstock) transplants. In their contributions, the authors reported that a treatment emitting 88/12% red/blue (12B) light and 12B including 5% far-red (12B + FR) improved the vegetative growth. The latter light treatment also accelerated flowering even compared to 12B which only lacked 5% far-red radiation, an important wavelength related to flowering. Following, these light treatments along with monochromatic red also accelerated the fruit production, which was eventually similar under all light treatments when the rest caught up after a few days. Monochromatic blue had a similar response to the control (fluorescence lamps: FL), showing decreased vegetative growth and slower flowering and fruit production. Fruit morphological and biochemical properties were also not affected by the different light qualities during healing. Overall, red, blue, and far-red wavelengths during the healing of grafted watermelon seedlings improved the growth and accelerated the flowering and fruit production of watermelon crops.

Another important crop which is widely established with the use of grafted seedlings is tomato (*Solanum lycopersicum* L.). In their contribution, Melissas et al. [4] studied the effect of light quality during the healing of grafted tomato (cv. Kabrera F1 scion grafted onto cv. Emperador F1 rootstock) seedlings, on seedling adaptation to transplant shock. Monochromatic blue light reaffirmed its inhibitory effect on plant growth by showing poor results both before and after the seedlings were transplanted in pots. Similar but less pronounced effects were reported for monochromatic red light. The control (FL) treatment also decelerated the seedling's growth due to inferior spectral distribution compared to light recipes containing proper red–blue amounts and wavelengths. On the other hand, light treatments including large amounts of red and 11–24% blue (i.e., 88/12% red–blue, 76/24% red–blue, and white with 71/18/11% red–green–blue) enhanced the seedling quality before and after transplanting in pots. This is also supported by the greater antioxidant activity and overall adaptation to transplant shock. Finally, the white treatment is preferable to achieve better light conditions for the human eye.

In another contribution involving grafted tomato (*Lycopersicon esculentum* Mill.) transplants, Zheng et al. [5] compared different photon flux densities during the healing of grafted seedlings, as well as broad-spectrum light qualities for the production of scions (Cv. Dongfeng No. 1) and rootstock (cv. Zhezhan No. 1), and during the post-grafting stage. The authors examined the growth and energy-use efficiency under the abovementioned treatments. During the experiment, the LEDs showed a 110% greater electrical energy saving compared to FL lights. The addition of red light to white LEDs enhances the dry matter accumulation and improves plant compactness and leaf thickness. However, red light should be added with caution since excessive amounts may lead to negative responses on seedling quality and could possibly increase the operation costs. Nevertheless, a higher photon flux density during healing improved the seedling growth and quality without increasing the energy consumption. Overall, a red/blue ratio of 1.2 and red/far-red ratio of 16, and a photon flux density of 150 μmol $m^{-2}$ $s^{-1}$ were suggested for the production of grafted tomato transplants, taking into account both the seedling quality and the energy consumption.

Ultraviolet (UV) radiation is known to affect the yield and quality of red lettuce (*Lactuca sativa* L.). In their contribution, Lycoskoufis et al. [6] aimed to enhance red lettuce (var. Redino Lollo rosso) yield and quality through a UV management system in a greenhouse. Specifically, the authors created a new cultivation system by combining polyethene film, which blocks UV radiation in a greenhouse (UV-block), and supplemental UV light, and compared with a UV-open greenhouse. It was reported that the latter showed decreased red lettuce growth (i.e., head weight) and quality (i.e., total phenols, flavonoids, and antioxidant capacity) compared to UV-block greenhouse including supplemental UV light.

Even in the Mediterranean region, climatic conditions and especially light quantity during winter are suboptimal for the efficient production of leafy vegetables. Nowadays, PFALs offer the option to cultivate such products all year round while maintaining crop yield and quality regardless of outside conditions. Voutsinos et al. [7] evaluated the commercial benefits arising from the cultivation of butterhead lettuce (*Lactuca sativa* L. cv. Glory) using only artificial lighting indoors compared to a glasshouse in the Mediterranean region during wintertime. High light intensity (310 μmol $m^{-2}$ $s^{-1}$) in the vertical facility resulted in greater photosynthetic capacity, biomass accumulation, and quality characteristics. Only in the vertical farming system, nitrate content was within the limits set for the human consumption of 100 g of fresh lettuce regardless of light intensity.

In another study involving "Elizium" romaine lettuce (*Lactuca sativa* L. var. longifolia), Matysiak et al. [8] analysed the impact of the light quality of the crops using two different nutrient solutions in a controlled environment. The authors reported that a RGB 70/18/12% combination enhanced the leaf biomass and inhibited the accumulation of potassium and magnesium, while showing low nitrogen balance index and high flavonol index. A 25% increase of mineral concentration in the nutrient solution (EC 2.0 mS $m^{-2}$ $s^{-1}$) negatively affected the food quality displayed by the nitrogen balance index, flavonol index, nitrate content, and tipburn.

PFALs can also be the production centre of less pronounced and less studied leafy vegetables such as watercress (*Nasturtium officinale* L.). In their contribution, Lam et al. [9] determined the optimal photoperiod and light intensity for the production of watercress with a view to increase its plant growth and glucosinolate content in a deep flow system. It is suggested that the 20 h photoperiod and 160 μmol $m^{-2}$ $s^{-1}$ light intensity promotes the plant growth and glucosinolate content of watercress. The decrease in light intensity and increase in the photoperiod led to a gradual decrease in net photosynthesis and stomatal conductance.

Cuttings of ornamental plant are among the plant materials that can be cultivated in a closed system with controlled conditions. In the case of chrysanthemums (*Chrysanthemum grandiflorum* Ramat./Kitam) cv. "Nova Lime", Schroeter-Zakrzewska and Pradita [10] investigated the rooting in cuttings and the subsequent plant growth as affected by light colour. White and monochromatic blue lights promoted rooting compared to white+blue and red+blue treatments. Blue also enhanced the plant height, while red+blue promoted the leaf number. Overall, the authors concluded that chrysanthemum showing good quality can be cultivated under low light intensity, leading to reduced energy costs.

A valuable aspect characterizing controlled environment agriculture is the potential to grow plants all year round. This is particularly important for regions facing harsh climate conditions for large part of the year such as the Nordics (Sweden, Denmark, Norway, Finland, and Iceland). Velasco [11] analysed the meteorological satellite data of these regions to evaluate the potential of technological advancements to be used for efficient year-round plant production. For that purpose, LEDs, greenhouse-integrated photovoltaics, and adaptive lighting control were jointly studied. It is reported that such a concept is indeed a feasible option even in the climatic conditions of the Nordics. In that region, natural temperatures and daily light integral are only sufficient in the summer months. Greenhouses can be used to extend the growth period further in the spring and autumn months, while transmittance levels of natural light affect the supplementary lighting used for the plants. Among the options tested, LEDs in combination with adaptive lighting control have the potential to save the most energy. In addition, greenhouses with integrated photovoltaics offer the possibility to use the abundant summer sunshine and decrease some electricity during the darker period of the year. Closed production systems provide a stable environment for plant production during the winter months.

This Special Issue provides a fraction of the recent research findings related to light effects on horticultural crops with respect to their yield, nutritional value, physiological responses, and overall production and development. Additional research efforts could be made towards the sustainability of the closed production systems, taking into account running costs for heating and cooling along with lighting. Moreover, large-scale experiments are necessary to evaluate the usefulness of such systems for the year-round production of vegetables compared to the conventional methods applied throughout the world. Finally, several underutilized leafy and fruit crops could be tested under different light configurations with a view to increase their overall production and quality. Furthermore, biodiversity is still an important aspect of plant production, and studying the underlying mechanisms of light on several genotypes is a long-lasting effort.

**Funding:** This research received no external funding.

**Acknowledgments:** We gratefully acknowledge all the authors that participated in this Special Issues.

**Conflicts of Interest:** The authors declare no conflict of interest.

## References

1. Foschi, M.L.; Juan, M.; Pascual, B.; Pascual-Seva, N. Influence of Lighting and Laser Irradiation on the Germination of Caper Seeds. *Agriculture* **2022**, *12*, 1612. [CrossRef]
2. Yan, Z.; Wang, C.; Wang, L.; Li, X.; Wang, G.; Yang, Y. The Combinations of White, Blue, and UV-A Light Provided by Supplementary Light-Emitting Diodes Promoted the Quality of Greenhouse-Grown Cucumber Seedlings. *Agriculture* **2022**, *12*, 1593. [CrossRef]
3. Bantis, F.; Gkotzamani, A.; Dangitsis, C.; Koukounaras, A. A Light Recipe including Far-Red Wavelength during Healing of Grafted Watermelon Seedlings Enhances the Floral Development and Yield Earliness. *Agriculture* **2022**, *12*, 982. [CrossRef]
4. Melissas, C.; Bantis, F.; Dangitsis, C.; Kostas, S.; Koukounaras, A. Proposed Light Wavelengths during Healing of Grafted Tomato Seedlings Enhance Their Adaptation to Transplant Shock. *Agriculture* **2022**, *12*, 797. [CrossRef]
5. Zheng, J.; Gan, P.; Ji, F.; He, D.; Yang, P. Growth and Energy Use Efficiency of Grafted Tomato Transplants as Affected by LED Light Quality and Photon Flux Density. *Agriculture* **2021**, *11*, 816. [CrossRef]
6. Lycoskoufis, I.; Kavga, A.; Koubouris, G.; Karamousantas, D. Ultraviolet Radiation Management in Greenhouse to Improve Red Lettuce Quality and Yield. *Agriculture* **2022**, *12*, 1620. [CrossRef]
7. Voutsinos, O.; Mastoraki, M.; Ntatsi, G.; Liakopoulos, G.; Savvas, D. Comparative Assessment of Hydroponic Lettuce Production Either under Artificial Lighting, or in a Mediterranean Greenhouse during Wintertime. *Agriculture* **2021**, *11*, 503. [CrossRef]
8. Matysiak, B.; Kaniszewski, S.; Dyśko, J.; Kowalczyk, W.; Kowalski, A.; Grzegorzewska, M. The Impact of LED Light Spectrum on the Growth, Morphological Traits, and Nutritional Status of 'Elizium' Romaine Lettuce Grown in an Indoor Controlled Environment. *Agriculture* **2021**, *11*, 1133. [CrossRef]
9. Lam, V.P.; Choi, J.; Park, J. Enhancing Growth and Glucosinolate Accumulation in Watercress (*Nasturtium officinale* L.) by Regulating Light Intensity and Photoperiod in Plant Factories. *Agriculture* **2021**, *11*, 723. [CrossRef]
10. Schroeter-Zakrzewska, A.; Pradita, F.A. Effect of Colour of Light on Rooting Cuttings and Subsequent Growth of Chrysanthemum (*Chrysanthemum* × *grandiflorum* Ramat./Kitam.). *Agriculture* **2021**, *11*, 671. [CrossRef]
11. Hernandez Velasco, M. Enabling Year-round Cultivation in the Nordics-Agrivoltaics and Adaptive LED Lighting Control of Daily Light Integral. *Agriculture* **2021**, *11*, 1255. [CrossRef]

agriculture

MDPI

Article

# Ultraviolet Radiation Management in Greenhouse to Improve Red Lettuce Quality and Yield

**Ioannis Lycoskoufis [1], Angeliki Kavga [2,*], Georgios Koubouris [3] and Dimitrios Karamousantas [1]**

[1] Department of Agriculture, Kalamata Campus, University of Peloponnese, Antikalamos, GR-24100 Kalamta, Greece
[2] Department of Agriculture, University of Patras, GR-26504 Patras, Greece
[3] ELGO-DIMITRA, Institute of Olive Tree, Subtropical Crops and Viticulture, Leoforos Karamanli 167, GR-73134 Chania, Greece
* Correspondence: akavga@upatras.gr

**Abstract:** The intensity of ultraviolet (UV) radiation affects the yield and quality of red lettuce. The current study aimed to develop a UV management system in a greenhouse to achieve high yield and quality in red lettuce production. The study consisted of two experiments. In the first experiment, the effects of the different UV transparencies of the plastic materials covering the greenhouse on plant growth and the concentration of antioxidants in red lettuce were studied. For this purpose, two greenhouses were covered with polyethene of different transparencies to UV radiation. One greenhouse was covered with a common type of polyethene transparent in a large spectrum of UV radiation (UV-open), while the second greenhouse was covered with polyethene untransparent to ultraviolet radiation (UV-block). The plants were grown in a deep flotation hydroponic system. At the end of the cultivation, plant growth measurements, leaf colour measurements, and the determination of antioxidant components' concentration were carried out. Red lettuce plants harvested 42 days after planting had an average head weight 42% greater in the UV-block greenhouse compared to plants grown in the UV-open greenhouse. However, the red leaf colour of plants in the UV-block greenhouse lagged significantly compared to that in the UV-open greenhouse. Moreover, the total phenolic content, the total flavonoid content, and the antioxidant capacity of the lettuce leaves in the UV-block greenhouse were significantly lower compared to the corresponding values of the plants in the UV-open greenhouse. During the second experiment, a new cultivation system of red lettuce, which combined a UV-block polyethene film as a greenhouse cover and a pre-harvested supplemental UV light, was tested. For this purpose, various doses of supplemental UV lighting were tested in the UV-block greenhouse for ten days prior to harvest. From these tests, it emerged that applying supplemental UV lighting with a dose of 425 kJ $m^{-2}$ $d^{-1}$ for ten days before harvest produces red lettuces of the same quality as those produced in a UV-open greenhouse. This technique of growing red lettuce increases its yield by 30% without a negative effect on the quality of the product.

**Keywords:** greenhouse cover; leaf colour; UV radiation; antioxidant capacity; supplemental light

**Citation:** Lycoskoufis, I.; Kavga, A.; Koubouris, G.; Karamousantas, D. Ultraviolet Radiation Management in Greenhouse to Improve Red Lettuce Quality and Yield. *Agriculture* **2022**, *12*, 1620. https://doi.org/10.3390/agriculture12101620

Academic Editors: Athanasios Koukounaras and Filippos Bantis

Received: 18 August 2022
Accepted: 4 October 2022
Published: 6 October 2022



## 1. Introduction

Growing plants in a greenhouse provides a more favourable environment compared to an open field due to the control and management of different environmental conditions, such as radiation, temperature, humidity, carbon dioxide, and more precise availability of water and nutrients. Consequently, crops have an increased vegetative development, are of higher quality (colour, firmness, taste, etc.), and have earlier harvests.

Lettuce can be grown both in the open field and in a greenhouse. However, in recent decades, greenhouse lettuce cultivation has become more popular as it extends seasonal availability and improves quality by producing cleaner lettuce and an earlier yield.

Nevertheless, the different materials covering a greenhouse affect the amount and spectral quality of light that passes inside the greenhouse and reaches the plant growth area. Greenhouse plastic cover materials are sensitive to UV radiation, and their exposure to UV radiation for a long period reduces their durability and transparency in the light. Protective agents are added to plastic covers to protect them from senescence. These agents expand the use life of the plastic covers; however, they reduce their transparency in UV radiation [1]. Consequently, in the commonly used greenhouse plastic covers, transparency in UV radiation is much lower than transparency in photosynthetically active radiation (PAR).

Preventing the entry of ultraviolet radiation into the greenhouse is achieved by choosing covers that are not transparent to this radiation. These covers are defined as UV-block, while the covers that allow UV radiation to enter the greenhouse are characterized as UV-transparent or UV-open. Most plastic greenhouse covers, as well as glass, do not transmit UV-B radiation (280–315 nm). The glass transmits only a part of UV-A radiation (315–400 nm). In the Mediterranean region, plastic materials, and mostly the polyethene variant types, have prevailed as greenhouse-covering materials. Photoselective sheets, which block UV radiation, have also been used as greenhouse covers to control diseases and pests in greenhouses. The absence of ultraviolet radiation prevents the sporulation of several phytopathogenic fungi, such as Botrytis and Alternaria, and reduces the populations of pests that need ultraviolet radiation for their orientation [2,3]. Moreover, in recent decades, types of polyethene covers with low to zero UV transmittance (UV-block polyethene) have been developed to control pest populations [4] and greenhouse crop diseases [5].

On the other hand, UV-block polyethene films can have a detrimental effect on the accumulation of antioxidant compounds in lettuce [6]. Ultraviolet radiation and the high density of photosynthetically active radiation increase the concentration of flavonoids in lettuce [7]. In addition, lettuce grown in an open field contains higher concentrations of flavonols than lettuce grown in a polycarbonate greenhouse [8] due to the different radiation densities and different wavelengths entering the greenhouse. Polycarbonate sheets absorb a small percentage of photosynthetically active radiation but most of the ultraviolet radiation. Moreover, exposure of red lettuce cultivars to high levels of ultraviolet radiation during cultivation causes reddening of the leaves and increases the concentration of total phenols and main flavonoids (quercetin and cyanidin), as well as phenolic acids [9]. Consequently, the antioxidant content of the plants is responsible for the intensity of the colouring of various fruits, leaves, and flowers, so ultraviolet radiation is required to initiate the violet colour in some ornamental and vegetable crops, such as eggplant and red-pigmented lettuce [10].

On the other hand, blocking the entry of ultraviolet solar radiation into the greenhouse favours the faster growth of red lettuce plants [11]. A greater explanation for the growth inhibition of red lettuce under ambient levels of UV radiation may be the high metabolic cost of photoprotection, such that the plants divert energy produced by photosynthesis to synthesize phenolic compounds [12]. In a similar experiment, eggplant plants grown in greenhouses with UV-block polyethene achieved higher height, larger leaf size, and higher yields compared to plants grown in a greenhouse with UV-transparent polyethene [10].

In order to achieve high yield and antioxidant content in red lettuce, Tsormpatsidis et al. [12] proposed initial cultivation in a UV-block greenhouse and transfer to a UV-transparent greenhouse a few days before harvest. However, transferring soil-grown red lettuce plants shortly before harvest is impossible, while transferring soilless-grown plants is difficult and requires a lot of labour or an expensive automated installation.

Therefore, there is a need to develop a technique that will allow the use of UV-block polyethene sheets with simultaneous ultraviolet radiation availability to plants at their critical growth stages (in which ultraviolet radiation is necessary). Ultraviolet radiation is not currently used in horticulture, but its effects on plant development and secondary metabolism could be implemented (especially UV-A in small amounts) for the production of high-quality compact plants [13]. However, little research has been completed on the effect of pre-harvest supplemental lighting containing the UV-A spectrum on red lettuce in a greenhouse production system [14]. The present study aimed to investigate the use of UV-block greenhouse covers with the possibility of artificially adding UV radiation before the harvest of red lettuce to achieve high nutritional quality.

## 2. Materials and Methods

Two experiments were performed in two greenhouses located at the experimental field of the Agriculture Department of the University of Peloponnese, Kalamata, 5 m above sea level (latitude 37°03′40″ N, longitude 22°03′41″ E). Each greenhouse had an area of 108 $m^2$ (6 × 9 m), with a height of 1.7 and 3 m in the gutter and ridge, respectively. During hot days, both greenhouses were ventilated and cooled by fan-pad system [15]. The two greenhouses were covered with polyethene films of different UV transmittance. One greenhouse was covered by a common polyethene film (TUV3965), which allowed a large part (26%) of UV-A radiation to pass through, while the second greenhouse was covered with a UV-block polyethene film (TUV3957) that allowed only 4.5% of outside UV-A radiation to transmit through it. Figure 1 shows the spectral transmission in UV radiation of each polyethene film used in the current study accordingly to film supplier (plastikakritis S.A.). This transmittance difference to UV radiation of the used polyethene films was confirmed by measurements with portable instruments (HD2102.1 and LP 471 UV-A probe, Delta-OHM, ITALY). According to the polyethene film supplier's laboratory, the transparency of UV-open and UV-block film at PAR was 86.5% and 85%, respectively.

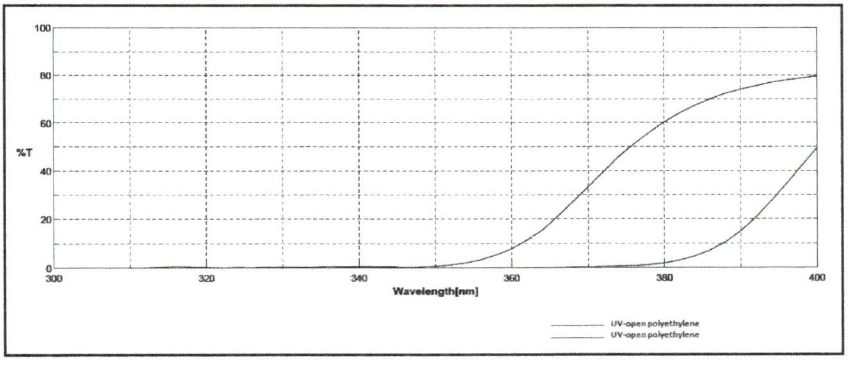


**Figure 1.** Spectral transmission in UV-A radiation of the UV-open and UV-block polyethene.

### 2.1. *Plant Material and Growing Conditions of 1st Experiment*

In the first experiment, the effect of different levels of ultraviolet radiation on the growth and quality characteristics (leaf colour, antioxidant accumulation) of red lettuce was studied. Thus, the experiment consisted of two treatments; the first treatment included the red lettuce plants grown in the UV-open greenhouse and the second treatment included the plants grown in the UV-block greenhouse.

Red lettuce plants (Redino Lollo rosso from Geoponiki S.A.) were grown in a deep floating hydroponic system. Tanks (105 cm × 52 cm × 40 cm) were used, which were sealed by a black–white polyethene sheet. Each tank was filled with 60 L of nutrient solution suitable for lettuce. EC and pH values of the nutrient solution in the solution tanks were adjusted to 2.1 dS $m^{-1}$ and 5.6, respectively. The composition of the lettuce nutrient solution was as follows: 8.47 mM $K^+$, 3.81 mM $Ca^{2+}$, 1.27 mM $Mg^{2+}$, 0.85 mM $NH_4^+$, 16.10 mM $NO_3^-$, 1.00 mM $H_2PO_4^-$, 1.19 mM $SO_4^{2-}$, 35 μM Fe, 20 μM Mn, 4 μM Zn, 5 μM B, 0.50 μM Cu, and 0.50 μM Mo.

Moreover, an air pump was placed in each tank to oxygenate the nutrient solution. A polystyrene plate (100 cm × 50 cm × 2 cm) with ten planting sites was placed on the surface of the nutrient solution in each tank, in which ten lettuce plants were transplanted, respectively.

The transplantation of lettuce seedlings took place on 17 March 2020, in the stage of 3–4 leaves, and 7 tanks with growing plants were placed in each greenhouse. Two plants from each greenhouse and tank (14 plants from each greenhouse) were randomly sampled 42 days after transplanting for growth measurements (plant weight, number of leaves per plant), leaf colour determination, antioxidants analysis, and nutritional analyses.

On the same date, three lettuce planting tanks were moved from the UV-block greenhouse to the UV-open greenhouse to determine whether a change in UV intensity could trigger the production of antioxidants and plant colouration. Seven days later, ten plants were randomly sampled from the three removed tanks in the UV-open greenhouse and from the tanks that were in the UV-open greenhouse from the beginning of the experiment.

#### 2.1.1. Leaf Colour Determination

To determine the leaf colour of red lettuce, two intermediate leaves of each sample plant (14 plants from each greenhouse) were used, and the measurement was made 1–2 cm from the margins of the leaf circumference. Colour was measured using a compact colorimeter Minolta CR-400 (Minolta, Osaka, Japan). The hue angle $[\tan^{-1}(b/a)]$ and chroma index $[(a^2 + b^2)^{1/2}]$ were determined from parameters L, a, and b. The hue angle was represented on a 360° polar chart, where 0° and 360° represent red colour, while 90°, 180°, and 270° represent yellow, green, and blue, respectively [16].

#### 2.1.2. Extraction of Phytochemicals for Measurements of Total Phenolics (TP), Total Flavonoids (TF), and Total Antioxidant Capacity (TAC)

Leaf tissues, which were used for antioxidants analysis, were frozen immediately and stored in a freezer (−25 °C) until analysis. The extraction of phenolic compounds was carried out according to [17] with some modifications. A sample of approximately 1 g of frozen tissue was homogenized in an Ultra-Turrax (T25, IKA Labortechnik, Germany) with 10 mL of cold 80% (*v*/*v*) methanol. The crude extract was placed in a supersonic bath (Elma, Transsonic 420) at 4 °C for 20 min, centrifuged at 4000 rpm for 6 min, and the supernatant was collected. The extraction was repeated twice, and the extracts were collected. After centrifugation, the supernatant was assessed for total phenols (TP), total flavonoids (TF), and total antioxidant capacity (TAC) determination, as described below.

Total phenols content was determined by the Folin–Ciocalteu assay method according to [18]. To 3.95 mL distilled water, 50 μL of extract was added and agitated thoroughly. After that, 250 μL of Folin–Ciocalteu reagent and 750 μL of 20% *w*/*v* $Na_2CO_3$ were added and thoroughly mixed. The intensity of blue colour developed was recorded on a spectrophotometer (Heλios γ, Unicam, UK) after 2 h at 760 nm. The results were expressed as milligrams of gallic acid equivalents per gram of fresh weight.

Total flavonoid content was determined using aluminium chloride ($AlCl_3$) according to [17]. To 2 mL distilled water 500 μL of extract (1:1 diluted) was added and 150 μL 5% *w*/*v* $NaNO_2$. After 5 min 150 μL of 10% *w*/*v* $AlCl_3$ was added. After a further 6 min in the reaction mixture, 1 mL of 1 N NaOH was added. Finally, the reaction mixture was diluted to 1.2 mL with water, and the absorbance was measured at 510 nm on a spectrophotometer (Heλios γ, Unicam, UK). The results were expressed as mg of catechin equivalents per gram of fresh weight.

The antioxidant capacity of lettuce leaves was evaluated in the supernatant produced by the extraction of phenolic compounds using the DPPH assay based on the method described by [19]. For the DPPH assay, 2 mL of 0.1 mM DPPH freshly prepared solution was added to 0.1 mL of the methanolic extract, and the absorbance was measured at 517 nm after a 60 min period in the dark. The scavenging capacity of the extracts was expressed in μmol Trolox equivalents per g fresh weight (μmol TE $g^{-1}$).

#### 2.1.3. Nutrient Status

Leaf samples were oven-dried at 70 °C to constant weight and were grounded for mineral analysis after ashing at 500 °C and extraction with 1N HCl solution. The concentrations of $K^+$, $Ca^{2+}$, and $Mg^{2+}$ were determined by atomic absorption spectrophotometry (Varian, SpectrAA-200, Australia). Total nitrogen was determined by means of Kjeldahl digestion using a Gerhard Vapodest 30 apparatus [20].

### 2.2. Plant Material and Growing Conditions of 2nd Experiment

Red lettuce seedlings (Redino Lollo rosso from Geoponiki S.A.), in the stage of 2–3 true leaves, were transplanted on 25 August 2020 in a deep floating hydroponic system, similar to the 1st experiment. In the UV-block greenhouse, 14 growth plant tanks were installed, and each tank had 10 lettuce plants. Four similar plant tanks were installed in the UV-open greenhouse. On the red lettuce plants grown in the UV-block greenhouse, supplemental UV lighting treatments started 26 days after the transplanting and lasted 10 days. Red lettuce plants were sampled randomly before and after supplemental UV lighting to determine the head weight of the plants.

Ultraviolet lamps (Philips, UVA, TLK40W/10R) were used to apply supplemental lighting. These lamps emit in the 350 to 400 nm waveband, with peak at 370 nm and 18.5% efficiency in UV-A radiation. Thus, each lamp emitted 7.4 W of UV-A radiation according to the manufacturer. The UV emission intensity of the lamps was confirmed by measurements with portable instruments (HD2102.1 and LP 471 UV-A probe, Delta-OHM, ITALY). Figure 2 shows the spectral power distribution of the lamps used.

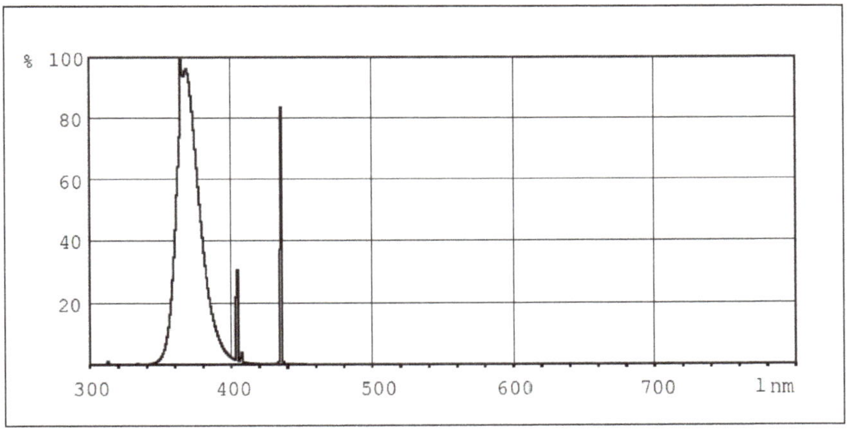


**Figure 2.** Spectral power distribution of UV-A lamps used in the second experiment in the UV-block greenhouse.

During supplemental lighting with UV radiation, care must be taken so that the applied dose does not exceed 30 W $m^{-2}$ because there is a risk of causing burns to the lettuce leaves. The plants in the UV-block greenhouse were divided into seven different groups (two plant growth tanks per group). Each group received a different treatment with supplemental ultraviolet radiation. With different combinations of operating times and bulb density, each group of plants received 0 kJ $m^{-2}$ $d^{-1}$, 210 kJ $m^{-2}$ $d^{-1}$, 265 kJ $m^{-2}$ $d^{-1}$, 320 kJ $m^{-2}$ $d^{-1}$, 425 kJ $m^{-2}$ $d^{-1}$, 530 kJ $m^{-2}$ $d^{-1}$, or 640 kJ $m^{-2}$ $d^{-1}$ of UV-A radiation daily. In Table 1, details are given about calculation of the doses of UV supplemental lighting in kJ $m^{-2}$ $d^{-1}$ and mmol $m^{-2}$ $d^{-1}$, number of used lamps per $m^2$, and the duration of the lighting in each treatment.

**Table 1.** Calculation of the dose of UV supplemental lighting in kJ $m^{-2}$ $d^{-1}$ and mol $m^{-2}$ $d^{-1}$ from lamps' UV emission, number of lamps per $m^2$, and the duration of the lighting in each treatment.

| Lamp Wattage W | Lamp UV Emission W | Number of Lamps Per $m^2$ | Intensity of UV Supplemental Lighting W $m^{-2}$ | Photon Flux of UV Supplemental Lighting μmol $m^{-2}$ $s^{-1}$ | Duration of Supplemental Lighting h | Dose of UV Supplemental Lighting kJ $m^{-2}$ $d^{-1}$ | Dose of UV Supplemental Lighting mmol $m^{-2}$ $d^{-1}$ |
|---|---|---|---|---|---|---|---|
| 40 | 7.4 | 1 | 7.4 | 33.6 | 8 | 210 | 960 |
| 40 | 7.4 | 1 | 7.4 | 33.6 | 10 | 265 | 1210 |
| 40 | 7.4 | 2 | 14.8 | 67.3 | 6 | 320 | 1450 |
| 40 | 7.4 | 2 | 14.8 | 67.3 | 8 | 425 | 1940 |
| 40 | 7.4 | 2 | 14.8 | 67.3 | 10 | 530 | 2420 |
| 40 | 7.4 | 4 | 29.6 | 134.5 | 6 | 640 | 2900 |

After 10 days of supplemental UV lighting, the irradiated plants were compared to the non-irradiated (0 kJ $m^{-2}$ $d^{-1}$) control plants. At the end of the supplemental ultraviolet lighting treatments, five plants per treatment were randomly sampled for growth, colour, and antioxidant content measurements. These parameters were determined as in the first experiment.

### 2.3. Statistical Analyses

The data were subjected to single-factor analyses of variance using the STATISTICA software version 7.0 (StatSoft Inc., Tulsa, OK, USA), and, when a significant F-test was obtained, means were separated using LSD test ($p < 0.05$).

## 3. Results

### 3.1. Results of the First Experiment

Head weight in red lettuce plants harvested 42 days after planting was 42% greater in the UV-block greenhouse compared to plants grown in the UV-open greenhouse. On the other hand, plant root weight was not affected by the different levels of UV radiation inside the two greenhouses. Thus, the total plant weight in the UV-block greenhouse was 34.7% higher than that of plants in the UV-open greenhouse (Table 2). The difference in red lettuce growth between the two greenhouses decreased as the harvest time extended to 49 days after transplanting. The head of red lettuces grown in the UV-block greenhouse weighed 36% more than that of the lettuces in the UV-open greenhouse. When red lettuce plants were transferred from the UV-block greenhouse to the UV-open greenhouse and grown there for seven days, their growth rate decreased in comparison to red lettuce plants, which continued to grow in the UV-block greenhouse. However, the head weight of these lettuces was significantly higher (23%) compared to that of lettuces grown for the entire growing season in the UV-open greenhouse.

**Table 2.** Effects of greenhouse cover transparency in UV-A radiation on growth parameters of red lettuce (Redino Lollo rosso) 42 and 49 days after transplanting.

| | Greenhouse -Treatments | Head Weight g | Root Weight g | Total Plant Weight g |
|---|---|---|---|---|
| 42 days after transplanting | UV-open | 265.3 a | 64.7 a | 330.0 a |
| | UV-block | 378.3 b | 66.1 a | 444.4 b |
| 49 days after transplanting | UV-open | 333.9 a | | |
| | UV-block | 454.5 c | | |
| | UV-block to UV-open | 411.6 b | | |

Means within the same column followed by the same letter do not differ significantly based on LSD test at α = 0.05.

Table 3 shows the results related to the effects of polyethene films on the colour parameters of red lettuce. The different intensities of UV radiation in the two greenhouses significantly affected the colouration of red lettuce leaves. The lettuces grown in the UV-open greenhouse had a more intense colouring than those in the UV-block greenhouse. Angle h° is the main colour parameter representing colour development in red lettuce leaves. Leaves that developed a more intense violet colour had lower values in the h angle parameter than greenish leaves.

**Table 3.** Effects of greenhouse cover transparency in UV-A radiation on colour parameters (L, a, b, Chroma, and hue angle) of red lettuce (Redino Lollo rosso) as measured by colorimeter Minolta CR-400, 42 days after transplanting.

| Greenhouse -Treatments | Colour Parameters | | | | |
|---|---|---|---|---|---|
| | **L** | **a** | **b** | **Chroma** | **Hue Angle** |
| UV-open | 34.60 a | −2.38 a | 19.39 a | 19.76 a | 93.57 a |
| UV-block | 44.05 b | −13.14 b | 27.76 b | 30.84 b | 113.86 b |

Means within the same column followed by the same letter do not differ significantly based on LSD test at $\alpha = 0.05$.

When red lettuce plants were transferred from the UV-block greenhouse to the UV-open greenhouse and grown there for seven days, their leaves changed from green to violet. However, their colour was less intensive than the colour of plants that were grown for the entire season in the UV-open greenhouse (data not shown).

The analysis of phytochemicals showed that higher antioxidant content, both total phenolics and flavonoids, was found in the lettuces grown in the UV-open greenhouse in comparison to that in the lettuces grown in the UV-block greenhouse (Table 4). The total phenolic content decreased by 42%, and the total flavonoid content decreased by 47% in the UV-block greenhouse in comparison to the UV-open greenhouse. In addition, the total antioxidant capacity (TAC) of the red lettuce leaves in the UV-open greenhouse significantly increased compared to that of the lettuce leaves in the UV-block greenhouse (Table 4). When red lettuce plants were transferred from the UV-block greenhouse to the UV-open greenhouse and grown there for seven days, the antioxidant content and antioxidant capacity in their leaves raised significantly in comparison to the plants which remained in the UV-block greenhouse until the end of the first experiment. However, the antioxidant content and antioxidant capacity in the leaves of these plants were significantly less in comparison to that of plants that were grown for the entire growing season in the UV-open greenhouse.

**Table 4.** Effects of greenhouse cover transparency in UV-A radiation on total phenols (TP), total flavonoids (TF), and antioxidant capacity (DPPH) of red lettuce (Redino Lollo rosso) 42 and 49 days after transplanting.

| | Greenhouse Treatments | TP (mg $g^{-1}$ FW) | TF (mg $g^{-1}$ FW) | DPPH (μmol TE $g^{-1}$) |
|---|---|---|---|---|
| 42 days after transplanting | UV-open | 6.39 a | 2.51 a | 38.15 a |
| | UV-block | 3.50 b | 1.33 b | 17.86 b |
| 49 days after transplanting | UV-open | 6.81 a | 2.46 a | 42.92 a |
| | UV-block | 3.27 c | 1.22 c | 16.20 c |
| | UV-block to UV-open | 4.25 b | 1.71 b | 28.78 b |

Means within the same column followed by the same letter do not differ significantly based on LSD test at $\alpha = 0.05$.

As shown in Table 5, no statistically significant differences were found in the concentration of the primary nutrients in the leaves of the red lettuce plants cultivated in the UV-open greenhouse or the UV-block greenhouse.

**Table 5.** Concentrations (mmol $g^{-1}$ dry weight) of total N, K, Ca, and Mg leaves of red lettuce (Redino Lollo rosso) as influenced by the greenhouse cover 42 days after transplanting.

| Greenhouse Treatments | Macrontrients (mmol $g^{-1}$ D.W.) | | | |
|---|---|---|---|---|
| | N | K | Ca | Mg |
| UV-open | 3.74 a | 1.73 a | 0.19 a | 0.10 a |
| UV-block | 3.65 a | 1.49 a | 0.20 a | 0.10 a |

Means within the same column followed by the same letter do not differ significantly based on LSD test at $\alpha = 0.05$.

### 3.2. Results of 2nd Experiment

Table 6 shows the effect of different levels of UV radiation created by the two different plastic greenhouse covers on the head weight and leaf number per red lettuce plant 25 days after transplanting. The lettuces grown in the UV-block greenhouse gained a 34% increase in head weight compared to those grown in the UV-open greenhouse. Additionally, lettuces grown in the UV-block greenhouse produced 10% more leaves than plants grown in the UV-open greenhouse.

**Table 6.** The effect of greenhouse cover transparency in UV-A radiation on head weight and leaf number per plant in red lettuce (Redino Lollo rosso), 25 days after transplanting, in the second experiment.

| Greenhouse | Head Weight g | Leaf Number per Plant |
|---|---|---|
| UV-open | 162.2 a | 19.2 a |
| UV-block | 217.6 b | 21.2 b |

Means within the same column followed by the same letter do not differ significantly based on LSD test at $\alpha = 0.05$.

Red lettuces growth was affected negatively by the high doses of supplemental UV radiation (425 to 640 kJ $m^{-2}$ $d^{-1}$) on the harvested day. However, their head weight was significantly higher in comparison to that of lettuces grown in the UV-open greenhouse. Supplemental UV light up to 320 kJ $m^{-2}$ $d^{-1}$ did not affect the head weight in comparison to that of untreated plants.

When supplemental UV light was applied to the red lettuces cultivated in the UV-block greenhouse, leaf colour measurements (h° angle) showed that a higher applied dose of supplemental UV light (640 kJ $m^{-2}$ $d^{-1}$) caused more intense colouration in plant leaves compared to plants cultivated in the UV-open greenhouse (Table 7). This dose resulted in the leaves of the plants grown in the UV-block greenhouse having the same total phenolic and total flavonoid concentrations and higher antioxidant capacity compared to the red lettuce plants grown in the UV-open greenhouse. Doses of supplemental UV light from 210 to 265 kJ $m^{-2}$ $d^{-1}$ had no influence on the colour and the antioxidant content of red lettuces in comparison to lettuces grown without supplemental UV lighting (Table 7). The dose of 320 kJ $m^{-2}$ $d^{-1}$ caused milder colouration, lower total phenolics, and total flavonoid content, as well as lower antioxidant capacity, compared to plants grown in the UV-open greenhouse. Supplemental UV light doses of 425 and 530 kJ $m^{-2}$ $d^{-1}$ caused similar colouration, total phenolic, and total flavonoid concentrations, as well as the same antioxidant capacity compared to those of plants in the UV-open greenhouse.

**Table 7.** Effects of 10-day pre-harvest supplemental UV lighting on red lettuce (Redino Lollo rosso) head weight, leaf colour (hue angle), total phenols (TP), total flavonoids (TF), and antioxidant capacity (DPPH).

| Supplemental UV Dose/Treatments | Head Weight (g) | Hue Angle | TP (mg $g^{-1}$ FW) | TF (mg $g^{-1}$ FW) | DPPH (μmol TE $g^{-1}$) |
|---|---|---|---|---|---|
| 0 kJ $m^{-2}$ $d^{-1}$ | 366 a | 116.6 a | 3.69 a | 1.27 a | 24.60 a |
| 210 kJ $m^{-2}$ $d^{-1}$ | 351 a | 113.6 a | 3.88 a | 1.37 a | 25.00 a |
| 265 kJ $m^{-2}$ $d^{-1}$ | 380 a | 111.6 a | 3.97 a | 1.49 a | 26.40 a |
| 320 kJ $m^{-2}$ $d^{-1}$ | 319 ab | 113.4 a | 4.71 b | 1.84 b | 28.80 b |
| 425 kJ $m^{-2}$ $d^{-1}$ | 298 b | 104.9 b | 5.80 c | 2.28 c | 37.25 c |
| 530 kJ $m^{-2}$ $d^{-1}$ | 299 b | 103.6 b | 6.03 cd | 2.36 c | 39.50 cd |
| 640 kJ $m^{-2}$ $d^{-1}$ | 279 b | 94.8 c | 6.39 d | 2.57 c | 43.75 d |
| UV-open | 230 c | 102.9 b | 6.11 cd | 2.33 c | 40.50 cd |

Means within the same column followed by the same letter do not differ significantly based on LSD test at $\alpha = 0.05$.

## 4. Discussion

In the present study, red lettuces were grown in two greenhouses with different transparency in UV-A radiation, and both greenhouses were not transparent to UV-B radiation. Recent research suggests that different UV transparencies of greenhouse cover do not affect the yield of red lettuce [21]. However, our results (Tables 2 and 6) agree with earlier research that found a significant increase in the yield of red lettuce in the absence of UV radiation or low UV intensity [9,11]. The effects of UV-A on biomass production depend on further environmental factors, as well as on the species or even the genotype, as different responses to UV-A are observed within particular studies [22]. In the UV-block greenhouse, the red lettuce plants developed more leaves (Table 6) and gained more weight (Tables 2 and 6) compared to the plants in the UV-open greenhouse. However, their root size was not affected (Table 2). UV intensity probably affects the distribution of assimilates only in the upper part of the red lettuce plant, and root size can supply water and nutrients to plants with a larger head. Differential partitioning of biomass between shoots and roots in response to UV-A has been reported in various species [22]. For instance, in all four cultivars of Cucumis sativus studied, UV-A decreased the amount of shoot biomass, although there was no effect on root biomass [23].

Previous research demonstrated the detrimental effect of the greenhouse covered with UV-block polyethene film in the antioxidant production of red lettuce [6] and the potential of using UV-transparent covers to increase the beneficial flavonoid content of red leaf lettuce when the crop is grown in greenhouses [9]. Similar conclusions emerge from the results of our first experiment. The cultivation of red lettuce plants in the UV-block greenhouse significantly reduced the content of total flavonoids and total phenolics, causing a corresponding significant reduction in antioxidant capacity (Table 4). On the other hand, growing red lettuce plants in a greenhouse with a transparent cover in UV-A radiation allowed them to synthesize flavonoids and phenolics and thus acquire their natural violet colour. The plants that presented high concentrations of the above secondary compounds (Table 4) also developed more intense violet colouration (Table 3). The violet colour of red lettuce origins mainly from the composition of phenolic substances, flavonoids, and anthocyanins. It has been suggested that when lettuce is grown under a high level of ultraviolet radiation, it will contain more phenolic compounds, which are produced by the plant as protective agents [11].

In the current study, from leaf analysis of red lettuces grown in UV-open or UV-block greenhouse, no differences in concentrations of four macronutrients were found (Table 5). On the other hand, Lee et al. [24], in an indoor experiment, found higher levels of calcium and magnesium in the leaves of red lettuces that received supplemental UV lighting. This finding indicates that the effect of UV radiation intensity on red lettuce nutrition may be cultivar-dependent, or different effects may be caused by different conditions in the greenhouse and growth chamber. However, little research has been conducted on the effect of UV intensity on the nutritional status of red lettuce, and further research is needed. The authors of [25] reported that the intensity of UV radiation causes a different response in stomatal conductance among the various plant species. In order to draw safe conclusions about the effect of UV intensity, simultaneous transpiration measurements, plant tissue analyses, and nutrient solution analyses are required.

Red lettuce yield is high in UV-block greenhouses, which is interesting for farmers, but its antioxidant capacity is low. Plant secondary metabolites that are affected by UV light are essential due to their health-promoting properties, and this is important for consumers. Tsormpatsidis et al. [12] supported that growing plants continuously under a UV-blocking film and then transferring them to a UV transparent film six days before the final harvest showed that high yields and high phytochemical content could be achieved complementarily. In the current study, the transfer of red lettuce plants from UV-block to UV-open greenhouse seven days pre-harvest caused a significant increase in their antioxidant content (total phenols, total flavonoids). However, this increase was not enough to reach the quality of the red lettuces grown in the UV-open greenhouse for the whole period (Table 4). Probably, more time in a UV-open greenhouse was required than a period of seven days. Ordidge et al. [26] showed that in the red lettuce Lollo Rosso, total phenolics, anthocyanin, luteolin, and quercetin levels were all raised by changing from a UV-blocking film to a film of low UV transparency and to a film of high UV transparency. Nevertheless, this finding confirms that a pre-harvest supplemental UV lighting treatment of red lettuces grown in a UV-block greenhouse can produce red lettuces with high yield and high quality. Light manipulation is a key environmental control method to increase the functional phytochemical concentrations of plants under controlled environments such as greenhouses and plant factories [27].

The second experiment of the present study sought to answer the question: what is the appropriate pre-harvest dose of additional supplemental UV light to produce red lettuces in a UV-block greenhouse with equal antioxidant content to those produced in a UV-open greenhouse?

According to our results in Table 7, the high intensity of supplemental UV radiation significantly decreased the red lettuces' head weight. However, in an experiment with supplemental UV treatments in a UV-open greenhouse, Lee et al. [24] reported that UV-A treatments did not affect the fresh or dry shoot biomass of lettuce varieties. However, plant physiological processes are variably affected by light, and the responses are species- and cultivar-dependent [28]. This indicates that plant response to supplemental UV-A may be cultivar-dependent or affected by pre-treatment acclimatization to UV radiation. In the present study, red lettuces before the starting of supplemental UV lighting were grown in a UV-block greenhouse, while in the study of [24], they were grown in a UV-transparent greenhouse.

The results of the second experiment indicated that supplemental UV-A lighting stimulated the accumulation of total flavonoids and total phenols. Red lettuce plants which received 210 and 265 kJ $m^{-2}$ $d^{-1}$ of supplemental UV lighting had no differences in yield, colour, total phenols content, total flavonoids content, and antioxidation capacity in comparison to plants grown without supplemental UV light in the UV-block greenhouse. Our findings are in agreement with that of [29], who used 11 μmol $m^{-2}$ $s^{-1}$ of supplemental UV light with a photoperiod duration of 20 h per day, that is, approximately equal to 172 kJ $m^{-2}$ $d^{-1}$, to improve the quality of red lettuce in a plant growth chamber. They reported that no treatment differences were found for total phenolics, anthocyanins content, and antioxidant capacity. Obviously, a higher dose of UV radiation is required to express the photomorphogenetic effect.

The higher dose of supplemental UV lighting (640 kJ $m^{-2}$ $d^{-1}$) induced more intense leaf colouration and a higher accumulation of antioxidants in the leaves of red lettuces in comparison to lettuces grown in the UV-open greenhouse. This indicates that this dose can produce the desired effect in a shorter time than 10 days. Lee et al. [24] improved the quality (antioxidant content) of red lettuces grown in a UV-open greenhouse using 700 kJ $m^{-2}$ $d^{-1}$ of supplemental UV-A radiation for 5–6 days pre-harvest.

The doses of 425 and 530 kJ $m^{-2}$ $d^{-1}$ of supplemental UV lighting for 10 days prior to the harvest produced red lettuces of similar quality to that in the UV-open greenhouse. Moreover, the yield in these treatments was 30% higher in comparison to that of red lettuces grown in the UV-open greenhouse. Gómez and Jiménez [29] supported that end-of-production radiation is a cost-effective, pre-harvest practice that can allow growers to manipulate product quality and thus increase the market value of lettuce without negatively affecting plant growth. From this research, we concluded that adding UV light 10 days prior to harvest is effective for the production of functional phytochemical-rich lettuce. Consequently, we can suggest the cultivation of the red lettuces in a UV-block greenhouse and adding 425 kJ $m^{-2}$ $d^{-1}$ of supplemental UV-A lighting 10 days prior to harvest. LED lights have great potential to provide supplemental light more efficiently than traditional lights, and their spectrum can be adjusted based on plant growth requirements [30]. Additionally, LEDs allow potential control of both irradiance and spectra and, when used in fully enclosed environments, photoperiod [13], so the results of the current study can be applied to both greenhouse and plant factory cultivations. Real-time electricity prices and crop value should be considered in the economic evaluation of the effectiveness of supplemental UV lighting.

In the present study, the addition of supplementary UV lighting was carried out during the light period, that is, during the working hours of the staff in the greenhouse. Although the above-recommended UV dose is not higher than the usual outside UV intensity, care must be taken for the safety of greenhouse staff with supplemental UV lighting. Night-time supplemental lighting in red lettuce was more effective than daytime supplemental lighting, as it resulted in better crop quality [31]. In addition, the application of short-duration supplemental blue LED light may become more effective in the pigmentation of red lettuce if applied during night breaks [32]. So, in future research, it is worth studying whether adding the same doses of supplemental UV lighting during the night produces the same results.

## 5. Conclusions

We designed and created a new cultivation system for red lettuce based on the combination of UV-block polyethene film as a greenhouse cover and supplemental UV light. Overall, the second experiment showed that pre-harvest supplemental UV light treatments in a UV-block greenhouse could increase red lettuce growth and quality compared to that in a UV-open greenhouse. Regarding phytonutrient content, our results showed that supplemental UV-A light could increase beneficial antioxidants such as flavonoids and phenolic compounds in lettuce. More research is needed to further understand the effects of pre-harvest supplemental UV lighting and its intensity on plant growth and quality.

**Author Contributions:** Conceptualization, methodology, experimental measurements, and writing—original draft: I.L.; investigation: I.L. and D.K.; resources: I.L., D.K. and G.K.; data analysis: I.L., D.K. and G.K.; writing—review and editing: I.L., A.K. and G.K.; supervision: I.L. and A.K. All authors have read and agreed to the published version of the manuscript.

**Funding:** The paper was funded in the frame of "Development of Intelligent and Energy-autonomous Greenhouse Using Innovative Technologies to Improve Productivity and Product quality" (IEnGreen) program, MIS:5045455 Smart Specialization Strategy-RIS3.

**Institutional Review Board Statement:** Not applicable.

**Data Availability Statement:** Not applicable.

**Acknowledgments:** We would like to thank Vasileios Thomopoulos for technical support.

**Conflicts of Interest:** The authors declare no conflict of interest.

## References


1. Dilara, P.A.; Briassoulis, D. Degradation and Stabilization of Low-density Polyethylene Films used as Greenhouse Covering Materials. *J. Agric. Eng. Res.* **2000**, *76*, 309–321. [CrossRef]
2. Krizek, D.T.; Clark, H.D.; Mirecki, R.M. Spectral properties of selected UV-blocking and UV-transmitting covering materials with application for production of high-value crops in high tunnels. *Photochem. Photobiol.* **2005**, *81*, 1047–1051. [CrossRef] [PubMed]
3. Paul, N.D.; Jacobson, R.J.; Taylor, A.; Wargent, J.J.; Moore, J.P. The use of wavelength selective plastic cladding materials in horticulture: Understanding of crop and fungal responses through the assessment of biological spectral weighting functions. *Photochem. Photobiol.* **2005**, *81*, 1052–1060. [CrossRef] [PubMed]
4. Doukas, D.; Payne, C.C. Greenhouse whitefly (Homoptera: Aleyrodidae) dispersal under different UV light environments. *J. Econ. Entomol.* **2007**, *100*, 389–397. [CrossRef]
5. West, J.S.; Pearson, S.; Hadley, P.; Wheldon, A.E.; Davis, F.J.; Gilbert, A.; Henbest, R.G.C. Spectral filters for the control of Botrytis cinerea. *Ann. Appl. Biol.* **2000**, *136*, 115–120. [CrossRef]
6. Zapata-Vahos, I.C.; Rojas-Rodas, F.; David, D.; Gutierrez-Monsalve, J.A.; Castro-Restrepo, D. Comparison of antioxidant contents of green and red leaf lettuce cultivated in hydroponic systems in greenhouses and conventional soil cultivation. *Rev. Fac. Nac. Agron. Medellín* **2020**, *73*, 9077–9088. [CrossRef]
7. Oh, M.M.; Carey, E.E.; Rajashekar, C.B. Environmental stresses induce health-promoting phytochemicals in lettuce. *Plant Physiol. Biochem.* **2009**, *47*, 578–583. [CrossRef]
8. Romani, A.; Pinelli, P.; Galardi, C.; Sani, G.; Cimato, A.; Heimler, D. Polyphenols in greenhouse and open-air grown lettuce. *Food Chem.* **2002**, *79*, 337–342. [CrossRef]
9. García-Macías, P.; Ordidge, M.; Vysini, E.; Waroonphan, S.; Battey, N.; Gordon, M.; Hadley, P.; John, P.; Lovegrove, J.; Wagstaffe, A. Changes in the Flavonoid and Phenolic Acid Contents and Antioxidant Activity of Red Leaf Lettuce (Lollo Rosso) Due to Cultivation under Plastic Films Varying in Ultraviolet Transparency. *J. Agric. Food Chem.* **2007**, *55*, 10168–10172. [CrossRef]
10. Kittas, C.; Tchamitchian, M.; Katsoulas, N.; Karaiskou, P.; Papaioannou, C. Effect of two UV-absorbing greenhouse-covering films on growth and yield of an eggplant soilless crop. *Sci. Hortic.* **2006**, *110*, 30–37. [CrossRef]
11. Tsormpatsidis, E.; Henbest, R.G.C.; Davis, F.J.; Battey, N.H.; Hadley, P.; Wagstae, A. UV irradiance as a major influence on growth, development and secondary products of commercial importance in Lollo Rosso lettuce 'Revolution' grown under polyethylene films. *Environ. Exp. Bot.* **2008**, *63*, 232–239. [CrossRef]
12. Tsormpatsidis, E.; Henbest, R.G.C.; Battey, N.H.; Hadley, P. The influence of ultraviolet radiation on growth, photosynthesis and phenolic levels of green and red lettuce: Potential for exploiting effects of ultraviolet radiation in a production system. *Ann. Appl. Biol.* **2010**, *156*, 357–366. [CrossRef]
13. Bantis, F.; Smirnakou, S.; Ouzounis, T.; Koukounaras, A. Current status and recent achievements in the field of horticulture with the use of lightemitting diodes (LEDs). *Sci. Horticult.* **2018**, *235*, 437–451. [CrossRef]
14. Hooks, T.; Masabni, J.; Sun, L.; Niu, G. Effect of Pre-Harvest Supplemental UV-A/Blue and Red/Blue LED Lighting on Lettuce Growth and Nutritional Quality. *Horticulturae* **2021**, *7*, 80. [CrossRef]
15. Ghoulem, M.; Moueddeb, K.; Nehdi, E.; Boukhanouf, R.; Calautit, J.K. Greenhouse design and cooling technologies for sustainable food cultivation in hot climates: Review of current practice and future status. *Biosyst. Eng.* **2018**, *183*, 121–150. [CrossRef]
16. Marin, A.; Ferreres, F.; Barberá, G.G.; Gil, M.I. Weather Variability Influences Color and Phenolic Content of Pigmented Baby Leaf Lettuces throughout the Season. *J. Agric. Food Chem.* **2015**, *63*, 1673–1681. [CrossRef] [PubMed]
17. Denaxa, N.-K.; Roussos, P.; Vemmos, S. Assigning a role to the endogenous phenolic compounds on adventitious root formation of olive stem cuttings. *J. Plant Growth Regul.* **2020**, *39*, 411–421. [CrossRef]
18. Roussos, P.; Pontikis, C.A. Phenolic compounds in olive explants and their contribution to browning during the establishment stage in vitro. *Gartenbauwissenschaft* **2001**, *66*, 298–303.
19. Roussos, P.; Denaxa, N.K.; Ntanos, E.; Tsafouros, A.; Mavrikou, S.; Kintzios, S. Organoleptic, nutritional and anticarcinogenic characteristics of the fruit and rooting performance of cuttings of black mulberry (*Morus nigra* L.) genotypes. *J. Berry Res.* **2020**, *10*, 77–93. [CrossRef]
20. Lycoskoufis, I.; Mavrogianopoulos, G. NDT, a new soilless growing system without substrate suitable for Mediterranean conditions. *Not. Bot. Horti Agrobot. Cluj-Napoca* **2020**, *48*, 1292–1305. [CrossRef]
21. Quintero-Arias, D.G.; Acuña-Caita, J.F.; Asensio, C.; Valenzuela, J.L. Ultraviolet Transparency of Plastic Films Determines the Quality of Lettuce (*Lactuca sativa* L.) Grown in a Greenhouse. *Agronomy* **2021**, *11*, 358. [CrossRef]
22. Verdaguer, D.; Jansen, M.A.; Llorens, L.; Morales, L.O.; Neugart, S. UV-A radiation effects on higher plants: Exploring the known unknown. *Plant Sci.* **2017**, *255*, 72–81. [CrossRef]
23. Krizek, D.T.; Mirecki, R.M.; Britz, S.J. Inhibitory effects of ambient levels of solar UV-A and UV-B radiation on growth of cucumber. *Physiol. Plant.* **1997**, *100*, 886–893. [CrossRef]
24. Lee, M.; Rivard, C.; Pliakoni, E.; Wang, W.Q.; Rajashekar, C.B. Supplemental UV-A and UV-B Affect the Nutritional Quality of Lettuce and Tomato: Health-Promoting Phytochemicals and Essential Nutrients. *Am. J. Plant Sci.* **2021**, *12*, 104–126. [CrossRef]
25. Katsoulas, N.; Bari, A.; Papaioannou, C. Plant Responses to UV Blocking Greenhouse Covering Materials: A Review. *Agronomy* **2020**, *10*, 1021. [CrossRef]

26. Ordidge, M.; García-Macías, P.; Battey, N.H.; Gordon, M.H.; Hadley, P.; John, P.; Lovegrove, J.A.; Vysini, E.; Wagstaffe, A. Phenolic contents of lettuce, strawberry, raspberry, and blueberry crops cultivated under plastic films varying in ultraviolet transparency. *Food Chem.* **2010**, *119*, 1224–1227. [CrossRef]
27. Goto, E.; Hayashi, K.; Furuyama, S.; Hikosaka, S.; Ishigami, Y. Effect of UV light on phytochemical accumulation and expression of anthocyanin biosynthesis genes in red leaf lettuce. *Acta Hortic.* **2016**, *1134*, 179–186. [CrossRef]
28. Ouzounis, T.; Rosenqvist, E.; Ottosen, C.-O. Spectral effects of artificial light on plant physiology and secondary metabolism. *Hortscience* **2015**, *50*, 1128–1135. [CrossRef]
29. Gómez, C.; Jiménez, J. Effect of End-of-production Highenergy Radiation on Nutritional Quality of Indoor-grown Red-leaf Lettuce. *Hortscience* **2020**, *55*, 1055–1060. [CrossRef]
30. Matysiak, B.; Kaniszewski, S.; Dyśko, J.; Kowalczyk, W.; Kowalski, A.; Grzegorzewska, M. The Impact of LED Light Spectrum on the Growth, Morphological Traits, and Nutritional Status of 'Elizium' Romaine Lettuce Grown in an Indoor Controlled Environment. *Agriculture* **2021**, *11*, 1133. [CrossRef]
31. Hooks, T.; Sun, L.; Kong, Y.; Masabni, J.; Niu, G. Short-Term Pre-Harvest Supplemental Lighting with Different Light Emitting Diodes Improves Greenhouse Lettuce Quality. *Horticulturae* **2022**, *8*, 435. [CrossRef]
32. Cammarisano, L.; Donnison, I.S.; Robson, P.R.H. Producing Enhanced Yield and Nutritional Pigmentation in Lollo Rosso through Manipulating the Irradiance, Duration, and Periodicity of LEDs in the Visible Region of Light. *Front. Plant Sci.* **2020**, *11*, 598082. [CrossRef] [PubMed]

agriculture

MDPI

Article

# Influence of Lighting and Laser Irradiation on the Germination of Caper Seeds

**María Laura Foschi [1,2], Mariano Juan [1], Bernardo Pascual [1] and Nuria Pascual-Seva [1,*]**

1 Plant Production Department, Universitat Politècnica de València, 46022 Valencia, Spain
2 Horticulture and Floriculture, Agriculture Faculty, National University of Cuyo, Mendoza M5528AHB, Argentina
* Correspondence: nupasse@upv.es

**Abstract:** Caper seeds present difficulties in their germination, which has been studied by several research teams. It is known that light can release dormancy in some seeds, but its effect on caper seed germination has not yet been deeply studied. The main aim of this study was to analyze the response of caper seeds germination to light exposure. The study analyzed the germination response of seeds to lighting with different wavelengths (white, red, blue, red + blue and darkness) and to the He-Ne laser light, using both dry seeds and seeds that had been previously soaked in water. Overall, it could be stated that caper seeds are insensitive to light during the germination process. Thus, germination could be carried out in lightness or darkness, so germination in nurseries could be carried out in the darkness, leading to substantial energy savings. Caper seed irradiation with a He-Ne laser during short exposure times improved the germination percentage for the seeds previously soaked in water, germinating all viable seeds. However, applying a solution of gibberellic acid was always required in all the cases studied.

**Keywords:** *Capparis spinosa* L.; darkness; gibberellic acid; light-emitting diodes (LEDs); light wavelength; soaking

**Citation:** Foschi, M.L.; Juan, M.; Pascual, B.; Pascual-Seva, N. Influence of Lighting and Laser Irradiation on the Germination of Caper Seeds. *Agriculture* **2022**, *12*, 1612. https://doi.org/10.3390/agriculture12101612

Academic Editors: Athanasios Koukounaras and Filippos Bantis

Received: 8 September 2022
Accepted: 30 September 2022
Published: 4 October 2022



## 1. Introduction

The caper (*Capparis spinosa* L.) is a deciduous creeping shrub native to Asia, which spreads throughout the Mediterranean basin, where it grows in dry lands. It is currently cultivated in Spain, Italy, France, Greece, and North Africa, as well as in South America [1,2]. It is mainly cultivated for the floral buds, called capers; however, their fruits and, to a lesser extent, their vegetative shoots are also consumed pickled or salted. As Shahrajabian et al. [3] have stated, different parts of the plant are rich sources of antioxidants and bioactive compounds beneficial to health. Furthermore, the flowers have a high ornamental value; thus, caper plants are included in gardening, particularly in xeriscape [4,5].

In a recent review article, Sottile et al. [1] stated that, as a crop, caper should not be considered a difficult crop to propagate. Pascual et al. [6] reported acceptable percentages of success in the rooting of cuttings, and Foschi et al. [7–9] obtained high germination percentages of dry seeds using gibberellic acid and without the need for any treatment in fresh seeds. These germination tests were carried out in a growth chamber under a photoperiod of 12 h.

Light is one of the main environmental signals for plants [10], being an important factor in breaking seed dormancy [11]. To respond to environmental signals, plants have developed several families of photoreceptors, which are photosensitive pigments capable of being activated by photons of specific wavelengths and, in turn, activatingsignal translation pathways, providing the ability to respond to light stimuli [12]. These photoreceptors include the following: (i) Phytochromes, which are red (600–700 nm) and far red (700–800 nm) light photoreceptors; (ii) Cryptochromes, which are photoreceptors for blue (400–500 nm) and ultraviolet A (320–400 nm) light; (iii) Phototropins, another group of blue and UV-A light photoreceptors [10,13].

Butler et al. [14] obtained a photo-reversible pigment from etiolated shoots of maize and named it phytochrome. Later, in studies carried out on *Arabidopsis thaliana* (L.) Heynh., Sharrock and Quail [15] identified sequences that showed that small families of genes encode phytochromes. Specifically, these authors hypothesized that the minimum number of phytochrome genes present in higher plants could be determined in studies in *A. thaliana*. They stated that phytochromes in this plant were encoded by a small gene family consisting of at least three genes and probably four or five. Subsequent studies have shown that *A. thaliana* contains five different phytochromes (*phyA, phyB, phyC, phyD* and *phyE*) [10] that are encoded by five genes (*PHYA, PHYB, PHYC, PHYD* and *PHYE*). Cryptochromes were first identified in *A. thaliana* [16]. Different organisms have different numbers of cryptochromes; plants have at least two types of cryptochromes, and this number can range up to six, as in soybean (*Glycine max* L.) [16].

According to sensitivity to white light, seeds have been classified into three categories [17]: (i) positive photoblastic (the seeds that germinate only under white light); (ii) negative photoblastic (the seeds that germinate only in the dark); (iii) light insensitive (the seeds that germinate both under white light and in darkness). A Photoblastic Index (PI; Equation (1)) has been utilized to evaluate the photoblastic responses [18]:

$$PI = (GD - GL)/(GD + GL) \quad (1)$$

GD is the germination (%) in darkness, and GL is the germination (%) under light. This index ranges from 1 (negative photoblastism) to $-1$ (positive photoblastism); PI = 0 indicates that germination is not dependent on light.

Another index expressing a light requirement is the Relative Light Germination (RLG; Equation (2) [19]):

$$RLG = (GL)/(GD + GL) \quad (2)$$

GD is the germination (%) in darkness, and GL is the germination (%) under light. The RLG values vary from 0 (only seeds exposed to darkness germinate) to 1 (only seeds exposed to light germinate), with values close to 0.5 in seeds germinating both in light and darkness.

Takaki [20] proposed to replace the term photoblastism with forms of phytochrome that control germination, proposing a classification based on three mechanisms, depending on the level of fluence (total energy received by a seed in a period of time, J $m^{-2}$ [21]) to saturate the responses [20,22]:

Low Fluency Responses (LFRs), which represent the classic reversible red/far red responses, in which Pfr (the active form of phytochrome, which absorbs light of 735 nm) production promotes plant responses and removal of Pfr reverses the response. These seeds have *phyB* controlling the germination process through LFR. The saturation of the response frequently occurs at low levels of Pfr/Ptotal ($10^{-2}$–0.87 Pfr/Ptotal) and intermediate fluences (1–1000 μmol $m^{-2}$ $s^{-1}$).

High Irradiance Responses (HIRs), which represent responses produced by prolonged high irradiation, which do not show reciprocity or reversibility. These seeds have *phyA* controlling germination through HIR, and the maximum reaction generally occurs at wavelengths that maintain low Pfr levels for long periods of time, such as occurs in far-red-rich environments.

Very low fluence responses (VLFRs), which represent the saturation of responses by very low fluences, with reciprocity but no reversibility because the photo-equilibrium maintained by far-red light (or even safe dim green light used in experiments) produces enough Pfr to saturate these responses, which occur at low levels of Pfr ($10^{-6}$–$10^{-3}$ Pfr/Ptotal). These seeds have *phyA* controlling the germination process through VLFR.

LED (Light Emitting Diode) lights are an alternative to incandescent lamps and cold white fluorescent tubes for growing plants. They have several advantages due to their small size, long lifespan, low emission temperature, high efficiency in energy conversion, and the possibility of selecting specific wavelengths [23,24]. In the last two decades, several studies have shown that LEDs of different wavelengths can modify the germination, growth, and development of seedlings in many species [11,25–30]. In this type of light, which is mainly monochromatic, it is essential to know the optimal light spectrum and the intensity required by the different species at each phenological stage to optimize yield and quality [28].

The irradiation of seeds with laser light (Light Amplification by Stimulated Emission of Radiation) can also be an alternative to improve the germination or growth and development of seedlings of various species based on the bio-stimulant effect of laser light [31–33]. Laser irradiation effects depend on many laser parameters, such as wavelength, irradiation duration, power, dose, and method (constant/pulse) [31]. However, seed properties are also important, particularly their genetic traits and physiological properties (health status, seed quality), and even their orientation during irradiation [31,32].

Among the different lasers used in agriculture, the helium-neon laser (He-Ne) is the most used [31], considering that its wavelength of 632.4 nm corresponds to the red light which is responsible for phytochrome activation [32]. The improvement and acceleration of germination have been related to an induction of the enzymatic activities, a change of thermodynamic parameters, and an acceleration of physiological and biochemical metabolism of seeds, increasing, in some cases, the levels of gemination-promoting growth hormones. such as gibberellic acid ($GA_3$). and decreasing inhibitors, such as abscisic acid [34–39].

Light and gibberellins can release dormancy in some seeds [40], specifically in those that present coat dormancy, promoting their germination [41]. Caper seeds have a non-deep physiological dormancy, specifically a coat-imposed dormancy due to a mechanical characteristic, which can be released by adding $GA_3$ to the germination substrate [7]. As far as is known, the effect of light on caper seed germination has not yet been deeply studied. The main aim of this study was to analyze the response of caper seeds germination to light exposure. Particularly, the study analyzes the germination response of seeds to lighting with different wavelengths (white, red, blue, red + blue and darkness) and to the He-Ne laser light.

## 2. Materials and Methods

### 2.1. Plant Material

Caper (*Capparis spinosa* L.) seeds were extracted from ripe fruits produced by adult plants grown in Llíria, Valencia, Spain (39°38′54.2″ N, 0°37′3.5″ W). The fruit collection was carried out over the first fortnight of September 2019, 2020, and 2021, each constituting a different lot.

After the extraction, mature seeds were selected using the flotation method [42] with tap water. The seeds were disinfected by soaking them in a 25% sodium hypochlorite solution and then rinsing with tap water. They were dried for 15 days at room temperature in the shade and kept in hermetically sealed glass containers at 7 ± 0.5 °C in a domestic refrigerator (Beko, Beko Electronics España, Barcelona, Spain). At the beginning of each germination test, the three seed lots were within the recommended storage period to not affect their viability [7,43].

### 2.2. Viability and Germination Tests

The seed lots' viability was determined by the tetrazolium test, as Foschi et al. [7] reported (four replicates of fifty seeds each), according to the International Rules for Seed Testing [44].

Germination tests were carried out with the Between Paper method, placing 100 seeds per Petri dish of 9 cm diameter [45]. In all cases, four replicates were performed per treatment. Ultrapure water (Wasserlab G.R type II analytical grade water system; from now on referred to as water) or a solution of 500 mg $L^{-1}$ of $GA_3$ (Semefil L, Nufarm L.) were used to wet the substrate. The Petri dishes were then placed under controlled conditions in a growth chamber (model Zimbueze, Seville, Spain) at 30 ± 1/20 ± 1 °C, 85 ± 1% relative humidity for a photoperiod of 12 h (cold white fluorescent tubes Philips TL-D 36W/54), providing a photosynthetic photon flux density (PPFD) of 81.1 ± 1.7 μmol $m^{-2}$ $s^{-1}$, unless stated otherwise.

The germination test lasted 120 days, and germinated seeds were counted and removed periodically. Seeds were considered germinated when the radicle protruded from the testa and the micropylar endosperm, reaching a length of approximately 2 mm. Results of germination tests were fitted to the logistic function [46,47], defined as a particular case of Richards' function ([48] Equation (3)):

$$G = A/1 + e^{(\beta - kt)} \quad (3)$$

$G$ is the cumulative germination (%), $A$ represents the final germination percentage, t is the germination time (d), and $\beta$ and $k$ are function parameters used to determine the time required to reach 50% of $G$ ($Gt_{50} = \beta/k$; d) and the mean relative cumulative germination rate ($k/2$; $d^{-1}$).

## 2.3. Experiment 1

This experiment evaluated the effect of light and darkness on the germination of the two caper seed lots (corresponding to seeds produced in 2019 and 2020). The experiment started in March 2021. The seeds were placed in the germination chamber with a photoperiod of 12 h of white light or continuous darkness. Light exposure was provided by cool white fluorescent tubes, as previously stated, and seeds that germinated in the dark were placed in closed opaque boxes. The Photoblastic Index (PI) and the Relative Light Germination (RLG) were determined using Equations (1) and (2).

Eight combinations of the three sources of variation were tested: 2 seed lots, light-darkness and 2 wetting solutions.

## 2.4. Experiment 2

The second experiment evaluated the effect of lighting with different wavelengths on caper seed germination. It started in July 2021. In light of the results obtained in the first experiment, tests were performed under white, red, blue, and red + blue lights. Color lighting was achieved with LED lights (AMZLAB GmbH), consisting of 80 LEDs (52 red, wavelength range 600–700 nm, and 28 blue, wavelength range 400–500 nm) placed 25 cm above the Petri dishes. The maximum power, when using the full spectrum of lights, was 30 W. Four types of light were tested: white was provided by fluorescent tubes (Philips TL-D 36W/54) with a power of 36 W, a distribution of the color spectrum ranging from wavelengths of 300 to 800 nm and with photosynthetically active radiation of 81.1 μmol $m^{-2}$ $s^{-1}$; red (52 red LEDs; 102.5 μmol $m^{-2}$ $s^{-1}$), blue (28 blue LEDs; 80.1 μmol $m^{-2}$ $s^{-1}$) and red + blue (52 red LEDs + 28 blue LEDs; 125.4 μmol $m^{-2}$ $s^{-1}$). Darkness was applied as in Experiment 1, placing the seeds in closed opaque boxes.

Table 1 shows the percentages and wavelengths for the color spectra of the lamps, measured with a Thorlabs spectrometer, model CCS200/M. The rates of each type of wavelength in each light were calculated using the ImageJ program [49], measuring the areas between the desired ranges of the wavelengths of the color spectrum.

**Table 1.** Relative percentage distribution of the different wavelengths for white light (cold white fluorescent lamp) and blue, red, and red + blue LED lights.

| Wavelength | White Light | Blue LED | Red LED | Red + Blue LED |
|---|---|---|---|---|
| Violet 300–400 nm | 1.51% | 0.00% | 0.00% | 0.00% |
| Blue 400–500 nm | 30.33% | 97.48% | 2.98% | 30.25% |
| Green 500–600 nm | 43.32% | 1.05% | 7.88% | 5.27% |
| Red 600–700 nm | 22.27% | 1.47% | 80.55% | 58.41% |
| Far Red 700–800 nm | 2.56% | 0.00% | 8.58% | 6.06% |
| R/FR | 8.69 | 0.00 | 9.39 | 9.63 |

This experiment analyzed 20 combinations of the three sources of variation: 2 seed lots, 5 types of lighting and 2 wetting solutions.

### 2.5. Experiment 3

The third experiment evaluated the influence of He-Ne laser irradiation, with different exposure times, on germination of caper seeds. Seeds from the lot harvested in 2021 were used in this experiment, which started in December 2021.

The seeds were irradiated with a He-Ne laser (JDS Uniphase model 1145), with an output emission power of 22.5 mW and wavelength of 632.8 nm, belonging to the red band of the spectrum. It had a circular beam of 0.7 mm in diameter.

To define the exposure times (and therefore energies) to be analyzed in this experiment, a preliminary study was carried out to analyze the effect on seed viability of the following exposure times: 0, 1, 5, 15, 30, 60, 120 and 180 s, for which 40 seeds were irradiated during each of the times. These assayed exposure times expanded the range Juan [50] tested with the same laser. Neither of the analyzed exposure times decreased ($p \leq 0.05$) the seed viability (data not shown), thus, maximum and minimum thresholds were included in the experiment, as well as three intermediate levels: 0, 1, 15, 60 and 180 s, corresponding to 0, 22.5, 337.5, 1350, 4050 mJ applied energy levels, respectively. Intact seeds were irradiated one by one. After irradiation, the germination test was performed as previously indicated.

A total of 10 combinations of two sources of variation were tested: 5 exposure times and 2 wetting solutions.

### 2.6. Experiment 4

The fourth experiment evaluated the effect on germination of soaking the seeds in water before irradiation with the He-Ne laser with two timings of exposure, 1 and 15 s, which led to the best results in Experiment 3. As in the previous experiment, seeds of the 2021 lot were used and the experiment started in February 2022.

Caper seeds were soaked in water for four days before laser irradiation by a He-Ne laser for 0, 1 and 15 s. The germination test was performed as previously indicated. The germination substrate was only wetted with the $GA_3$ solution, as this experiment was not analyzing the effect of the wetting solution, since it haf already been stated that the $GA_3$ solution was required to get an acceptable germination percentage that allowed adjusting the logistic model.

This experiment tested six combinations of two sources of variation: 3 exposure times and soaked/dry seeds.

### 2.7. Data Analysis

All the tests verified that the tolerance required by the ISTA standards [45] was met, either between the replicates or between the germination tests. The statistical analysis program Statgraphics [51] was used to perform multi-way analyses of variance (ANOVA; $p \leq 0.05$) and verify the normality of the data. Mean separations were performed where appropriate, using Fisher's smallest significance difference (LSD test) at $p \leq 0.05$.

## 3. Results and Discussion

### 3.1. Experiment 1

The viability of the two seed lots was very high (82.5 ± 2.5% and 90.0 ± 3.1% in the seeds produced in 2019 and 2020, respectively), and, as a consequence, among other factors of careful harvesting, cleaning, and drying of the seeds, and according to that reported by Foschi et al. [9], there were no differences ($p < 0.01$) between both values.

Germination percentages obtained when water was used to wet the germination substrate were very low (specifically 9.5% in light and 6.8% in darkness, on average, for the two lots). These values were significantly lower ($p \leq 0.05$) than those obtained with the addition of $GA_3$ (81.5% in light and 81.2% in darkness, on average for the two lots). The low germination percentages obtained with water did not allow adjusting the logistic model;

thus, the statistical analysis (Table 2) was only carried out for the seeds wetted with the $GA_3$ solution. The high germination values obtained with the $GA_3$ addition in the two lots are worth noting, accounting for 92% and 97% of the viable seeds. The 16 curves were adjusted to the germination logistic model with a determination coefficient higher than 99.6%, which allowed using the variable $A$ (instead of $G$), as well as the other variables derived from the logistic function, as occurred in previous studies of caper seed germination carried out by Pascual et al. [52] and Foschi et al. [8] It applied to all the experiments reported in this manuscript. Figure 1 shows the logistic model adjusted to the average curves of accumulated germination of caper seeds from this experiment.

**Table 2.** Effect of the seed lot and the exposure to light–darkness on the germination parameters: final germination percentage ($A$, %), time required to reach 50% of final germination ($Gt_{50}$, d), and average germination rate ($k/2$; $d^{-1}$); average values from Experiment 1. A 500 mg $L^{-1}$ $GA_3$ solution was used to wet the substrate.

| | *A* | *Gt*$_{50}$ | *k*/2 |
|---|---|---|---|
| Seed lot (L) | | | |
| 2019 | 79.87 | 24.33 | 0.098 |
| 2020 | 82.88 | 25.24 | 0.088 |
| Exposure to (E) | | | |
| Light | 81.50 | 23.91 | 0.094 |
| Darkness | 81.25 | 25.68 | 0.092 |
| | Analysis of Variance | | |
| Factors (degrees of freedom) | | % Sum of squares | |
| L (1) | 16.30 NS | 5.41 NS | 10.18 NS |
| E (1) | 0.12 NS | 20.32 NS | 0.61 NS |
| L × E (1) | 2.49 NS | 3.37 NS | 5.14 NS |
| Residual (12) | 81.09 | 70.90 | 84.08 |
| Standard deviation (+) | 3.88 | 1.92 | 0.02 |

NS: Not significant differences ($p \leq 0.05$) according to the LSD test. (+) The standard deviation has been calculated as the square root of the residual mean square.

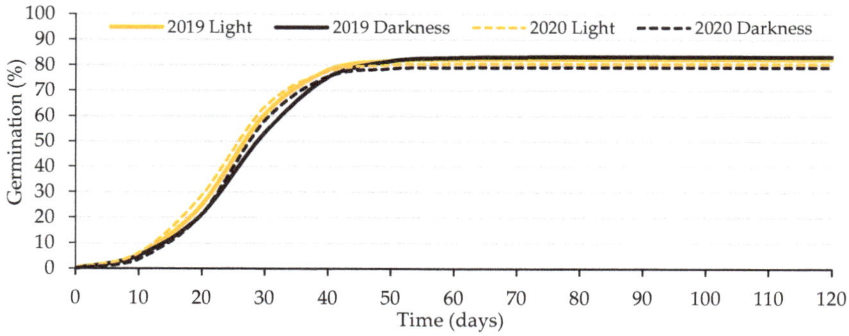


**Figure 1.** Logistic model adjusted to the curves of accumulated germination of caper seeds from Experiment 1. Average values of the combination of the seed lot (2019 and 2020) and the exposure to light-darkness of the seeds. A 500 mg $L^{-1}$ $GA_3$ solution was used to wet the substrate.

Neither the seed lot and the seeds' exposure to light–darkness, nor their interaction, influenced ($p \leq 0.05$) any of the determined germination parameters. These results coincided with those obtained by Germanà and Chiancone [53], in the sense that no significant difference was obtained when incubating mechanically scarified caper seeds in lightness and darkness. The non-statistical significance between the mean values of $A$ obtained in light (81.5%) and darkness (81.2%) indicated that the caper seeds germinated equally in light and darkness, which was corroborated by the values of the photoblastism (PI = −0.002) and Relative Light Germination (RLG = 0.5) indices. PI and RLG were calculated only for seeds germinated in substrate wetted with $GA_3$ solution and not with water to exclude calculations of both indices based on a small number of seeds, as Milberg et al. [19] did.

As Takaki [20] reported, light insensitive seeds present *phyA*, corresponding to the so-called very low fluence responses (VLFR [21]). They saturate the responses by very low fluences because the photoequilibrium maintained by far-red light or extremely low light fluences in most regions of the visible spectrum produces enough Pfr to saturate this response [20,22].

The second experiment was set up with the aim of assessing d the response of seeds exposed to light with different wavelengths, particularly blue light, of which the cryptochromes are photoreceptors [10,13].

### 3.2. Experiment 2

With the low germination percentages obtained with the seeds in water (on average, 3%, data not shown), it was not possible to fit the logistic model, so these data were not included in the subsequent statistical analysis. The germination percentage obtained with the $GA_3$ addition did significantly ($p \leq 0.01$) fit the logistic function, obtaining determination coefficients greater than 98.6%.

The seed lot did not affect any germination parameters ($p \leq 0.05$; Table 3); thus, in Figure 2, to facilitate its interpretation, average values for both lots are presented. As seen in Table 3 and Figure 2, the light wavelength did not influence ($p \leq 0.05$) the germination of the seeds, not differing from permanent darkness. In all cases, these germination percentages were high, comparable to those obtained in the first experiment; the difference between the germination values obtained in this and the previous experiment (both for white light and darkness) did not exceed the tolerance level established by ISTA Rules [45]. Neither the $Gt_{50}$ nor the $k/2$ was affected by the different wavelengths. The interaction between the seed lot and the exposure to different wavelengths did not affect ($p \leq 0.05$) any of the germination parameters.

**Table 3.** Effect of the seed lot and the exposure to white, red, blue or red + blue lights and to darkness on the germination parameters: final germination percentage ($A$, %), time required to reach 50% of final germination ($Gt_{50}$; d), and average germination rate ($k/2$; $d^{-1}$); average values from Experiment 2. A 500 mg $L^{-1}$ $GA_3$ solution was used to wet the substrate.

| | $A$ | $Gt_{50}$ | $k/2$ |
|---|---|---|---|
| Seed Lot (L) | | | |
| 2019 | 78.0 | 28.1 | 0.077 |
| 2020 | 77.7 | 26.1 | 0.080 |
| Exposure to Lighting (EL) | | | |
| White | 80.3 | 29.6 | 0.073 |
| Red | 76.3 | 27.7 | 0.083 |
| Blue | 75.7 | 28.5 | 0.085 |
| Red + Blue | 76.8 | 26.2 | 0.081 |
| Darkness | 80.1 | 23.4 | 0.073 |
| Analysis of variance | | | |
| Factors (degrees of freedom) | % Sum of squares | | |
| L (1) | 0.1 NS | 5.2 NS | 0.7 NS |
| EL (4) | 12.3 NS | 23.2 NS | 13.0 NS |
| L × EL (4) | 2.5 NS | 4.5 NS | 9.5 NS |
| Residual (30) | 85.2 | 67.1 | 76.7 |
| Standard deviation (+) | 5.9 | 4.2 | 0.01 |

NS: Not significant differences ($p \leq 0.05$) according to the LSD test. (+) The standard deviation has been calculated as the square root of the residual mean square.

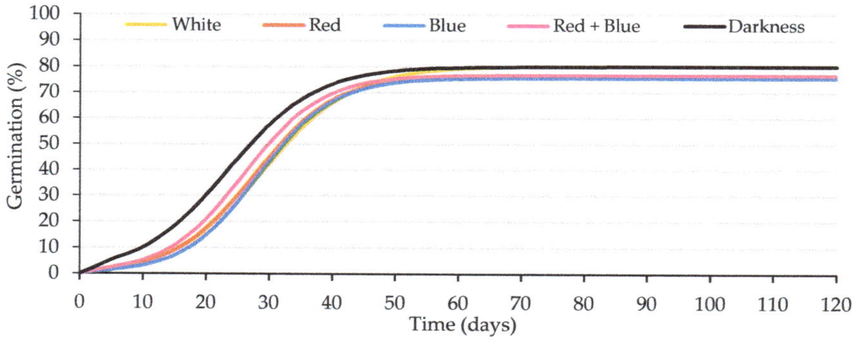


**Figure 2.** Logistic model adjusted to the curves of accumulated germination of caper seeds from Experiment 2. Average values for the seed exposure to white, red, blue or red + blue lights and to darkness. A 500 mg $L^{-1}$ $GA_3$ solution was used to wet the substrate.

No significant differences were obtained between the *A* values obtained in white light (80.3%) and in darkness (80.1%), indicating that the caper seeds germinated equally in white light and darkness, which was corroborated by the values of the Photoblastic Index (PI = −0.001) and Relative Light Germination (RLG = 0.5). These results confirmed those obtained in the previous experiment, that caper seeds are insensitive to white light; thus, nurseries could save the energy needed to illuminate the seeds during germination.

Caper seeds were also insensitive to blue light, even though all plants had cryptochromes [16], photoreceptors of blue light (400–500 nm; [13]). They were also insensitive to red light, even though phytochromes are photoreceptors of red light (600–700 nm).

Neither Marín [54], nor Moreno et al. [55], obtained significant differences ($p \leq 0.05$) when applying white, red, and blue light in the Serrano variety of pepper (*Capssicumm annuum* L.) nor orchids *Encyclia*, respectively. Paniagua et al. [27] and Cho et al. [26] analyzed the effect of LED lights of different wavelengths on broccoli (*Brassica oleracea* Plenck var. *italica*), with none of them obtaining any statistical difference ($p \leq 0.05$) in the final germination percentage. Aguado and Álvarez [25] obtained no differences in the germination and emergence of lettuce (*Lactuca sativa* L.), basil (*Ocimum basilicum* L.), and tomato (*Solanum lycopersicum* L.) seedlings exposed to LED lights with spectra differing on the proportions of red and blue lights.

However, Enache and Livadariu [29], with the use of red LED lights in *Artemisia dracunculus* L.. obtained a 10% greater germination percentage than with white light, which was, in turn, higher than those obtained under blue and green LEDs. However, the authors did not statistically compare these results.

### 3.3. Experiment 3

The viability of the 2021 lot was 85 ± 4%, according to what was obtained for the 2019 and 2020 lots. The germination obtained in the seeds wetted with water was very low, on average 6.5% (6.2% in irradiated and 8% in unirradiated seeds). The low germination data did not adjust to the logistic model; thus, they were not included in the analysis of variance, as in Experiments 1 and 2.

Germination data for seeds wetted with the $GA_3$ solution was fitted ($p \leq 0.01$) to the logistic function, presenting coefficients of determination for the 20 curves greater than 98.7%. Figure 3 shows the cumulative germination curves fitted to the logistic model obtained for the average values of each irradiation duration, the final germination percentages ranging between 69% and 82%. The tested laser irradiation durations (0, 1, 15, 60 and 180 s) did not significantly affect ($p \leq 0.05$) *A* or $k/2$, but an increase in $Gt_{50}$ was observed in seeds irradiated for 180 s in relation to the control seeds (7 days delay; Table 4) and to those irradiated for 1 and 15 s (up to 10 days delay).

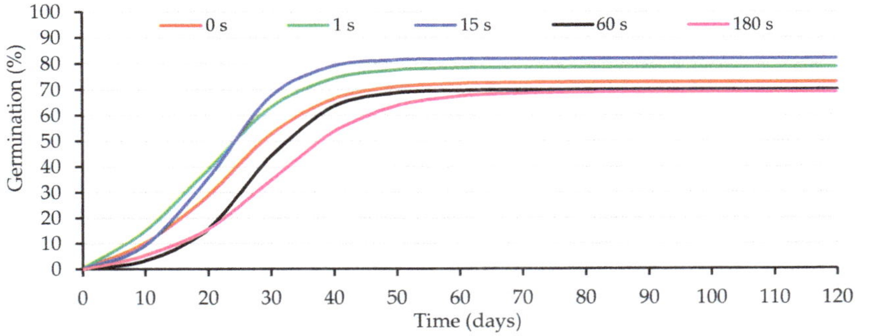


**Figure 3.** Logistic model adjusted to the curves of accumulated germination of caper seeds from Experiment 3. Average values of the He-Ne laser irradiation during 0, 1, 15, 60 and 180 s. A 500 mg $L^{-1}$ $GA_3$ solution was used to wet the substrate.

**Table 4.** Effect of the He-Ne laser irradiation during 0, 1, 15, 60 and 180 s on the germination parameters: final germination percentage (*A*, %), time required to reach 50% of final germination ($Gt_{50}$; d), and average germination rate ($k/2$; $d^{-1}$); average values from Experiment 3. A 500 mg $L^{-1}$ $GA_3$ solution was used to wet the substrate.

| | *A* | $Gt_{50}$ | *k*/2 |
|---|---|---|---|
| Time (T) | | | |
| 0 s | 72.81 | 23.05 bc | 0.069 |
| 1 s | 78.69 | 20.43 c | 0.071 |
| 15 s | 81.96 | 22.37 c | 0.089 |
| 60 s | 69.86 | 27.41 ab | 0.088 |
| 180 s | 69.10 | 30.68 a | 0.061 |
| | Analysis of variance | | |
| Factors (degrees of freedom) | % Sum of squares | | |
| T (4) | 35.58 NS | 66.05 ** | 26.31 NS |
| Residual (15) | 64.42 | 33.95 | 73.69 |
| Standard deviation (+) | 7.82 | 3.08 | 0.021 |

Different letters in the same column within each factor indicate significant differences ($p \leq 0.05$) according to the LSD test. **: significance level $p \leq 0.01$; NS: Not significant. (+) The standard deviation has been calculated as the square root of the residual mean square.

The results coincided with those that Juan, [50], obtained with the same type of laser used, in the sense that laser irradiation did not improve the germination of caper seeds. Similar results were obtained by Álvarez et al. [56] in tomato seeds irradiated with a He-Ne laser, which did not improve the germination percentages.

On the other hand, other studies on many species found that irradiation with laser light improved germination parameters, including wheat (*Triticum aestivum* L. [36]), lupine (*Lupinus albus* L.) and bean (*Vicia faba* L.) [38], radish (*Raphanus sativus* L. [57]), soybean (*Glycine max* L. [58]), safflower (*Carthamus tinctorius* L. [59]), sunflower (*Helianthus annuus* L. [37]) and Chinese woad (*Isatis indigotica* L. [34]). The laser light can affect the thermodynamic parameters of the seeds by increasing their internal energy, affecting the enzymatic activity (mainly of amylases, proteases and glucosidases), which may positively influence the germination [36]. Another He-Ne laser effect is the acceleration of seed metabolism through increased levels of germination-promoting hormones, such as $GA_3$, and decreased inhibitors, such as abscisic acid, as stated by Soliman and Harith [39] in *Acacia farnesiana* L. and by Swathy et al. [33] in eggplant (*Solanum melongena* L.).

The success of bio-stimulation caused by monochromatic laser light depends on the wavelength, irradiation duration, power, dose and method (constant or pulse), but also on the seed physiological properties and even the seed position during the laser irradiation [31]. Krawiec et al. [32] related the greatest response of seeds irradiated with a laser beam to the fact that these seeds had been previously soaked in water; thus, it was decided to analyze the effect of soaking the seeds in water before their irradiation, as presented in Experiment 4.

### 3.4. Experiment 4

Germination data were fitted ($p \leq 0.01$) to the logistic function, presenting coefficients of determination for the 24 curves greater than 98.1%. Figure 4 shows the germination curves adjusted to the logistic model in which germination was higher in seeds irradiated when they had been previously soaked in water compared to dry seeds and control. These differences ($p \leq 0.05$) are shown in Table 5, where it can also be seen that germination was affected ($p \leq 0.01$) by the irradiation duration. Soaking the seeds in water before irradiation represented 48% of the variation in the data, while the irradiation duration represented 26% of the variability. Figure 5 presents the significant interaction ($p \leq 0.01$) between the irradiation duration and the seed soaking before the irradiation. Germination percentages were comparable to those obtained in Experiment 3 with the same seed lot; the difference between the germination values obtained in this and the previous experiment for non-irradiated seeds, did not exceed the tolerance level established by ISTA Rules [45]. It can be seen that soaking the seeds before laser irradiation significantly increased ($p \leq 0.05$) the germination percentage, germinating all viable seeds. Neither $Gt_{50}$ nor $k/2$ was affected by the analyzed ($p \leq 0.05$) factors.

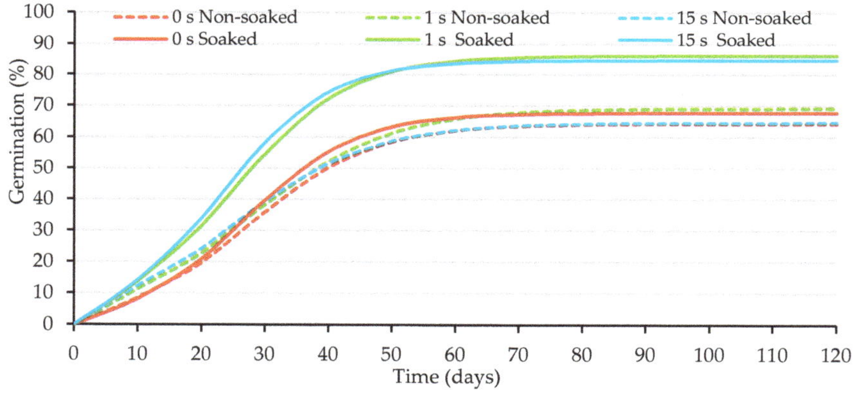


**Figure 4.** Logistic model adjusted to the curves of accumulated germination of caper seeds from Experiment 4. Average values corresponding to the combination of the exposure time of the seeds to the He-Ne laser irradiation (0, 1 or 15 s) after being soaked in water or not. A 500 mg $L^{-1}$ $GA_3$ solution was used to wet the substrate.

**Table 5.** Effect of He-Ne laser irradiation during 0, 1 and 15 s applied to dry and soaking the seeds in water, for 4 days, on the germination parameters: final germination percentage (*A*, %), time required to reach 50% of final germination ($Gt_{50}$; d), and average germination rate ($k/2$; $d^{-1}$); average values from Experiment 4. A 500 mg $L^{-1}$ $GA_3$ solution was used to wet the substrate.

| | *A* | $Gt_{50}$ | *k*/2 |
|---|---|---|---|
| Time (T) | | | |
| 0 s | 66.0 b | 23.8 | 0.055 |
| 1 s | 77.6 a | 22.9 | 0.051 |
| 15 s | 74.6 a | 21.8 | 0.054 |
| Soaking (S) | | | |
| Soaked seeds | 79.4 a | 21.6 | 0.058 |
| Dry seeds | 65.9 b | 24.1 | 0.049 |
| | Analysis of variance | | |
| Factors (degrees of freedom) | % Sum of squares | | |
| T (2) | 25.6 ** | 3.4 NS | 1.9 NS |
| S (1) | 48.2 ** | 8.2 NS | 11.3 NS |
| T × S (2) | 13.8 ** | 3.8 NS | 1.3 NS |
| Residual (18) | 12.4 | 84.6 | 85.5 |
| Standard deviation (+) | 4.0 | 4.6 | 0.014 |

Different letters in the same column within each factor indicate significant differences ($p \leq 0.05$) according to the LSD test. **: significance level $p \leq 0.01$, NS: Not significant. (+) The standard deviation has been calculated as the square root of the residual mean square.

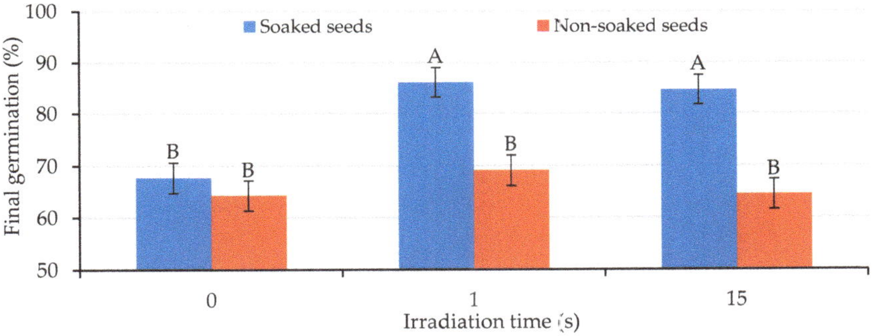


**Figure 5.** Analysis of the significant interactions of the analysis of variance in Table 5 between irradiation time and seed soaking prior to irradiation on the final germination. Average values of four replicates. Different letters indicate significant differences according to the LSD test. Error bars represent the LSD ($p \leq 0.05$).

As already mentioned in Experiment 3, the laser light can affect the thermodynamic parameters of the seeds [36]. It can also increase seed metabolism through increased levels of germination-promoting hormones ($GA_3$) and decreased inhibitors (abscisic acid), positively affecting germination [33]. In this experiment, irradiating the seeds once the germination metabolism had started, due to the prior seed soaking, improved the effectiveness of the laser irradiation in relation to that obtained in Experiment 3, resulting in significant differences ($p \leq 0.05$). This was in accordance with that reported by Perveen et al. [37,59] for sunflower and safflower, respectively, who obtained good germination results irradiating seeds that had been previously soaked in water with a He-Ne laser.

Future research will focus on studying the photoreceptors (phytochromes, cryptochromes and phototropins) present in caper seeds, as well as on analyzing the enzymatic activities and the levels of germination promoting and inhibiting hormones in seed irradiated with laser light.

## 4. Conclusions

Caper seeds are insensitive to exposure to white, red, blue, and red + blue lights during the germination process, not showing differences between the germination response in lightness in relation to darkness. Thus, germination can be carried out in lightness or darkness, and, therefore, germination in nurseries could be carried out in darkness, leading to important energy savings. Caper seed irradiation with a He-Ne laser during short exposure times improved the germination percentage when the seeds had been previously soaked in water, germinating all viable seeds. However, applying a solution of gibberellic acid was always required in all the cases studied.

**Author Contributions:** Conceptualization, M.J. and B.P.; methodology, M.L.F.; software, M.L.F.; validation, M.L.F., M.J., B.P. and N.P.-S.; formal analysis, M.L.F., M.J., B.P. and N.P.-S.; investigation, M.L.F. and N.P.-S.; resources, M.J.; data curation, M.L.F., B.P. and N.P.-S.; writing—original draft preparation, M.L.F., B.P. and N.P.-S.; writing—review and editing, M.L.F., M.J., B.P. and N.P.-S.; visualization, M.L.F., M.J., B.P. and N.P.-S.; supervision, B.P. and N.P.-S. All authors have read and agreed to the published version of the manuscript.

**Funding:** This research received no external funding.

**Data Availability Statement:** Not applicable.

**Acknowledgments:** Authors thank Vicente Ferrando, from the Applied Physics Department (UPV), for the technical support in the measurements with the spectrometer for the wavelength of the lamps used in this research.

**Conflicts of Interest:** The authors declare no conflict of interest.

## References


1. Sottile, F.; Caltagirone, C.; Peano, C.; Del Signore, M.B.; Barone, E. Can the caper (*Capparis spinosa* L.) still be considered a difficult-to-propagate crop? *Horticulturae* **2021**, *7*, 316. [CrossRef]
2. Rhizopoulou, S.; Ioannidi, E.; Alexandredes, N.; Argiropoulos, A. A study on functional and structural traits of the nocturnal flowers of *Capparis spinosa* L. *J. Arid Environ.* **2006**, *66*, 635–647. [CrossRef]
3. Shahrajabian, M.H.; Sun, W.; Cheng, Q. Plant of the millennium, caper (*Capparis spinosa* L.), chemical composition and medicinal uses. *Bull. Natl. Res. Cent.* **2021**, *45*, 131. [CrossRef]
4. Rivera, D.; Inocencio, C.; Obón, C.; Alcaraz, F. Review of food and medicinal uses of *Capparis* L. subgenus *Capparis* (Capparidaceae). *Econ. Bot.* **2003**, *57*, 515–534. [CrossRef]
5. Fici, S. A taxonomic revision of the *Capparis spinosa* group (Capparaceae) from the Mediterranean to Central Asia. *Phytotaxa* **2014**, *174*, 1–24. [CrossRef]
6. Pascual, B.; San Bautista, A.; Pascual-Seva, N.; García-Molina, R.; López Galarza, S.; Maroto, J.V. Estudio del enraizamiento de estaquillas de madera suave en alcaparra (*Capparis spinosa* L.). *Actas Hortic.* **2008**, *50*, 89–93.
7. Foschi, M.L.; Juan, M.; Pascual, B.; Pascual-Seva, N. The imbibition, viability, and germination of caper seeds (*Capparis spinosa* L.) in the first year of storage. *Plants* **2022**, *11*, 202. [CrossRef]
8. Foschi, M.L.; Juan, M.; Pascual, B.; Pascual-Seva, N. Water uptake and germination of caper (*Capparis spinosa* L.) seeds. *Agronomy* **2020**, *10*, 838. [CrossRef]
9. Foschi, M.L.; Juan, M.; Pascual, B.; Pascual-Seva, N. Collection guidelines to achieve a viable caper commercial propagation. *Agronomy* **2022**, *12*, 74. [CrossRef]
10. Takano, M.; Inagaki, N.; Xie, X.; Kiyota, S.; Baba-Kasai, A.; Tanabata, T.; Shinomura, T. Phytochromes are the sole photoreceptors for perceiving red/far-red light in rice. *Proc. Natl. Acad. Sci. USA* **2009**, *106*, 14705–14710. [CrossRef]
11. Bewley, J.D.; Bradford, K.J.; Hilhorst, H.W.M.; Nonogaki, H. *Seeds: Physiology of Development, Germination and Dormancy*, 3rd ed.; Springer: New York, NY, USA, 2013; ISBN 978-1-4614-4692-7.
12. Meisel, L.A.; Urbina, D.C.; Pinto, M.E. Fotorreceptores y respuestas de plantas a señales lumínicas. *Fisiol. Veg.* **2011**, *18*, 1–10.
13. Casal, J.J. Fotomorfogénesis: La luz como factor regulador del crecimiento. In *Fundamentos de Fisiología Vegetal*; Azcón-Bieto, J., Talón, M., Eds.; McGraw-Hill–Interamericana, S.L.: Madrid, Spain; Publicacions i Edicions de la Universitat De Barcelona: Barcelona, Spain, 2013; pp. 467–481. ISBN 978-84-481-5168-3.
14. Butler, W.L.; Norris, K.H.; Siegelman, H.W.; Hendricks, S.B. Detection, assay, and preliminary purification of the pigment controlling photoresponsive development of plants. *Proc. Natl. Acad. Sci. USA* **1959**, *45*, 1703–1708. [CrossRef]
15. Sharrock, R.A.; Quail, P.H. Novel phytochrome sequences in *Arabidopsis thaliana*: Structure, evolution, and differential expression of a plant regulatory photoreceptor family. *Genes Dev.* **1989**, *3*, 1745–1757. [CrossRef]
16. Wang, Q.; Lin, C. Mechanisms of cryptochrome-mediated photoresponses in plants. *Annu. Rev. Plant Biol.* **2020**, *71*, 103–129. [CrossRef]

17. Orozco-Segovia, A. Fisiología y ecología del fitocromo: Su función en las semillas. *Bot. Sci.* **1989**, *49*, 71–84. [CrossRef]
18. Castillo, M.L.C.; Bustamante, R.O.; Peña-Gómez, F.T.; Gutiérrez, V.L.; Reyes, C.A.; Arredondo-Núñez, A.; Marey, M. Negative photoblastism in the invasive species *Eschscholzia californica* Cham. (Papaveraceae): Patterns of altitudinal variation in native and invasive range. *Gayana Bot.* **2013**, *70*, 331–337. [CrossRef]
19. Milberg, P.; Andersson, L.; Thompson, K. Large-seeded spices are less dependent on light for germination than small-seeded ones. *Seed Sci. Res.* **2000**, *10*, 99–104. [CrossRef]
20. Takaki, M. New proposal of classification of seeds based on forms of phytochrome instead of photoblastism. *Rev. Bras. Fisiol. Veg.* **2001**, *13*, 104–108. [CrossRef]
21. Baskin, C.C.; Baskin, J.M. *Seeds: Ecology, Biogeography, and Evolution of Dormancy and Germination*, 2nd ed.; Academic Press: San Diego, CA, USA, 2014; ISBN 978-0-12-416683-7.
22. Mathews, S. Phytochrome-mediated development in land plants: Red light sensing evolves to meet the challenges of changing light environments. *Mol. Ecol.* **2006**, *15*, 3483–3503. [CrossRef]
23. Lejeune, P.; Fratamico, A.; Bouché, F.; Huerga-Fernández, S.; Tocquin, P.; Périlleux, C. LED color gradient as a new screening tool for rapid phenotyping of plant responses to light quality. *Gigascience* **2022**, *11*, 1–13. [CrossRef]
24. He, D.; Kozai, T.; Niu, G.; Zhang, X. Light-Emitting Diodes for horticulture. In *Light-Emitting Diodes, Solid State Lighting Technology and Application*; Li, J., Zhang, G.Q., Eds.; Springer International Publishing AG, Part of Springer Nature: Cham, Germany, 2019; pp. 513–547. ISBN 978-3-319-99211-2.
25. Aguado, G.D.; Álvarez, M. INTA Mendoza Realiza Ensayos Sobre los Efectos de Iluminación LED en Plantines Hortícolas. Available online: https://inta.gob.ar/noticias/inta-mendoza-realiza-ensayos-sobre-los-efectos-de-iluminacion-led-en-plantines-horticolas-0 (accessed on 6 September 2022).
26. Cho, J.Y.; Son, D.M.; Kim, J.M.; Seo, B.S.; Yang, S.Y.; Bae, J.H.; Heo, B.G. Effect of LED as light quality on the germination, growth and physiological activities of broccoli sprouts. *J. Bio-Environ. Control* **2008**, *17*, 116–123.
27. Paniagua, G.; Hernández, C.; Rico, F.; Domínguez, F.A.; Martínez, E.; Martínez, C.L. Effect of high intensity LED light on the germination and growth of broccoli seedlings (*Brassica oleracea* L.). *Polibotánica* **2015**, *40*, 199–212. [CrossRef]
28. Paradiso, R.; Proietti, S. Light-quality manipulation to control plant growth and photomorphogenesis in greenhouse horticulture: The state of the art and the opportunities of modern LED systems. *J. Plant Growth Regul.* **2022**, *41*, 742–780. [CrossRef]
29. Enache, I.M.; Livadariu, O. Preliminary results regarding the testing of treatments with Light-Emitting Diode (LED) on the seed germination of *Artemisia dracunculus* L. *Sci. Bull. Ser. F Biotechnol.* **2016**, *20*, 51–56.
30. Tang, D.S.; Hamayun, M.; Khan, A.L.; Jan, S.; Nawaz, Y.; Irshad, M.; Na, Y.E.; Lee, I.J. Exposure to red light, temperature and exogenous gibberellins influenced germination of some winter weeds. *Afr. J. Biotechnol.* **2012**, *11*, 273–279. [CrossRef]
31. Hernandez, A.C.; Dominguez, P.A.; Cruz, O.A.; Ivanov, R.; Carballo, C.A.; Zepeda, B.R. Laser in agriculture. *Int. Agrophysics* **2010**, *24*, 407–422.
32. Krawiec, M.; Dziwulska-Hunek, A.; Kornarzyński, K. The use of physical factors for seed quality improvement of horticultural plants. *J. Hortic. Res.* **2018**, *26*, 81–94. [CrossRef]
33. Swathy, P.S.; Kiran, K.R.; Joshi, M.B.; Mahato, K.K.; Muthusamy, A. He–Ne laser accelerates seed germination by modulating growth hormones and reprogramming metabolism in brinjal. *Sci. Rep.* **2021**, *11*, 7948. [CrossRef]
34. Chen, Y.P.; Liu, Y.J.; Wang, X.L.; Ren, Z.Y.; Yue, M. Effect of microwave and He-Ne laser on enzyme activity and biophoton emission of *Isatis indigotica* Fort. *J. Integr. Plant Biol.* **2005**, *47*, 849–855. [CrossRef]
35. Chen, Y.P.; Yue, M.; Wang, X.-L. Influence of He–Ne laser irradiation on seeds thermodynamic parameters and seedlings growth of *Isatis indogotica*. *Plant Sci.* **2005**, *168*, 601–606. [CrossRef]
36. Jamil, Y.; Perveen, R.; Ashraf, M.; Ali, Q.; Iqbal, M.; Ahmad, M.R. He–Ne laser-induced changes in germination, thermodynamic parameters, internal energy, enzyme activities and physiological attributes of wheat during germination and early growth. *Laser Phys. Lett.* **2013**, *10*, 045606. [CrossRef]
37. Perveen, R.; Ali, Q.; Ashraf, M.; Al-Qurainy, F.; Jamil, Y.; Raza Ahmad, M. Effects of different doses of low power continuous wave He-Ne laser radiation on some seed thermodynamic and germination parameters, and potential enzymes involved in seed germination of sunflower (*Helianthus annuus* L.). *Photochem. Photobiol.* **2010**, *86*, 1050–1055. [CrossRef] [PubMed]
38. Podleśny, J.; Stochmal, A.; Podleśna, A.; Misiak, L.E. Effect of laser light treatment on some biochemical and physiological processes in seeds and seedlings of white lupine and faba bean. *Plant Growth Regul.* **2012**, *67*, 227–233. [CrossRef]
39. Soliman, A.S.; Harith, M.A. Effects of laser biostimulation on germination of *Acacia farnesiana* (L.) Willd. *Acta Hortic.* **2009**, *854*, 41–50. [CrossRef]
40. Demotes-Mainard, S.; Péron, T.; Corot, A.; Bertheloot, J.; Le Gourrierec, J.; Pelleschi-Travier, S.; Crespel, L.; Morel, P.; Huché-Thélier, L.; Boumaza, R.; et al. Plant responses to red and far-red lights, applications in horticulture. *Environ. Exp. Bot.* **2016**, *121*, 4–21. [CrossRef]
41. Finch-Savage, W.E.; Leubner-Metzger, G. Seed dormancy and the control of germination. *New Phytol.* **2006**, *171*, 501–523. [CrossRef]
42. Davies, F.T.; Geneve, R.L.; Wilson, S.B. *Hartmann and Kester's Plant Propagation Principles and Practices*, 9th ed.; Pearson Education: New York, NY, USA, 2018; ISBN 13: 978-0-13-448089-3.
43. Pascual, B.; San Bautista, A.; López-Galarza, S.; Alagarda, J.; Maroto, J.V. Germination behavior after storage of caper seeds. *Seed Sci. Technol.* **2006**, *34*, 151–159. [CrossRef]

44. ISTA. *ISTA Working Sheets on Tetrazolium Testing. Tree and Shrub Species*, 1st ed.; The International Seed Testing Association (ISTA): Bassersdorf, Switzerland, 2003; Volume II, ISBN 3-906549-41-1.
45. ISTA. *International Seed Testing Association (ISTA)*; ISTA Rules; ISTA: Bassersdorf, Switzerland, 2018; ISBN 3906549275. ISSN 0251-3952.
46. Pascual, B.; San Bautista, A.; Ferreros, N.; López-Galarza, S.; Maroto, J.V. Analysis of germination of caper seeds as influenced by the position of fruit on the mother plant, fruit maturation stage and fruit weight. *J. Hortic. Sci. Biotechnol.* **2003**, *78*, 73–78. [CrossRef]
47. Torres, M.; Frutos, G. Logistic function analysis of germination behaviour of aged fennel seeds. *Environ. Exp. Bot.* **1990**, *30*, 383–390. [CrossRef]
48. Causton, D.R.; Venus, J.C. Single leaf growth and the Richards function: Methodology. In *The Biometry of Plant Growth*; Causton, D.R., Venus, J.C., Eds.; Edward Arnold: London, UK, 1981; pp. 87–143. ISBN 978-0-7131-2812-3.
49. Rasband, W.; Ferreira, T. *ImageJ. Programa de Procesamiento de Imágenes Diseñado para Imágenes Científicas*; National Institutes of Health: Bethesda, MD, USA, 2016.
50. Juan, M. Estudio para la Mejora de las Técnicas de Propagación de la Alcaparra (*Capparis spinosa* L.). Ph.D. Thesis, Universitat Politècnica de València, València, España, 2017.
51. *Statgraphics Statistical Graphics Statgraphics Centurion XVIII*; Statistical Graphics Corporation: Rockville, MD, USA, 2018.
52. Pascual, B.; San Bautista, A.; Pascual Seva, N.; García Molina, R.; López-Galarza, S.; Maroto, J.V. Effects of soaking period and gibberellic acid addition on caper seed germination. *Seed Sci. Technol.* **2009**, *37*, 33–41. [CrossRef]
53. Germanà, M.A.; Chiancone, B. In vitro germination and seedling development of caper (*Capparis spinosa* L.) mature seeds. *Acta Hortic.* **2009**, *839*, 181–186. [CrossRef]
54. Marín, L. Estudio de la Influencia de Diferentes Longitudes de onda de luz LED en la Germinación de una Orquídea *Encyclia* sp. Bachelor's Thesis, Universidad Federal de Integración Latino-Americana, Foz do Iguaçu, Brasil, 2016.
55. Moreno, A.; Loza, S.; Ortiz, M. Efecto de luz LED sobre semillas de *Capsicum annuum* L. var. serrano. *Biotecnol. Veg.* **2017**, *17*, 145–151.
56. Álvarez, A.; Ramírez, R.; Chávez, L.; Camejo, Y.; Licea, L.; Porras, E.; García, B. Efecto del tratamiento de semillas con láser de baja potencia, sobre el crecimiento y rendimiento en plantas de tomate (*Solanum lycopersicum* L.). *ITEA Inf. Tec. Econ. Agrar.* **2011**, *107*, 290–299.
57. Krawiec, M.; Dziwulska-Hunek, A.; Kornarzyński, K.; Palonka, S. Wpływ wybranych czynników fizycznych na kiełkowanie nasion rzodkiewki (*Raphanus sativus* L.). *Acta Agrophysica* **2012**, *19*, 737–748.
58. Klimek-Kopyra, A.; Dłużniewska, J.; Ślizowska, A.; Dobrowolski, J.W. Impact of coherent laser irradiation on germination and mycoflora of soybean seeds—Innovative and prospective seed quality management. *Agriculture* **2020**, *10*, 314. [CrossRef]
59. Perveen, R.; Wang, X.; Jamil, Y.; Ali, Q.; Ali, S.; Zakaria, M.Q.; Afzaal, M.; Kasana, R.A.; Saleem, M.H.; Fiaz, S. Quantitative determination of the effects of He–Ne laser irradiation on seed thermodynamics, germination attributes and metabolites of safflower (*Carthamus tinctorius* L.) in relation with the activities of germination enzymes. *Agronomy* **2021**, *11*, 1411. [CrossRef]

*agriculture* MDPI

*Article*

# The Combinations of White, Blue, and UV-A Light Provided by Supplementary Light-Emitting Diodes Promoted the Quality of Greenhouse-Grown Cucumber Seedlings

**Zhengnan Yan [1,†], Chunling Wang [2,†], Long Wang [1], Xin Li [1], Guanjie Wang [3] and Yanjie Yang [1,*]**

[1] College of Horticulture, Qingdao Agricultural University, Qingdao 266109, China
[2] College of Water Resources and Architecture Engineering, Tarim University, Aral 843300, China
[3] Shandong Qicai Manor Vegetable Food Base Co., Ltd., Shouguang 262700, China
* Correspondence: 200401170@qau.edu.cn
† These authors contributed equally to this work.

**Citation:** Yan, Z.; Wang, C.; Wang, L.; Li, X.; Wang, G.; Yang, Y. The Combinations of White, Blue, and UV-A Light Provided by Supplementary Light-Emitting Diodes Promoted the Quality of Greenhouse-Grown Cucumber Seedlings. *Agriculture* **2022**, *12*, 1593. https://doi.org/10.3390/agriculture12101593

Academic Editors: Athanasios Koukounaras and Filippos Bantis

Received: 12 September 2022
Accepted: 29 September 2022
Published: 2 October 2022



**Abstract:** Insufficient solar light in winter inside the greenhouse may lead to a lower quality of vegetable seedlings, and supplemental light is an effective technique to solve this problem. This study evaluated the impacts of supplementary white (W)-light-emitting diodes (LEDs), ultraviolet A LEDs (UV-A), white and blue LEDs (WB), the combinations of white and UV-A LEDs (W-UVA), and white, blue, and UV-A LEDs (WB-UVA) on the leaf morphology, photosynthetic traits, biomass accumulation, root architecture, and hormone content of cucumber (*Cucumis sativus* L. cv. Tianjiao No. 5) seedlings grown in the greenhouse. The results indicated that supplementary LED lighting led to a decreased plant height, shorter hypocotyl length, bigger leaf area, and thicker leaf compared with those grown with solar light only, regardless of light quality. The shoot fresh weight, root fresh weight, and seedling quality index of cucumber seedlings grown under the combinations of white, blue, and UVA radiations increased by 30.8%, 3.2-fold, and 1.8-fold, respectively, compared with those grown with natural light only. However, no significant differences were exhibited in the biomass accumulation of greenhouse-grown cucumber seedlings between the control and the UVA treatment. The cellulose content and stem firmness of greenhouse-grown cucumber seedlings grown under the combinations of white, blue, and UVA radiations increased by 49.9% and 13.1%, respectively, compared with those grown under white light only. Additionally, the cytokinin content of cucumber seedlings was promoted by over 36.7% by applying supplementary light. In summary, the combinations of white, blue, and UVA radiations led to compact morphological characteristics, superior mechanical properties, and preferable growth performance, which could be applied as an available lighting strategy to obtain the desired morphological and quality properties of vegetable seedlings.

**Keywords:** cytokinin content; supplementary light; net photosynthetic rate; ultraviolet; stem firmness

## 1. Introduction

Cucumbers (*Cucumis sativus* L.) are widely cultivated worldwide as an important vegetable variety with a world production of 91.2 million tons in 2021 [1]. Prior studies have indicated that the quality of vegetable seedlings influenced the subsequent growth and yield of mature plants at harvest [2,3], and the annual demand of cucumber seedlings in China is 47 billion plants [4]. Therefore, producers focus on various environmental factors that affect the yield and quality of cucumber seedlings during the production process. In addition, the growing conditions in the protected horticulture (e.g., greenhouse or plant factory with artificial lighting) are superior compared with the open field as a result of the fact that growers could adjust the environmental elements based on the crop needs, which was beneficial for cultivation of high-quality vegetable seedlings.

Light is one of the most important variables for plant growth and development, and there are three dimensions of light to be noticed: light quality, light intensity, and photoperiod [5]. Previous studies have indicated that lower light intensity was not conducive to plant growth and development. For instance, Pennisi et al. [6] found that lower light intensity reduced the leaf functionality of lettuce, which resulted in reduced nutritional content, and lower antioxidant capacity, phenolics, and flavonoids concentrations. Similar results were also observed in sweet basil [7], broccoli microgreens [8], and dwarf tomato [9]. Recently, more researchers paid attention to the daily light integral (DLI) and observed that increasing the DLI within limits could improve the growth status of plants and promote the accumulation of nutrients in plants [10–12]. Additionally, increasing the DLI in seasons with insufficient light could increase the stem firmness of cucumber seedlings, which was beneficial for mechanized transplanting [13].

Light quality also has significant impacts on the formation and composition of plant organic matter [14,15]. For instance, red light is efficient in driving plant photosynthesis compared with other wavelengths [16]; blue light promoted the synthesis of chlorophyll in cucumber, wheat, and spinach [17,18]. Many researchers have reported that red plus blue LEDs are the most important parts of spectral regions for plant growth [19]. Therefore, red plus blue lights provided by light-emitting diodes (LEDs) were commonly applied or investigated by researchers in leafy vegetables [20,21], vegetable seedlings [22,23], and herbs [24,25] grown in the greenhouse or plant factory with artificial lighting. However, white LEDs exhibited similar or preferable influences on plant growth and energy use efficiency compared with red plus blue LEDs [26,27], and white LEDs created a "friendly" light environment to human eyes [9,28]. Thus, white LEDs or white LEDs combined with other wavelengths were investigated in lettuce [28], spinach [29], and grafted tomato transplants [30]. Generally, the hypocotyl of vegetable seedlings would elongate under low light conditions, and the reduction in blue light had a similar response in plants [31]. However, the circumstances of increased tomato plant compactness and reduced stem elongation occurred with the increase in blue light [18,32]. The suitable combinations of supplementary white and blue lights were investigated by Yan et al. [13] in greenhouse-grown cucumber seedlings and the results indicated that the plant height and hypocotyl length of cucumber seedlings decreased with the increased blue fraction of supplementary light; however, the stem diameter and leaf area of cucumber seedlings increased first and then decreased with the increased blue fraction, and similar trends were observed in the biomass accumulation of cucumber seedlings.

The impacts of ultraviolet (UV) radiation on the plant morphology and physiological characteristics of plants were investigated by researchers in recent years. The majority of the UV radiation part is UV-A radiation (315–400 nm), which accounts for 98–99% of the ultraviolet radiation reaching the Earth's surface [33,34]. The UV-A radiation affected the growth, photosynthesis, and specific substance contents of plants [34]. Zhang et al. [35] indicated that UV-A functions had similar effects in maintaining leaf photosynthetic function as compared with blue light. In addition, previous reports have shown that supplementing certain amounts of UV-A to visible radiation (400–700 nm) promoted plant biomass accumulation compared with those grown without supplements [36].

Our previous study investigated the proper proportion created by white and blue LEDs for growth of greenhouse-grown cucumber seedlings [37]; nevertheless, hardly any studies have reported the influences of supplementary white or white plus blue LEDs combined with UV-A on the leaf morphology, pigment content, photosynthetic traits, biomass accumulation, root architecture, and hormone contents of cucumber seedlings grown in the greenhouse. The results could provide guidelines for the regulation of supplemental light strategies for greenhouse-grown cucumber seedlings produced in winter, early spring, or seasons with insufficient sunlight.

## 2. Materials and Methods

### 2.1. Plant Materials

Cucumber seedlings (*Cucumis sativus* L. cv. Tianjiao No. 5) were grown in 72-cell plug trays containing a mixture of 60% vermiculite, 20% peat, and 20% perlite. One seed was used per cell, plug trays were placed in a Venlo-type greenhouse with a floor area of 2736 $m^2$ in Qingdao Agricultural University, Qingdao, Shandong Province, China, at a temperature of (24 ± 1) °C/(16 ± 1) °C during the day/night period, and the relative humidity was maintained at 60–70%, for 22 days. Hoagland's nutrient solution was used for cucumber seedlings during the experimental period, and the management of seedlings was reported by our previous study [13].

### 2.2. Treatment Design

The average daily light intensity of sunlight inside the Venlo-type greenhouse was 128 μmol $m^{-2}$ $s^{-1}$ during the experimental period (19 November–11 December 2021), with an average DLI of 5.0 mol $m^{-2}$ $d^{-1}$. Considering the suitable DLI and supplementary duration [13] for growth of cucumber seedlings, they were grown under supplemental DLI at 6.5 mol $m^{-2}$ $d^{-1}$ with a light intensity and photoperiod at 180 μmol $m^{-2}$ $s^{-1}$ and 10 h $d^{-1}$, respectively, created by white LEDs (W) (Weifang Hengxin Electric Appliance Co., Ltd., Weifang, China), ultraviolet A LEDs (UV-A) (Xiamen Lumigro Technology Co., Ltd., Xiamen, China), the combinations of white and blue LEDs (WB) (Weifang Hengxin Electric Appliance Co., Ltd., Weifang, China), the combinations of white and UV-A LEDs (W-UVA), and the combinations of white, blue, and UV-A LEDs (WB-UVA), respectively. The ratio of combinations of blue and white LEDs and the spectral distribution of the LEDs were applied based on our previous study [37] and the supplemental light intensity of UV-A light was 15 μmol $m^{-2}$ $s^{-1}$. Additionally, cucumber seedlings grown with natural light only was set as the control (DLI at 5.0 mol $m^{-2}$ $d^{-1}$). The experiment was arranged in a randomized complete block design with three replications, and 72 seedlings were applied in each replication in this experiment.

### 2.3. Growth Measurements

#### 2.3.1. Plant Morphology and Growth Traits

The plant height, hypocotyl length, and stem diameter of cucumber seedlings were measured using a ruler and digital caliper (Shanghai Tool Factory Co., Ltd., Shanghai, China). Fresh and dry weights, and the root architecture of cucumber seedlings were measured based on Yan et al. [13]. Seedling quality index and specific leaf area were calculated according to Han et al. [38] and Dou et al. [7].

#### 2.3.2. Determinations of Photosynthetic Performance

Photosynthetic performances of cucumber seedlings were determined by a portable photosynthetic measuring system (LI-6400XT, Li-Cor Inc., Lincoln, NE, USA) with a leaf chamber (6400-02B), according to Yan et al. [13]. The apparent mesophyll conductance ($g_m$) and stomatal limitation value ($L_s$) were calculated according to Wang et al. [20]. A chlorophyll meter (SPAD-502 Plus, Konica Minolta Inc., Tokyo, Japan) was applied to determine the relative chlorophyll contents of cucumber seedlings.

#### 2.3.3. Measurement of Root Activity, Stem Firmness, and Cellulose Content of Cucumber Seedling

The triphenyl tetrazolium chloride (TTC) method and the Updegraff method were applied to determine the root activity and cellulose content of cucumber seedlings, according to Li [39] and Updegraff [40], respectively. The stem firmness of cucumber stems was determined according to Yan et al. [13].

#### 2.3.4. Measurement of Hormone Content of Cucumber Seedlings

The fully expanded cucumber leaf was flash-frozen and ground in liquid nitrogen and then transferred to a freezer with −80 °C for storage. The cytokinin contents of cucumber seedlings were quantified by a competitive Enzyme-linked immunosorbent assay (ELISA) technique, according to Aguilar et al. [41].

2.3.5. Supplementary Light Use Efficiency

Supplementary light use efficiency was estimated based on the increase in fresh weight, according to Wei et al. [42] and Wang et al. [12].

### *2.4. Statistical Analysis*

Statistical analysis was conducted using the SPSS 18.0 software (IBM, Inc., Chicago, IL, USA). The least significant difference (LSD) test was performed across all treatments at $p < 0.05$. The data were exhibited as the mean ± standard deviation (SD) values. The heat map was performed using the TBtools (https://github.com/CJ-Chen/TBtools, accessed on 25 September 2022) based on Chen et al. [43]. We calculated the Euclidean distance among samples and the complete clustering method was applied according to Gao et al. [8].

## **3. Results**

### *3.1. Impacts of Supplementary Light on Morphological Performances of Greenhouse-Grown Cucumber Seedlings*

Morphological characteristics of greenhouse-grown cucumber seedlings were significantly impacted by supplementary light (Table 1). In general, cucumber seedlings grown with natural light only exhibited a higher plant height, longer hypocotyl, and smaller leaf area compared with those grown with supplementary light; however, no significant differences were found in stem diameter, leaf width, and leaf area of cucumber seedlings grown between the control and the UV-A treatments. Cucumber seedlings grown with combinations of supplementary white, blue, and UV-A light exhibited the shortest plant height and smallest specific leaf area, which decreased by 39.8% and 47.7%, respectively, compared with those grown with natural light only. The stem diameter and leaf area of greenhouse-grown cucumber seedlings grown in the WB-UVA treatment increased by 33.3% and 44.6% compared with those grown under the control treatment, respectively.

### *3.2. Impacts of Supplementary Light on Photosynthetic Performances of Cucumber Seedlings Cultivated in the Greenhouse*

The SPAD value, net photosynthetic rate, substomatal $CO_2$ concentration, transpiration rate, and apparent mesophyll conductance of cucumber seedlings cultivated with sunlight only were significantly lower than those cultivated with supplementary light (Table 2). However, no significant differences were found in these parameters of cucumber seedlings grown between WB and WB-UVA treatments. The SPAD value, net photosynthetic rate, and transpiration rate of cucumber seedlings grown in the WB-UVA treatment increased by 39.1%, about 1.2- and 2.7-fold compared with those cultivated without supplementary light, respectively.

### *3.3. Impacts of Supplementary Light on Growth Characteristics and Root Architecture on Greenhouse-Grown Cucumber Seedlings*

Fresh and dry weights of greenhouse-grown cucumber seedlings were significantly influenced by supplementary light (Table 3). Fresh and dry weights of cucumber seedlings were significantly increased by using white LEDs alone or the combinations of white and other LEDs. No significant differences were observed in these above parameters in cucumber seedlings grown between supplemented UVA light and the control. The shoot fresh weight, root fresh weight, and seedling quality index of cucumber seedlings grown in the WB-UVA treatment increased by 30.8%, about 3.2- and 1.8-fold, compared with those grown with natural light only, respectively.

**Table 1.** Morphological performances of greenhouse-grown cucumber seedlings cultivated under supplementary light provided by white (W), UV-A light-emitting diodes (LEDs), and various combinations of W, blue (B), and UV-A LEDs; cucumber seedlings grown without supplementary light were used as control.

| Treatments | Plant Height (cm) | Hypocotyl Length (cm) | Stem Diameter (mm) | Leaf Length (cm) | Leaf Width (cm) | Leaf Area ($cm^2$) | Specific Leaf Area ($cm^2$ $mg^{-1}$) |
|---|---|---|---|---|---|---|---|
| Control | 18.6 ± 2.2 a | 13.8 ± 1.4 a | 3.6 ± 0.4 b | 6.6 ± 0.6 c | 6.7 ± 0.6 c | 30.7 ± 3.1 c | 0.501 ± 0.035 a |
| W | 14.0 ± 1.0 bc | 8.5 ± 0.4 c | 4.7 ± 0.3 a | 7.4 ± 0.3 b | 7.9 ± 0.4 b | 38.8 ± 2.1 b | 0.305 ± 0.013 c |
| UVA | 15.8 ± 0.3 b | 12.1 ± 0.6 b | 3.7 ± 0.2 b | 7.3 ± 0.1 b | 6.7 ± 0.3 c | 33.5 ± 2.2 c | 0.415 ± 0.031 b |
| WB | 12.7 ± 1.0 cd | 8.6 ± 0.6 c | 4.8 ± 0.4 a | 7.7 ± 0.2 ab | 8.4 ± 0.3 a | 42.9 ± 3.6 ab | 0.271 ± 0.021 cd |
| W-UVA | 13.5 ± 1.3 c | 8.5 ± 0.7 c | 4.9 ± 0.2 a | 7.3 ± 0.4 b | 8.0 ± 0.4 ab | 39.4 ± 2.8 b | 0.278 ± 0.012 cd |
| WB-UVA | 11.2 ± 0.4 d | 7.1 ± 0.5 d | 4.8 ± 0.4 a | 8.0 ± 0.1 a | 8.4 ± 0.2 ab | 44.4 ± 2.4 a | 0.262 ± 0.013 d |

Note: Different letters within each parameter were significantly different tested by the least significant difference (LSD) test at $p < 0.05$.

**Table 2.** Photosynthetic characteristics of cucumber seedlings cultivated in the greenhouse under supplementary light provided by white (W), UV-A light-emitting diodes (LEDs), and various combinations of W, blue (B), and UV-A LEDs; cucumber seedlings cultivated without supplementary light were used as control.

| Treatments | SPAD Value | Net Photosynthetic Rate (μmol $m^{-2}$ $s^{-1}$) | Stomatal Conductance (mol $m^{-2}$ $s^{-1}$) | Substomatal $CO_2$ Concentration (μmol $mol^{-1}$) | Transpiration Rate (mmol $m^{-2}$ $s^{-1}$) | Apparent Mesophyll Conductance (mol $m^{-2}$ $s^{-1}$) | Stomatal Limitation Value |
|---|---|---|---|---|---|---|---|
| Control | 36.3 ± 2.1 d | 5.2 ± 0.2 d | 0.06 ± 0.01 d | 243 ± 24 b | 0.78 ± 0.09 d | 0.025 ± 0.004 d | 0.40 ± 0.06 a |
| W | 46.3 ± 1.8 b | 9.6 ± 0.6 b | 0.19 ± 0.02 b | 295 ± 8 a | 2.16 ± 0.04 b | 0.031 ± 0.002 c | 0.24 ± 0.02 c |
| UVA | 39.1 ± 3.5 c | 6.8 ± 0.3 c | 0.10 ± 0.01 c | 281 ± 25 a | 1.17 ± 0.15 c | 0.026 ± 0.002 d | 0.31 ± 0.03 b |
| WB | 50.5 ± 1.7 a | 10.8 ± 0.4 a | 0.24 ± 0.03 a | 300 ± 8 a | 2.65 ± 0.22 a | 0.036 ± 0.001 b | 0.24 ± 0.02 c |
| W-UVA | 47.9 ± 1.4 ab | 10.1 ± 0.6 b | 0.19 ± 0.03 b | 293 ± 14 a | 2.13 ± 0.32 b | 0.033 ± 0.002 bc | 0.24 ± 0.03 c |
| WB-UVA | 50.5 ± 1.5 a | 11.2 ± 0.8 a | 0.22 ± 0.02 ab | 295 ± 12 a | 2.39 ± 0.21 ab | 0.039 ± 0.004 a | 0.23 ± 0.02 c |

Note: Different letters within each parameter were significantly different tested by the least significant difference (LSD) test at $p < 0.05$.

**Table 3.** Fresh and dry weights of greenhouse-grown cucumber seedlings cultivated under supplementary light provided by white (W), UV-A light-emitting diodes (LEDs), and various combinations of W, blue (B), and UV-A LEDs; cucumber seedlings grown with natural light only were used as control.

| Treatments | Shoot Fresh Weight (g Per Plant) | Root Fresh Weight (g Per Plant) | Shoot Dry Weight (g Per Plant) | Root Dry Weight (g Per Plant) | Seedling Quality Index |
|---|---|---|---|---|---|
| Control | 2.92 ± 0.13 d | 0.45 ± 0.03 d | 0.198 ± 0.030 c | 0.016 ± 0.002 d | 0.025 ± 0.003 c |
| W | 3.49 ± 0.44 bc | 1.17 ± 0.09 c | 0.283 ± 0.020 b | 0.040 ± 0.005 bc | 0.055 ± 0.006 b |
| UVA | 3.05 ± 0.27 cd | 0.44 ± 0.01 d | 0.216 ± 0.026 c | 0.015 ± 0.001 d | 0.022 ± 0.002 c |
| WB | 4.09 ± 0.19 a | 1.42 ± 0.18 b | 0.316 ± 0.014 ab | 0.045 ± 0.006 b | 0.065 ± 0.009 ab |
| W-UVA | 4.13 ± 0.10 a | 1.09 ± 0.09 c | 0.339 ± 0.038 a | 0.037 ± 0.004 c | 0.057 ± 0.005 b |
| WB-UVA | 3.82 ± 0.45 ab | 1.90 ± 0.18 a | 0.348 ± 0.046 a | 0.059 ± 0.009 a | 0.069 ± 0.008 a |

Note: Different letters within each parameter were significantly different tested by the least significant difference (LSD) test at $p < 0.05$.

The root architecture of greenhouse-grown cucumber seedlings was remarkably influenced by supplementary light (Figure 1). The root length, root area, and root volume of cucumber seedlings grown with supplementary WB-UVA light increased by 184.9%, 219.0%, and 266.7% compared with those grown with natural light only, respectively. Similarly, no significant differences were found in these parameters in cucumber seedlings grown between the control and UVA treatment. The root activity of cucumber seedings exposed to WB and WB-UVA treatments was significantly higher compared with other treatments, which increased by 3.8 and 3.6 times compared with the control.

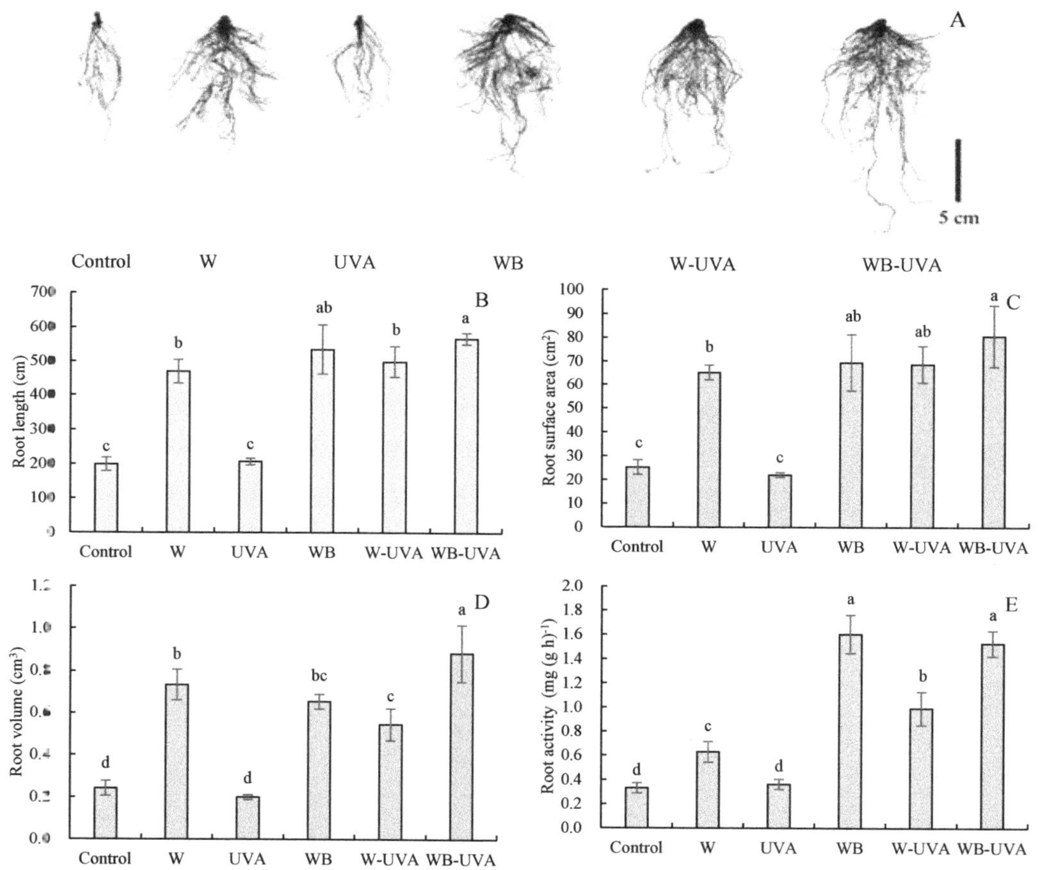


**Figure 1.** Effects of supplementary light provided by white (W), UV-A light-emitting diodes (LEDs), and various combinations of W, blue (B), and UV-A LEDs on root morphology (**A**), root length (**B**), root surface area (**C**), root volume (**D**), and root activity (**E**) of greenhouse-grown cucumber seedlings cultivated for 22 days after sowing; cucumber seedlings grown without supplementary light were used as control. Different letters within each parameter were significantly different tested by the least significant difference (LSD) test at $p < 0.05$.

### 3.4. *Influences of Supplementary Light on Stem Firmness and Cellulose Content of Greenhouse-Grown Cucumber Seedlings*

The stem firmness and cellulose content of greenhouse-grown cucumber seedlings cultivated with supplementary light were higher than those cultivated with sunlight only

(Figure 2). The stem firmness and cellulose content of cucumber seedlings grown under the combination of white, blue, and UVA light were 1.9- and 2.7-times higher compared with cucumber seedlings grown with supplementary white LEDs, respectively. In addition, UVA light also led to higher stem firmness and cellulose contents of greenhouse-grown cucumber seedlings.

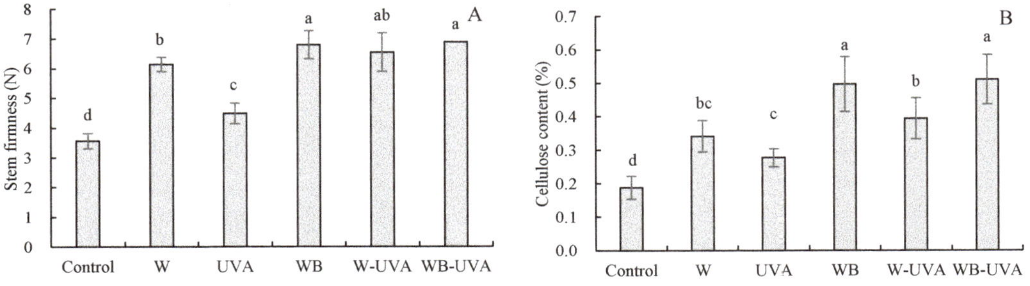


**Figure 2.** Effects of supplementary light provided by white (W), UV-A light-emitting diodes (LEDs), and various combinations of W, blue (B), and UV-A LEDs on stem firmness (**A**) and cellulose content (**B**) of greenhouse-grown cucumber seedlings cultivated for 22 days after sowing; cucumber seedlings grown without supplementary light were used as control. Different letters within each parameter were significantly different tested by the least significant difference (LSD) test at $p < 0.05$.

### *3.5. Hormone Content of Cucumber Seedlings Cultivated in the Greenhouse as Affected by Supplementary Light*

The hormone contents of cucumber seedlings grown in the greenhouse reacted differently by different supplementary light treatments (Figure 3). The cytokinin content of cucumber seedlings cultivated with supplemental light increased remarkably compared with those grown with natural light only. Clearly, UVA light was effective in promoting the cytokinin content of cucumber seedlings. Moreover, the cytokinin content of cucumber seedlings grown with WB-UVA increased by 1-fold compared with those grown under the control treatment.

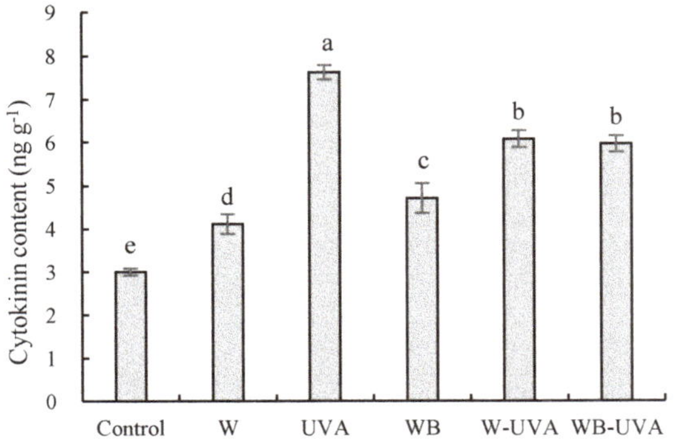


**Figure 3.** Impacts of supplementary light provided by white (W), UV-A light-emitting diodes (LEDs), and various combinations of W, blue (B), and UV-A LEDs on cytokinin content of greenhouse-grown cucumber seedlings cultivated for 22 days after sowing; cucumber seedlings grown without supplementary light were used as control. Different letters within each parameter were significantly different tested by the least significant difference (LSD) test at $p < 0.05$.

### 3.6. Supplementary Light Use Efficiency of Greenhouse-Grown Cucumber Seedlings

The supplementary light use efficiency of greenhouse-grown cucumber seedlings was significantly affected by supplementary LEDs (Figure 4). Supplementary white and UVA LEDs led to the lowest and highest supplementary light use efficiency in greenhouse-grown cucumber seedling production, respectively. In addition, the supplementary light use efficiency of cucumber seedlings cultivated under supplementary white LEDs increased by over 70% compared with those grown under the WB or WB-UVA treatments. No significant differences were found in supplementary light use efficiency in cucumber seedlings grown under the WB, W-UVA, and WB-UVA treatments.

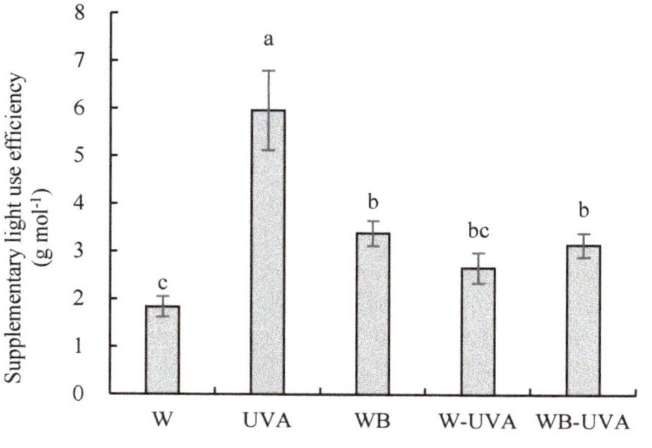


**Figure 4.** Supplementary light use efficiency of greenhouse-grown cucumber seedlings cultivated under supplementary white (W), UV-A light-emitting diodes (LEDs), and various combinations of W, blue (B), and UV-A LEDs at 22 days after sowing; cucumber seedlings grown without supplementary light were used as control. Different letters within each parameter were significantly different tested by the least significant difference (LSD) test at $p < 0.05$.

### 3.7. Heat Map Analysis

A heat map was applied to analyze the response among the tested parameters and exhibited a broad view of the influences of the supplementary lights on greenhouse-grown cucumber seedlings (Figure 5). The WB and the WB-UVA clusters were the closest to each other, the W and the W-UVA clusters were the closest to each other, and the control and the UVA clusters were the closest to each other. In addition, the control and WB-UVA treatment showed opposite responses in most of the measured parameters. Plant height and hypocotyl length were negatively related to the combinations of W-UVA, WB-UVA, and WB. However, the leaf area, root activity, root fresh weight, dry weight, and other indexes were positively related to the combinations of different lights. From the heat map, we could determine that different light combinations had different impacts on the growth indicators of cucumber seedlings. WB-UVA was characterized by a higher leaf area, shoot dry weight, gm, and biomass accumulation of the cucumber seedings. WB-UVA had a better performance than UVA, W, and W-UVA in the morphological characteristics, photosynthetic properties, growth characteristics, and root architecture of greenhouse-grown cucumber seedlings.

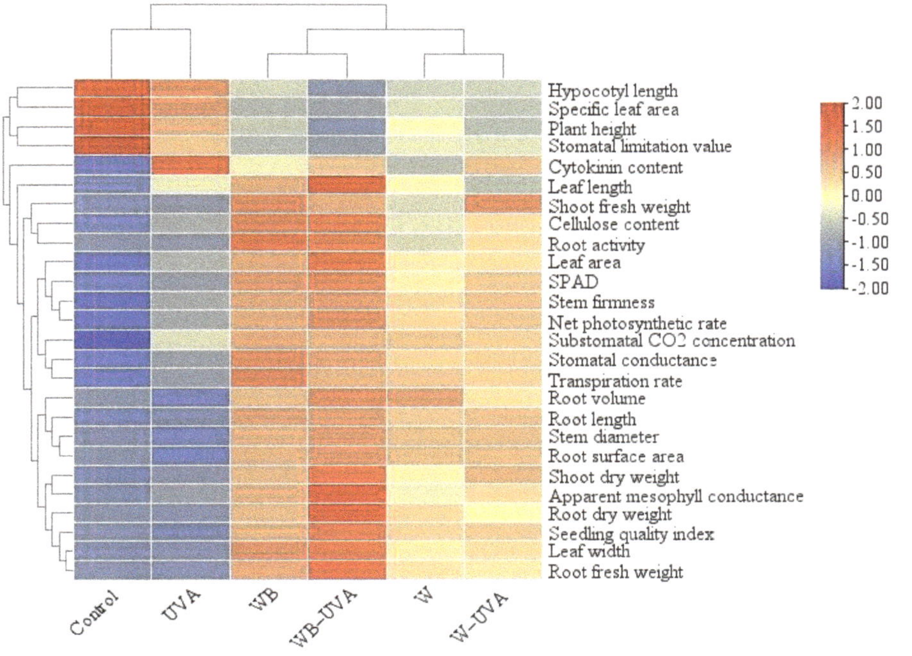


**Figure 5.** Cluster heat map analysis of greenhouse-grown cucumber seedlings for 22 days after sowing as influenced by white (W), UV-A light-emitting diodes (LEDs), and various combinations of W, blue (B), and UV-A LEDs. Blue and pure red indicated an increase and a decrease in the response parameters, respectively.

## 4. Discussion

Appling supplementary light in the greenhouse had become an indispensable method to improve the growth conditions and quality of plants grown in seasons or latitudes with insufficient solar light, and different light qualities had distinctly different biological effects on plants [42,44]. In this study, supplementary lighting led to a decreased plant height, shorter hypocotyl length, and bigger leaf area compared with those grown with solar light only. Moreover, the changes in morphology of plants were also related to the spectral composition of supplementary lights. The stem diameter and leaf area of greenhouse-grown cucumber seedlings cultivated in the WB-UVA treatment increased by 33.3% and 44.6% compared with those grown with natural light only, respectively. In addition, UVA radiation led to lower plant height, shorter hypocotyl length, and thicker leaves of cucumber seedlings compared with those cultivated with natural light only (Table 1). The results were similar with the previous studies reporting that UV radiation would lead to plants with larger internode diameters and shorter internodes [45,46]. In addition, Zhang et al. [35] observed that UVA radiation resulted in a bigger leaf area and smaller specific leaf area of tomato plants. Moreover, blue light significantly promoted the leaf expansion of plants [23]; when blue and UV lights were applied in the meantime, the elongation of the main stem of cucumber seedlings could be inhibited through activating cryptochromes [47], and the stem diameter and leaf area of plants could be promoted significantly. Similarly, Azad et al. [21] observed that the plant height of leaf lettuce decreased with the increase in blue light fraction, and a higher blue light fraction resulted in a compact leaf arrangement with green color and small petioles of lettuce, which was regulated by cryptochromes [31].

Different types of light sources, such as energy and signal sources, significantly affected plant photomorphogenesis [48]. It could be found that photosynthetic traits of the greenhouse-grown cucumber seedlings were changed remarkably when exposed to different combinations of supplementary light (Table 2). The results indicated that UVA radiation affected the plant chlorophyll contents significantly, and some studies had shown that UVA radiation could promote the increase in total chlorophyll content in lettuce, barley seedlings, and broccoli sprouts under greenhouse cultivation conditions [49–51]. Moreover, chlorophyll contents of plants directly influenced the photosynthesis process, and they were affected by the light quality [52]. In this study, the SPAD value, net photosynthetic rate, and transpiration rate of cucumber seedlings cultivated with supplementary light increased substantially compared with those cultivated with sunlight only, and supplementary lights with more blue light fraction led to a higher net photosynthetic rate and transpiration rate, indicating that increasing the DLI and blue light increased the photosynthetic characteristics of plants. Similar results were also found in lettuce [21] and grafted watermelon seedlings [53]. In addition, it could be observed that UVA radiation affected the photosynthetic process of cucumber seedlings between the control and UVA treatments, but not observed in W vs. W-UVA, and WB vs. WB-UVA, which may be related with the supplementary light intensity or background light. Gao et al. [54] indicated that UVA radiation improved the actual photochemical efficiency of photosystem II, thus increasing the growth of Chinese kale. Simultaneously, the enhancement of photosynthetic activity by UVA radiation was caused by the re-absorption of UVA-induced blue-green fluorescence [50]. However, Kang et al. [36] found that the net photosynthesis rate of tomato seedlings was unaffected by UVA radiation when supplemented with red plus blue lights. This difference may be caused by the supplementary light intensity, the background light quality, or the plant species.

The light spectral composition affected the biomass accumulation of plants, and different vegetable varieties responded differently to the light spectrum [55]. Supplemental LED lighting improved the quality of vegetable seedlings by increasing the DLI in the greenhouse, where a 1% increase in DLI may lead to a 1% increase in the yields of fruit vegetable, which may be due to the increased Cytochrome b6f complex and Rubisco contents in plants [56,57]. Our study indicated that supplementing with UVA radiation alone unaffected the biomass accumulation of cucumber seedlings, but the combination of UVA with white light, or white plus blue lights significantly increased the biomass accumulation of greenhouse-grown cucumber seedlings. Additionally, the seedling quality index of greenhouse-grown cucumber seedlings cultivated in the WB-UVA treatment was significantly higher compared with other treatments, except for the cucumber seedlings grown with supplementary white and blue LEDs, indicating that UVA light was beneficial to the growth of cucumber seedlings when it was used in conjunction with other LEDs. Root architecture is also a vital trait in evaluating seedling quality, as well-rooted vegetable seedlings are more appropriate to support transport conditions [44]. Little research has been conducted on the root architecture of greenhouse-grown cucumber seedlings cultivated under various supplementary lights. Yan et al. [13] found that the root growth of cucumber seedlings was increased by supplementary white LEDs, and similar trends were also found in the root length, root surface area, root volume, and root activity of cucumber seedlings. Moreover, our results showed that no remarkable differences were observed in the aforementioned parameters of greenhouse-grown cucumber seedlings grown between the control and UVA treatment, which may be caused by the supplementary DLI.

The stem firmness and cellulose content of plants indicated the mechanical properties of plants [58]. Vegetable seedlings with a higher cellulose content characterized the good lodging resistance of their stems, which was convenient for vegetable grafting and transplanting. It can be seen in this study that the stem firmness and cellulose content of greenhouse-grown cucumber seedlings cultivated with supplementary light were higher than those cultivated without supplementary light. This change could improve the mechanical properties of greenhouse-grown cucumber seedlings and was conducive for transplanting. Similar results could be found in the research of Yan et al. [13]. In addition, the stem firmness and cellulose content of cucumber seedlings were improved by supplementary blue light.

The hormone contents in the plant reflected the degree of the physiological process of the plant. Cytokinins are a major class of phytohormones, which play a vital role in the retardants of leaf senescence, flower bud differentiation, and root growth [59,60]. From the results of this study, it could be found that different light qualities had remarkable impacts on the hormone content of cucumber seedlings, and the cytokinin content of greenhouse-grown cucumber seedlings cultivated with supplementary light increased compared with those grown with sunlight only. Chory et al. [61] suggested that adding higher contents of cytokinin to dark-grown seedlings led to de-etiolation, suggesting that etiolation depended on low levels of cytokinin. In general, blue and UVA radiations promoted the synthesis of cytokinin of cucumber seedling. Similarly, Marchetti et al. [62] indicated that cytokinin reactivation was delayed in the absence of blue light.

Supplementary light use efficiency should be considered due to the increased electricity consumption [63], which could be applied to estimate the effectiveness of supplementary light sources for cultivating crops in the protected horticulture. Our results indicated that cucumber seedlings grown in the UVA light led to the highest supplementary light use efficiency as a result of the small amount of light used in the treatment. However, the increased lighting use efficiency should also be considered with the growth performances of plants [12,28]. From the perspective of energy saving, no remarkable differences were found in supplementary light use efficiency in greenhouse-grown cucumber seedlings cultivated in the WB, W-UVA, and WB-VUA treatments, which were significantly higher as compared with those grown in the W treatment.

From the results of the heat map, the effects of different combinations of supplementary LEDs on cucumber seedlings varied greatly, and the appropriate combinations of supplementary LEDs could promote the growth of plants. It showed that UVA could not promote the growth of cucumber seedlings when it was used alone (the control vs. the UVA). However, when it was combined with white light or white and blue lights, the growth status of cucumber seedlings had been significantly improved as compared with those grown with natural light only, including morphological characteristics, biomass, and photosynthetic characteristics. This result is consistent with previous studies, which believed that UVA can be used in conjunction with blue light to promote plant growth [36,64,65]. Moreover, the promotion effects were more obvious when cucumber seedlings were grown under the combinations of blue, white, and UVA lights, especially on the growth characteristics and photosynthetic properties.

## 5. Conclusions

Supplementary lighting remarkably affected the leaf morphology, growth, and physiological traits of greenhouse-grown cucumber seedlings. In addition, different light spectra had significant and diverse influences on the growth of cucumber seedlings. Combinations of white, blue, and UVA radiations led to compact morphological characteristics, superior mechanical properties, and preferable growth performance, which could be used as an effective tool to obtain the desired morphological and quality properties of targeted plants. It also promoted photosynthetic processes and hormone synthesis of greenhouse-grown cucumber seedlings, which was suitable as supplementary lighting sources in the seasons with insufficient light.

**Author Contributions:** Conceptualization, Y.Y. and Z.Y.; methodology, L.W., Z.Y., C.W., and Y.Y.; investigation, X.L., L.W., and G.W.; resources, Z.Y., C.W., L.W., and Y.Y.; writing—original draft preparation, Z.Y. and C.W.; writing—review and editing, Z.Y., C.W., and Y.Y.; visualization, Z.Y. and L.W.; supervision, Z.Y., G.W., and Y.Y.; project administration, Y.Y. and Z.Y.; funding acquisition, Y.Y. and Z.Y. All authors have read and agreed to the published version of the manuscript.

**Funding:** The research was supported by the Science and Technology of People-benefiting Project of Qingdao (21-1-4-ny-6-nsh), the Key Research and Development Program of Shandong Province (2021TZXD007-02), the Natural Science Foundation of Shandong Province (ZR2021QC174), and the National Key Research and Development Program of China (2019YFD1001901).

**Institutional Review Board Statement:** Not applicable.

**Data Availability Statement:** Data are reported in the Results section. Please contact the corresponding author for any additional information.

**Conflicts of Interest:** The authors declare no conflict of interest.

## References

1. FAO. 2022. Available online: http://www.fao.org/faostat/en/#data/QC (accessed on 7 July 2022).
2. Chang, C.L.; Chang, K.P. The growth response of leaf lettuce at different stages to multiple wavelength-band light-emitting diode lighting. *Sci. Hortic.* **2014**, *179*, 78–84. [CrossRef]
3. Nakano, A.; Ohtake, N.; Yoneda, T.; Shinoda, A. Effects of LED alternating irradiation on tomato seedling quality and root development after transplanting. *Root Res.* **2017**, *26*, 3–9. [CrossRef]
4. Ji, F.; Wei, S.Q.; Liu, N.; Xu, L.J.; Yang, P. Growth of cucumber seedlings in different varieties as affected by light environment. *Int. J. Agric. Biol. Eng.* **2020**, *13*, 73–78. [CrossRef]
5. Özdemir, Y.; Demirel, H.; Yildirim, Y.E. Developing of DLI (Daily Light Integral) and spectrum control systems for scientific cultivation in agriculture. *Int. J. Agric. For. Sci.* **2016**, *1*, 14–17.
6. Pennisi, G.; Pistillo, A.; Orsini, F.; Cellini, A.; Spinelli, F.; Nicola, S.; Juan, A.; Crepaldi, F.A.; Gianquinto, G.; Marcelis, L.F.M. Optimal light intensity for sustainable water and energy use in indoor cultivation of lettuce and basil under red and blue LEDs. *Sci. Hortic.* **2020**, *272*, 109508. [CrossRef]
7. Dou, H.J.; Niu, G.; Gu, M.M.; Masabni, J.G. Responses of sweet basil to different daily light integrals in photosynthesis, morphology, yield, and nutritional quality. *HortScience* **2018**, *53*, 496–503. [CrossRef]
8. Gao, M.; He, R.; Shi, R.; Zhang, Y.; Song, S.; Su, W.; Liu, H. Differential Effects of Low Light Intensity on Broccoli Microgreens Growth and Phytochemicals. *Agronomy* **2021**, *11*, 537. [CrossRef]
9. Ke, X.; Yoshida, H.; Hikosaka, S.; Goto, E. Optimization of photosynthetic photon flux density and light quality for increasing radiation-use efficiency in dwarf tomato under LED light at the vegetative growth stage. *Plants* **2022**, *11*, 121. [CrossRef]
10. Kelly, N.; Choe, D.; Meng, Q.W.; Runkle, E.S. Promotion of lettuce growth under an increasing daily light integral depends on the combination of the photosynthetic photon flux density and photoperiod. *Sci. Hortic.* **2020**, *272*, 109565. [CrossRef]
11. Cui, J.; Song, S.; Yu, J.; Liu, H. Effect of daily light integral on cucumber plug seedlings in artificial light plant factory. *Horticulturae* **2021**, *7*, 139. [CrossRef]
12. Wang, Y.; Chu, Y.; Wan, Z.; Zhang, G.; Liu, L.; Yan, Z. Root architecture, growth and photon yield of cucumber seedlings as influenced by daily light integral at different stages in the closed transplant production system. *Horticulturae* **2021**, *7*, 328. [CrossRef]
13. Yan, Z.N.; Wang, L.; Wang, Y.; Chu, Y.; Lin, D.; Yang, Y. Morphological and physiological properties of greenhouse-grown cucumber seedlings as influenced by supplementary light-emitting diodes with same daily light integral. *Horticulturae* **2021**, *7*, 361. [CrossRef]
14. Dyśko, J.; Kaniszewski, S. Effects of LED and HPS lighting on the growth, seedling morphology and yield of greenhouse tomatoes and cucumbers. *Hortic. Sci.* **2021**, *48*, 22–29. [CrossRef]
15. Sobczak, A.; Sujkowska-Rybkowska, M.; Gajc-Wolska, J.; Kowalczyk, W.; Borucki, W.; Kalaji, H.M.; Kowalczyk, K. Photosynthetic efficiency and anatomical structure of pepper leaf (*Capsicum annuum* L.) transplants grown under high-pressure sodium (HPS) and light-emitting diode (LED) supplementary lighting systems. *Plants* **2021**, *10*, 1975. [CrossRef] [PubMed]
16. Hogewoning, S.W.; Wientjes, E.; Douwstra, P.; Trouwborst, G.; Van Ieperen, W.; Croce, R.; Harbinson, J. Photosynthetic quantum yield dynamics: From photosystems to leaves. *Plant Cell* **2012**, *24*, 1921–1935. [CrossRef]
17. Hogewoning, S.W.; Trouwborst, G.; Maljaars, H.; Poorter, H.; van Ieperen, W.; Harbinson, J. Blue light dose-responses of leaf photosynthesis, morphology, and chemical composition of Cucumis sativus grown under different combinations of red and blue light. *J. Exp. Bot.* **2010**, *61*, 3107–3117. [CrossRef]
18. Hernandez, R.; Kubota, C. Physiological responses of cucumber seedlings under different blue and red photon flux ratios using LED. *Environ. Exp. Bot.* **2016**, *121*, 66–74. [CrossRef]
19. Clavijo-Herrera, J.; van Santen, E.; Gómez, C. Growth, Water-use efficiency, stomatal conductance, and nitrogen uptake of two lettuce cultivars grown under different percentages of blue and red light. *Horticulturae* **2018**, *16*, 3–13. [CrossRef]
20. Wang, J.; Wei, L.; Tong, Y.; Yang, Q.C. Leaf morphology, photosynthetic performance, chlorophyll fluorescence, stomatal development of lettuce (*Lactuca sativa* L.) exposed to different ratios of red light to blue light. *Front. Plant Sci.* **2016**, *7*, 250. [CrossRef]

21. Azad, M.O.K.; Kjaer, K.H.; Adnan, M.; Naznin, M.T.; Lim, J.D.; Sung, I.J.; Park, C.H.; Lim, Y.S. The evaluation of growth performance, photosynthetic capacity, and primary and secondary metabolite content of leaf lettuce grown under limited irradiation of blue and red LED light in an urban plant factory. *Agriculture* **2020**, *10*, 28. [CrossRef]
22. Yang, Z.C.; He, W.; Mou, S.T.; Wang, X.X.; Chen, D.Y.; Hu, X.T.; Chen, L.H.; Bai, J.Y. Plant growth and development of pepper seedlings under different photoperiods and photon flux ratios of red and blue LEDs. *Trans. CSAE* **2017**, *33*, 173–180.
23. Liang, Y.; Kang, C.Q.; Kaiser, E.; Kuang, Y.; Yang, Q.C.; Li, T. Red/blue light ratios induce morphology and physiology alterations differently in cucumber and tomato. *Sci. Hortic.* **2021**, *281*, 109995. [CrossRef]
24. Pennisi, G.; Blasioli, S.; Cellini, A.; Maia, L.; Crepaldi, A.; Braschi, I.; Spinelli, F.; Nicola, S.; Fernandez, J.A.; Stanghellini, C.; et al. Unraveling the role of red:blue LED lights on resource use efficiency and nutritional properties of indoor grown sweet basil. *Front. Plant Sci.* **2019**, *10*, 305. [CrossRef] [PubMed]
25. Hammock, H.A.; Kopsell, D.A.; Sams, C.E. Supplementary blue and red LED narrowband wavelengths improve biomass yield and nutrient uptake in hydroponically grown basil. *HortScience* **2020**, *55*, 1888–1897. [CrossRef]
26. Park, Y.; Runkle, E.S. Spectral effects of light-emitting diodes on plant growth, visual color quality, and photosynthetic photon efficacy: White versus blue plus red radiation. *PLoS ONE* **2018**, *13*, e0202386. [CrossRef] [PubMed]
27. Yan, Z.N.; He, D.X.; Niu, G.H.; Zhou, Q.; Qu, Y.H. Growth, nutritional quality, and energy use efficiency in two lettuce cultivars as influenced by white plus red versus red plus blue LEDs. *Int. J. Agric. Biol. Eng.* **2020**, *13*, 33–40. [CrossRef]
28. Yan, Z.N.; He, D.X.; Niu, G.H.; Zhou, Q.; Qu, Y.H. Growth, nutritional quality and energy use efficiency of hydroponic lettuce as influenced by daily light integrals exposed to white versus white plus red light-emitting diodes. *HortScience* **2019**, *54*, 1737–1744. [CrossRef]
29. Gao, W.; He, D.X.; Ji, F.; Zhang, S.; Zheng, J.F. Effects of daily light integral and LED spectrum on growth and nutritional quality of hydroponic spinach. *Agronomy* **2020**, *10*, 1082. [CrossRef]
30. Zheng, J.F.; Peidian, G.; Ji, F.; He, D.X.; Yang, P. Growth and energy use efficiency of grafted tomato transplants as affected by LED light quality and photon flux density. *Agriculture* **2021**, *11*, 816. [CrossRef]
31. Kalaitzoglou, P.; Taylor, C.; Calders, K.; Hogervorst, M.; Ieperen, W.V.; Harbinson, J.; Visser, P.D.; Nicole, C.C.S.; Marcelis, L.F.M. Unraveling the effects of blue light in an artificial solar background light on growth of tomato plants. *Environ. Exp. Bot.* **2021**, *184*, 104377. [CrossRef]
32. Kaiser, E.; Ouzounis, T.; Giday, H.; Schipper, R.; Heuvelink, E.; Marcelis, L.F.M. Adding blue to red supplemental light increases biomass and yield of greenhousegrown tomatoes, but only to an optimum. *Front. Plant Sci.* **2019**, *9*, 2002. [CrossRef] [PubMed]
33. Verdaguer, D.; Jansen, M.A.K.; Llorens, L.; Morales, L.O.; Neugart, S. UV-A radiation effects on higher plants: Exploring the known unknown. *Plant Sci.* **2017**, *255*, 72–81. [CrossRef] [PubMed]
34. Lee, J.W.; Park, S.Y.; Oh, M.M. Supplemental radiation of ultraviolet-A light-emitting diode improves growth, antioxidant phenolics, and sugar alcohols of ice plant. *Hortic. Environ. Biotechnol.* **2021**, *62*, 559–570. [CrossRef]
35. Zhang, Y.T.; Elias, K.; Zhang, Y.Q.; Zou, J.; Bian, Z.H.; Yang, Q.C.; Li, T. UVA radiation promotes tomato growth through morphological adaptation leading to increased light interception. *Environ. Exp. Bot.* **2020**, *176*, 104073. [CrossRef]
36. Kang, S.Y.; Zhang, Y.T.; Zhang, Y.Q.; Zou, J.; Yang, Q.C.; Li, T. Ultraviolet-A radiation stimulates growth of indoor cultivated tomato (Solanum lycopersicum) seedlings. *HortScience* **2018**, *53*, 1429–1433. [CrossRef]
37. Yan, Z.N.; Wang, L.; Cheng, J.; Lin, D.; Yang, Y.J. Morphology, Growth and physiological traits of greenhouse-grown cucumber seedlings as affected by supplementary white and blue light-emitting diodes. In Proceedings of the 2021 International Symposium on Efficient Production of Value-Added Plants, Online, 27–28 December 2021; pp. 57–70.
38. Han, S.Q.; Wang, X.F.; Wei, M.; Li, Y. Study of plug seedling index if sweet pepper and relationship between seedling index and characters. *J. Shandong Agric. Univ.* **2004**, *35*, 187–190.
39. Li, H.S. *Principle and Technology of Plant Physiological and Biochemical Experiments*; Higher Education Press: Beijing, China, 2003; pp. 119–120.
40. Updegraff, D.M. Semi-micro determination of cellulose in biological materials. *Anal. Biochem.* **1969**, *32*, 420–424. [CrossRef]
41. Aguilar, M.L.; Espadas, F.; Maust, B.; Sáenz, L. Endogenous cytokinin content in coconut palms affected by lethal yellowing. *J. Plant Pathol.* **2009**, *91*, 141–146.
42. Wei, H.; Wang, M.W.; Jeong, B.R. Effect of supplementary lighting duration on growth and activity of antioxidant enzymes in grafted watermelon seedlings. *Agronomy* **2020**, *10*, 337. [CrossRef]
43. Chen, C.; Chen, H.; Zhang, Y.; Thomas, H.R.; Frank, M.H.; He, Y.; Xia, R. TBtools: An integrative toolkit developed for interactive analyses of big biological data. *Mol. Plant.* **2020**, *13*, 1194–1202. [CrossRef]
44. Kobori, M.M.R.G.; Mello, S.D.C.; Freitas, I.S.D.; Silveira, F.F.; Alves, M.C.; Azevedo, R.A. Supplemental light with different blue and red ratios in the physiology, yield and quality of Impatiens. *Sci. Hortic.* **2022**, *306*, 111424. [CrossRef]
45. Robson, T.M.; Klem, K.; Urban, O.; Jansen, M.A.K. Re-interpreting plant morphological responses to UV-B radiation. *Plant Cell Environ.* **2015**, *38*, 856–866. [CrossRef] [PubMed]
46. Qian, M.J.; Rosenqvist, E.; Flygare, A.M.; Kalbina, I.; Teng, Y.W.; Marcel, A.K.; Jansen, A. UV-A light induces a robust and dwarfed phenotype in cucumber plants (*Cucumis sativus* L.) without affecting fruit yield. *Sci. Hortic.* **2020**, *263*, 109110. [CrossRef]
47. Liu, B.; Yang, Z.; Gomez, A.; Liu, B.; Lin, C.; Oka, Y. Signaling mechanisms of plant cryptochromes in *Arabidopsis thaliana*. *J. Plant Res.* **2016**, *129*, 137–148. [CrossRef] [PubMed]

48. Bantis, F.; Smirnakou, S.; Ouzounis, T.; Koukounaras, A.; Ntagkas, N.; Radaglou, K. Current status and recent achievements in the field of horticulture with the use of light-emitting diodes (LEDs). *Sci. Hortic.* **2018**, *235*, 437–451. [CrossRef]
49. Caldwell, C.R.; Britz, S.J. Effect of supplemental ultraviolet radiation on the carotenoid and chlorophyll composition of greenhouse-grown leaf lettuce (*Lactuca sativa* L.) cultivars. *J. Food Compos. Anal.* **2006**, *19*, 637–644. [CrossRef]
50. Štroch, M.; Meterová, Z.; Vrábl, D.; Karlický, V.; Šigut, L.; Nezval, J.; Špunda, V. Protective effect of UV-A radiation during acclimation of the photosynthetic apparatus to UV-B treatment. *Plant Physiol. Biochem.* **2015**, *96*, 90–96. [CrossRef]
51. Moreira-Rodriguez, M.; Nair, V.; Benavides, J.; Cisneros-Zevallos, L.; Jacobo-Velazquez, D.A. UVA, UVB light doses and harvesting time differentially tailor glucosinolate and phenolic profiles in broccoli sprouts. *Molecules* **2017**, *22*, 1065. [CrossRef]
52. Yang, X.; Xu, H.; Shao, L.; Li, T.; Wang, Y.; Wang, R. Response of photosynthetic capacity of tomato leaves to different LED light wavelength. *Environ. Exp. Bot.* **2018**, *150*, 161–171. [CrossRef]
53. Moosavi-Nezhad, M.; Salehi, R.; Aliniaeifard, S.; Tsaniklidis, G.; Woltering, E.J.; Fanourakis, D.; Zuk-Gołaszewska, K.; Kalaji, H.M. Blue light improves photosynthetic performance during healing and acclimatization of grafted watermelon seedlings. *Int. J. Mol. Sci.* **2021**, *22*, 8043. [CrossRef]
54. Gao, M.; Li, Y.; Jiang, H.; He, R.; Shi, R.; Song, S.; Liu, H. UVA-Radiation Exposure of Different Durations Promoted the Growth, Phytochemicals and Glucosinolate Biosynthesis of Chinese Kale. *Int. J. Mol. Sci.* **2022**, *23*, 7619. [CrossRef] [PubMed]
55. Barbara, F.; Szczak, B.F.; Monika, K.M. The preferences of different cultivars of lettuce seedlings (*Lactuca sativa* L.) for the spectral composition of light. *Agronomy* **2021**, *11*, 1211.
56. Paucek, I.; Pennisi, G.; Pistillo, A.; Appolloni, E.; Crepaldi, A.; Calegari, B.; Spinelli, F.; Cellini, A.; Gabarrell, X.; Orsini, F.; et al. Supplementary LED interlighting improves yield and precocity of greenhouse tomatoes in the Mediterranean. *Agronomy* **2020**, *10*, 1002. [CrossRef]
57. He, J.; Jawahir, N.K.B.; Lin, Q. Quantity of supplementary LED lightings regulates photosynthetic apparatus, improves photosynthetic capacity and enhances productivity of Cos lettuce grown in a tropical greenhouse. *Photosynth. Res.* **2021**, *149*, 187–199. [CrossRef] [PubMed]
58. Liu, W.G.; Deng, Y.C.; Hussain, S.; Zou, J.L.; Yuan, J.; Luo, L.; Yang, C.Y.; Yuan, X.Q.; Yang, W.Y. Relationship between cellulose accumulation and lodging resistance in the stem of relay intercropped soybean [*Glycine max* (L.) men.]. *Field Crop. Res.* **2016**, *196*, 261–267. [CrossRef]
59. Rowe, J.H.; Topping, J.F.; Liu, J.; Lindsey, K. Abscisic acid regulates root growth under osmotic stress conditions via an interacting hormonal network with cytokinin, ethylene and auxin. *New Phytol.* **2016**, *211*, 225–239. [CrossRef]
60. Zürcher, E.; Müller, B. Cytokinin synthesis, signaling, and function—Advances and new insights. *Int. Rev. Cell Mol. Biol.* **2016**, *324*, 1–38.
61. Chory, J.; Reinecke, D.; Sim, S.; Washburn, T.; Brenner, M. A role for cytokinins in de-etiolation in Arabidopsis (det mutants have an altered response to cytokinins). *Plant Physiol.* **1994**, *104*, 339–347. [CrossRef]
62. Marchetti, C.F.; Skrabisova, M.; Galuszka, P.; Novak, O.; Causin, H.F. Blue light suppression alters cytokinin homeostasis in wheat leaves senescing under shading stress. *Plant Physiol. Biochem.* **2018**, *130*, 647–657. [CrossRef]
63. Elkins, C.; van Iersel, M.W. Longer photoperiods with the same daily light integral improve growth of Rudbeckia seedlings in a greenhouse. *Hort. Sci.* **2020**, *55*, 1676–1682. [CrossRef]
64. Lee, M.; Rivard, C.; Pliakoni, E.; Wang, W.; Rajashekar, C.B. Supplemental UV-A and UV-B affect the nutritional quality of lettuce and tomato: Health-promoting phytochemicals and essential nutrients. *Am. J. Plant Sci.* **2021**, *12*, 104–126. [CrossRef]
65. Chen, Y.; Li, T.; Yang, Q.; Zhang, Y.; Zou, J.; Bian, Z.; Wen, X. UVA radiation is beneficial for yield and quality of indoor cultivated lettuce. *Front. Plant Sci.* **2019**, *10*, 1563. [CrossRef] [PubMed]

*agriculture*

Article

# A Light Recipe including Far-Red Wavelength during Healing of Grafted Watermelon Seedlings Enhances the Floral Development and Yield Earliness

Filippos Bantis [1,*], Anna Gkotzamani [1], Christodoulos Dangitsis [2] and Athanasios Koukounaras [1]

[1] Department of Horticulture, Aristotle University, 54124 Thessaloniki, Greece; agkotzam@agro.auth.gr (A.G.); thankou@agro.auth.gr (A.K.)
[2] Agris S.A., Kleidi, 59300 Imathia, Greece; cdaggitsis@agris.gr
* Correspondence: fbantis@agro.auth.gr

**Abstract:** Watermelon is widely propagated through grafting, after which seedlings are subjected to healing under controlled conditions including artificial lighting. Light wavelengths, such as blue, red, and far-red, impose considerable effects on seedlings, which possibly carry on to the mature plants. The aim of the present study is to examine whether different light wavelengths during healing of grafted watermelon seedlings impose variable effects during field cultivation. After grafting, seedlings were healed in an environmentally controlled healing chamber under fluorescent (FL) lamps and light-emitting diodes, providing 100% red (R), 100% blue (B), 88/12% R/B (12B), and 12B including 5% far-red (12B + FR). After acclimatization, seedlings were transplanted in the field. Vegetative growth until floral initiation was enhanced by 12B and 12B + FR, as shown by stem diameter and leaf number measurements. Flowering was mainly accelerated by 12B + FR and considerably decelerated by FL and B. The same pattern was followed by fruit yield, which was similar for all treatments at the end of the experiment. Nevertheless, fruit quality was not affected by any of the light treatments. It is concluded that a light recipe, including red, blue and far-red, wavelengths during healing of grafted seedlings enhances the overall growth, and flowering and yield earliness of watermelon crops.

**Keywords:** *Citrullus lanatus*; nursery; healing chamber; transplantation; photomorphogenesis; flowering; crop production; antioxidants

**Citation:** Bantis, F.; Gkotzamani, A.; Dangitsis, C.; Koukounaras, A. A Light Recipe including Far-Red Wavelength during Healing of Grafted Watermelon Seedlings Enhances the Floral Development and Yield Earliness. *Agriculture* **2022**, *12*, 982. https://doi.org/10.3390/agriculture12070982

Academic Editor: Maria Giordano

Received: 7 June 2022
Accepted: 6 July 2022
Published: 7 July 2022



## 1. Introduction

Watermelon (*Citrullus lanatus* L.) is one of the most cultivated species among the *Cucurbitaceae* family worldwide. It is also one of the most exported horticultural species, mainly due to early harvests in some regions of southwest Greece. On average, watermelon yield in Greece reached more than 45 ton/ha during 2016–2020 and its export value has risen to over 52 million euros. In 2020, the area harvested was 8770 ha reaching productivity of 49.1 t/ha. The importance of the crop is highlighted by the export profit in that period, which constituted 12% of the total watermelon exported in Europe (FAOSTAT, 2022). In 2020, the total export value in Greece was over 60 million euros.

Watermelon is susceptible to soilborne pathogens, and, thus, it is grafted mostly onto gourds (*Lagernaria siceraria* Standl.) or onto interspecific hybrids (*C. maxima* Duch. × *C. moschata* Duch.) [1]. Healing is the most delicate stage of grafted watermelon seedlings production. During healing, seedlings must be grown in an environmentally controlled space, ideally a growth (healing) chamber, where temperature, relative humidity, and light are fully adjusted.

Light is a key factor in plant growth and development. Light acts through its characteristics, such as photoperiod (duration of emission), quantity (intensity), and quality (wavelength). Photosynthetically active radiation (PAR, 400–700 nm) is the part of visible light which is utilized by the plants for photosynthesis. Blue (425–490 nm) and red light

(620–700 nm) in particular are crucial for plant growth and development, increased yield, and fruit quality [2]. Far-red light (700–780 nm) is also important in the plant's life as it affects biological functions besides photosynthesis, such as seed germination, phototropism, and flowering [3]. In a recent publication involving hydroponically grown lettuce, Zhen and Bugbee (2020) [4] reported that far-red photons are equally effective for photosynthesis when acting synergistically with PAR photons.

Furthermore, climate change is an inevitable process that is expected to alter the prevailing environmental conditions. According to the 6th IPCC report (2021) [5], heat waves and drought periods will increase in frequency, thus increasing the risk for open-field vegetable crops to suffer from heat stress while also increasing their need for water. Increased heat is expected to deteriorate the abiotic stress of vegetable crops leading to the introduction of additional agrochemicals to cope with pests and pathogens, while the yields will mostly decline, thus raising the cost of production. Moreover, the increased water requirements will also lead to a production cost raise. To this end, the application of cultivation methods that reduce the growing cycle (i.e., production time) even for a few days is crucial in economic terms. This includes the earlier coverage of market demand, leading to higher income for all participants in the supply chain, from growers to retailers.

There is complete lack of literature regarding the effect of different light-emitting diode (LED) light spectra on the flowering, yield, and quality of watermelon crops. Previous studies of our group [6,7] highlighted the importance of red and blue wavelengths for the production of vigorous, high-quality grafted watermelon seedlings, which points to why we used such wavelengths in our research. Moreover, cucumber transplants illuminated with various wavelengths exhibited an after-effect during flowering and harvest [8]. Our research hypothesis was that light quality influences vegetative growth during the first few weeks after transplantation, and possibly affects flowering since the first flower buds differentiate during the nursery growth. Therefore, our objective was to examine whether different light wavelengths during the healing of grafted watermelon seedlings impose variable effects during field cultivation. To this end, our efforts focused on evaluating the plant development, the flowering, the yield, and fruit quality with the aim of increasing the fruit earliness.

## 2. Materials and Methods

### 2.1. Plant Material and Grating

The grafted seedlings were produced in the facilities of Agris S.A. in Kleidi, Imathia, Greece. Watermelon hybrid scions (*Citrullus lanatus* L., Celine F1) and interspecific squash hybrid rootstocks (*Cucurbita maxima* × *C. moschata*, TZ-148) were cultivated according to standard commercial practices. For a detailed description of this stage of cultivation, please refer to Bantis et al. [6]. When the scion and rootstock seedlings achieved appropriate growth, they were grafted with the splice grafting technique, re-planted in 72-cell plug trays, and immediately transferred in a healing chamber.

### 2.2. Healing, Light Conditions, and Acclimatization

The healing chamber is basically a growth room where conditions are fully controlled. Specifically, the temperature was set at 25 °C, and relative humidity was initially set at 98%, gradually decreasing down to 89%, while fans ensured air circulation. The plug trays were placed on shelves irradiated with five different light wavelengths including FL (Fluora 58 W, Osram, GmbH, Munich, Germany) and LED light sources. The LEDs emitted were: (a) monochromatic red (R) with peak wavelength at 661 nm, (b) monochromatic blue (B) with peak wavelength at 450 nm, (c) an 88/12% red/blue combination (12B) which proved optimum for the healing of grafted watermelon seedlings in a recent publication of our group [7], and (d) 12B with additional 5% far-red (12B + FR) radiation with peak wavelength at 725 nm. All light treatments emitted 85 ± 5 µmol $m^{-2}$ $s^{-1}$ with a photoperiod of 18 h. Information about the light treatments, such as waveband percentages and the phytochrome photostationary state, were obtained with a spectroradiometer (HD 30.1, DeltaOhm Srl, Padova, Italy) and are provided in Table 1.

**Table 1.** Wavelength distribution and photobiological parameters of the tested light treatments tested. PPS: phytochrome photostationary state. PPS was calculated according to Sager et al. [9].

| Waveband | Light Treatment | | | | |
|---|---|---|---|---|---|
| | FL | R | B | 12B | 12B + FR |
| UV %; 380–399 nm | 0 | 0 | 0 | 0 | 0 |
| Blue %; 400–499 nm | 35 | 0 | 100 | 12 | 12 |
| Green %; 500–599 nm | 24 | 0 | 0 | 0 | 0 |
| Red %; 600–699 nm | 37 | 100 | 0 | 88 | 83 |
| Far-red %; 700–780 nm | 4 | 0 | 0 | 0 | 5 |
| PPS | 0.82 | 0.89 | 0.51 | 0.89 | 0.88 |

Following the successful healing stage which lasted six days, the grafted seedlings were moved in a greenhouse for a two-week period of acclimatization where the minimum temperature was set at 21.5 °C. At this stage, the seedlings were considered commercial product and were ready for transplantation in the field.

### 2.3. Field Cultivation

Field cultivation was conducted in the experimental farm of the Laboratory of Vegetable Crops, in Aristotle University of Thessaloniki, Greece (N 40.536; E 22.995), in 2021. Following a typical analysis, the soil was characterized as sandy clay loam (SCL), moderate to heavy type. Organic matter constituted 2.3%, the pH was 7.8, and the electrical conductivity was 0.80 mS/cm. Prior to transplantation, the farm soil was plowed and crumbled, while fertilization, irrigation, and control of weeds and pathogens were in accordance with local practices. Specifically, basal dressing was conducted using a fertilizer (500 kg per hectare) including 20-5-20 (nitrogen–phosphorus–potassium) + 3 units of magnesium. During the cultivation period, plants were also fertigated twice using potassium nitrate (13.5–0–46). Irrigation was conducted depending on temperature and precipitation. Typically, the plants were irrigated every two days since precipitation was very low and the temperature was high at all times. Weeds were regularly hoed for about a month until the vines expanded and made it difficult to walk through the plants without damaging them. Finally, a few proactive crop dustings were conducted for aphids and soilborne pathogens (i.e., fusarium).

Sixteen plants from every light treatment were transplanted on 2 June 2021 over four rows considered as replicate, with a row distance of 3 m. The distance between plants within a row was 1 m. The experimental design was a randomized complete block (RCBD) with four replicates (rows). Within each row, plants were arranged in groups of four consecutive plants per light treatment and each light treatment was represented once in each row.

### 2.4. Determinations

Vegetative growth was evaluated for the first two weeks after transplanting until flowering initiation. Every week leaf number was measured, while stem diameter was determined with a digital caliber.

Flowering initiated 19 days after transplanting (DAT), but only male flowers were detected at that time. When the first female flowers bloomed, they were numbered and labeled every two days (until DAT 36) in order to identify the treatment and flowering date of the produced fruits. Specifically, female flowers were recorded on DAT 24, 26, 28, 30, 32, 34, and 36. Total, average, and gradual sum of female flowers were calculated for each measuring date.

Watermelon requires about 40 days between flowering and fruit maturity depending on the environmental conditions. At the end of the experiment, fruits were harvested separately for each of the recorded flowering dates. For example, fruits labeled on DAT 30 were harvested after 40 days, i.e., on DAT 70. The same procedure was applied for each flowering date recorded. The few flowers that bloomed on DAT 24 did not leaf to produced fruits. Total yield and fruit number were calculated for each harvest date.

Three fruits per light treatment were evaluated regarding their biochemical content. A refractometer (PAL-α, Atago, Tokyo, Japan) was used for the determination of total soluble solids (°Brix). The Singleton and Rossi [10] method was used for the measurement of total phenolics. Antioxidant capacity was measured with the ferric reducing antioxidant power (FRAP) method according to Benzie and Strain [11]. Lycopene and total carotenoid contents were measured according to Luterotti et al. [12].

### *2.5. Statistical Analysis*

Data were statistically analyzed with analysis of variance (ANOVA) using the IBM SPSS software (SPSS 23.0, IBM Corp., Armonk, NY, USA). Mean comparisons were conducted with the Scott–Knott method [13], using the statistical software StatsDirect v.2.8.0. (StatsDirect, Ltd., Birkenhead, UK) at significance level a = 0.05. The unique characteristic of this method is that it does not present overlapping in its grouping results.

## **3. Results and Discussion**

Upon seedling transplantation in the field, vegetative growth was evaluated until initiation of flowering. From DAT 0, stems were significantly narrower under the influence of FL and B, an effect which carried on up to DAT 14 (Figure 1A). This is in accordance with a previous study of our group which showed that monochromatic B limited the stem diameter of acclimatized watermelon seedlings compared to red-containing LEDs [14], while FL also induced narrow stem development of the final product (unpublished observation). Stem diameter has been proposed and has widely been used as an indicator of seedling quality in vegetable species, such as tomato, pepper, eggplant [15], cucumber [16], and watermelon [17].

At the beginning of the experiment, all seedlings had an identical number of leaves (four). However, by DAT 14 12B and 12B + FR enhanced the leaf formation compared to FL, B, and R (Figure 1B). Four tomato genotypes exhibited greater leaf number under an 88/12% red/blue treatment, which is similar to our 12B, and it was concluded that the addition of blue light increases plant development and biomass production [18]. Furthermore, supplemental blue LED lighting with high-pressure sodium lamps increased fresh and dry weight and the leaf area of cucumber transplants and enhanced their development [19]. In three artichoke cultivars, blue light negatively affected the leaf number compared to red [20], while no effect was found in cucumber seedlings [21]. Tomato transplants treated with red or red–blue and red–white combinations and pepper transplants treated with high ratios of red light developed fewer leaves before the first cluster [22]. In a study with lettuce irradiated by different light sources at the seedling stage, the authors reported greater mature yield and quality when the seedlings were treated with a red/blue ratio of 2.2 compared to 1.2 or fluorescent lamps [23].

Regarding floral evaluation, male flowers started to bloom on DAT 19. On DAT 24, the first female flowers started to bloom mainly with 12B and 12B + FR and secondarily with R. With FL and B, the first female flowers bloomed on DAT 26. In the following days, flower number sharply increased up to a maximum which was on a different day for each light treatment. Specifically, with 12B and 12B + FR, flowering peaked on DAT 32, while with R, FL, and B flowering peaked on DAT 34 (Figure 2A). On every DAT, sum female flower number was significantly smaller with FL and B compared to the red-containing LEDs, with 12B + FR (mainly) and 12B (secondarily) showing the greatest values on almost every DAT (Figure 2B).

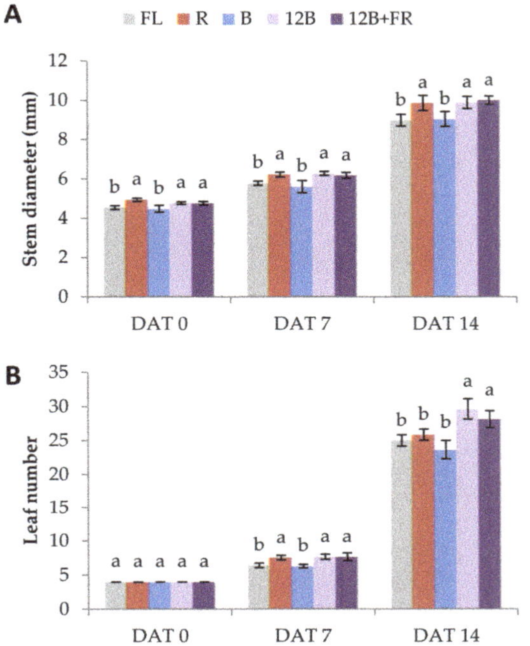


**Figure 1.** (**A**) Stem diameter and (**B**) leaf number of watermelon plants until 14 days after transplanting in the field. The seedlings were treated for six days in a healing chamber with five light treatments. Mean values ($n$ = 8; ±SE) within a row followed by different letters are significantly different ($\alpha < 0.05$).

Plants possess a suite of protein photoreceptors found in the model plant Arabidopsis, which are triggered by red and far-red (phytochromes), blue (cryptochromes, phototropins, and zeitlupe group), and ultraviolet (UVR8) wavelengths [24,25]. Photoreceptors are involved in flowering through a FLOWERING LOCUS T gene, whose expression is regulated by light quality [26,27]. Far-red in particular has been found to drive the expression of the *FvFT1* gene and trigger flowering in strawberry [28]. In petunia, flowering was also promoted by far-red light under two PPFDs (98 and 288 µmol m$^{-2}$ s$^{-1}$) [3], an effect also reported in other long-day plants [29]. In a study with cucumber, red + blue light generated higher biomass, growth rate, and average internode distance in comparison with red + blue + yellow light, while the latter light treatment led to increased sucrose content, which promoted the production of female flowers [30]. In our case, 12B + FR emits red–blue light including 5% far-red, a spectrum which obviously triggered floral development earlier than FL and the monochromatic R and B wavelengths. Strikingly, only 5% far-red of 12B + FR induced flowering to a greater extent compared to 12B, pointing to the considerable effect of red/far-red ratio.

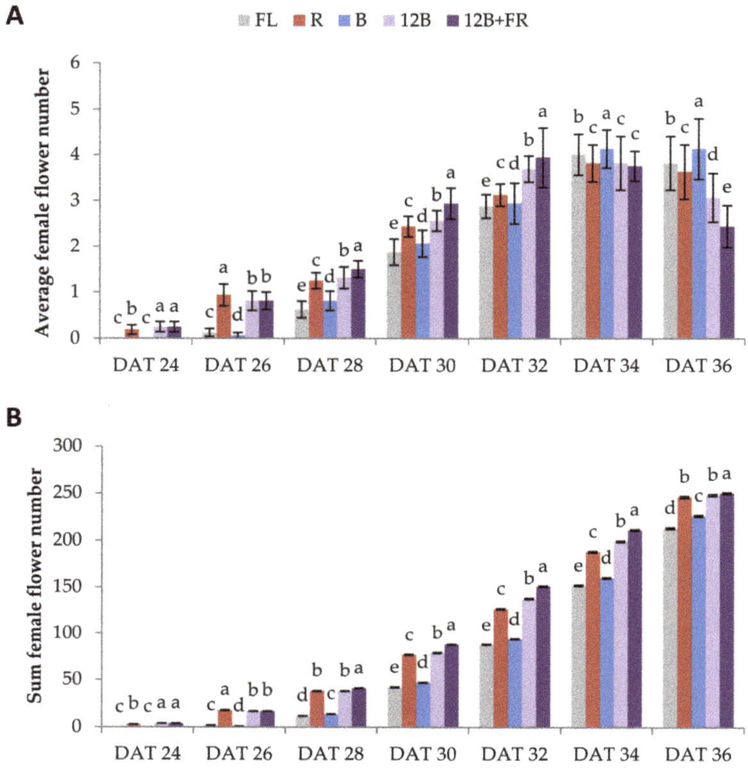


**Figure 2.** (**A**) Average female flower number on each date and (**B**) sum of female flower number of watermelon plants from the 24th to the 36th day after transplanting in the field. The seedlings were treated for six days in a healing chamber with five light treatments. Mean values ($n$ = 8; ±SE) within a row followed by different letters are significantly different ($\alpha < 0.05$).

As far as fruit production is concerned, flowers that bloomed on DAT 24 did not lead to fruits in any light treatment. The first fruits were produced with R, B, 12B, and 12B + FR from flowers that bloomed on DAT 26. From DAT 28 onward, sum yield and fruit number were significantly greater with R, 12B, and 12B + FR compared to FL and B. By DAT 34, total yield and fruit number were similar for every light treatment even though there was a tendency for reduced values with B and FL (Figure 3A,B). In general, fruit production followed the pattern of flowering, indicating that fruit set was similar in plants of all light treatments.

In other cucurbits, an Italian landrace of *Cucumis melo* L. called 'Carosello leccese', grown under red + blue + far red, and red + blue LEDs, resulted in higher growth rate in comparison to plants grown under natural light spectra. Moreover, the higher number of fruits harvested and the higher water content of the fruits resulted in a 27% higher yield in total for the plants grown under LEDs [31]. Another study showed that the yield of cucumber plants grown under LEDs was higher than those grown under high-pressure sodium lamps and those grown under the combination of the two [32]. Brazaitytė et al. [8] reported that cucumber transplants treated with various light wavelengths did not exhibit after-effects on yield, but the beginning of flowering and harvest were significantly affected. Tomato and pepper transplants showed greater rates of first cluster formation and first yield when treated with blue–red combinations [22].

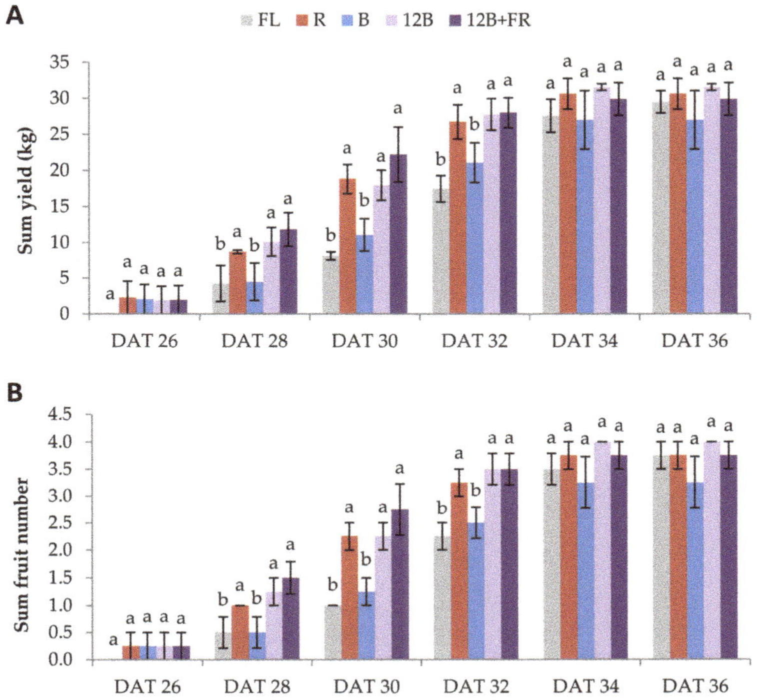


**Figure 3.** (**A**) Sum of watermelon fruit yield and (**B**) sum of fruit number derived from flowers bloomed from the 26th to the 36th day after transplanting in the field. The seedlings were treated for six days in a healing chamber with five light treatments. Mean values ($n$ = 4; ±SE) within a row followed by different letters are significantly different ($\alpha < 0.05$).

At the end of the experiment, fruits were evaluated regarding their morphological and biochemical attributes. Specifically, the fruit length, width, and rind thickness were similar under all light treatments (Table 2). The rind/mesocarp thickness was also similar in all cases (data not shown). The same observation was made for biochemical compounds, such as total soluble solids, total phenolics, total carotenoids, lycopene, and antioxidant compounds (FRAP), which were not significantly affected by the different light treatments (Table 2).

**Table 2.** Morphological and biochemical parameters of ripe watermelon fruits after field cultivation. The seedlings were treated for six days in a healing chamber with five light treatments. Mean values ($n$ = 3; ±SE) within a row followed by different letters are significantly different ($a < 0.05$).

| Parameters | Light Treatments | | | | |
|---|---|---|---|---|---|
| | FL | R | B | 12B | 12B + FR |
| Length (cm) | 23.50 ± 1.26 a | 34.83 ± 2.42 a | 30.17 ± 2.17 a | 34.67 ± 0.93 a | 30.00 ± 2.02 a |
| Width (cm) | 21.67 ± 0.33 a | 20.00 ± 0.29 a | 20.50 ± 0.29 a | 19.67 ± 0.67 a | 20.67 ± 0.73 a |
| Rind thick. (cm) | 0.80 ± 0.10 a | 0.73 ± 0.13 a | 0.93 ± 0.07 a | 0.87 ± 0.09 a | 0.77 ± 0.07 a |
| TSS (°Brix) | 11.53 ± 0.13 a | 11.87 ± 0.22 a | 11.50 ± 0.35 a | 11.27 ± 0.27 a | 11.20 ± 0.42 a |
| TPC (mg/g) | 0.20 ± 0.01 a | 0.21 ± 0.01 a | 0.20 ± 0.01 a | 0.23 ± 0.01 a | 0.20 ± 0.01 a |
| TCC (µg/g) | 24.45 ± 3.54 a | 24.51 ± 0.91 a | 24.20 ± 1.10 a | 25.77 ± 2.51 a | 26.88 ± 1.14 a |
| LC (µg/g) | 19.11 ± 4.74 a | 17.89 ± 1.87 a | 18.88 ± 1.47 a | 20.60 ± 3.00 a | 21.26 ± 0.92 a |
| FRAP (µg/g) | 87.34 ± 1.39 a | 81.46 ± 1.66 a | 76.19 ± 1.94 a | 87.88 ± 3.25 a | 79.39 ± 2.35 a |

Typically, watermelon fruits can be harvested about 40 days after flowering. Summer 2021 was exceptionally stressful for outfield crops in Greece, with very high temperatures for successive weeks. Even though there were not significant differences among our treatments, fruits exhibited considerable quality in terms of sweetness (total soluble solids above 11 °Brix) and antioxidant capacity. Javanmardi and Emami [22] reported greater total soluble solids in tomato fruits treated with blue light at the seedling stage [22], but this was not evident in our case. In another study of our group [14], watermelon transplants treated with varying spectral compositions of red and blue light, and then cultivated on the field in 2018, maintained fruit quality and high quality of important nutritive characteristics. Compared to 2018, our 2021 watermelons produced about 30% greater total phenolics (0.12–0.15 mg/g in 2018 versus 0.20–0.23 mg/g in 2021) and about 50% greater antioxidant capacity displayed by FRAP (35–40 µg/g in 2018 versus 76–87 µg/g in 2021), pointing to the harsh climatic conditions which acted positively towards watermelon fruit quality. In the same study in 2018, proper morphology, root development, photosynthesis, and high fruit quality were noted on grafted watermelon plants that were treated with blue + far-red LED light during the healing stage [14].

## 4. Conclusions

The varying light qualities obviously affected the seedlings during the critical stage of tissue healing leading to certain responses. In general, 12B and 12B + FR enhanced the vegetative growth, which was evaluated until flowering initiation. Flower buds differentiated during the nursery growth when the seedlings were irradiated in the healing chamber. The same light treatments induced earlier flowering compared to the rest of the treatments. A slight addition of only 5% far-red in 12B + FR enhanced flowering compared to 12B, pointing to the significant effect of this particular wavelength for flowering. Subsequently, yield and number of fruits of 12B and 12B + FR, along with R, peaked earlier compared to B and FL. Total yield and fruit number were similar for all treatments. Overall, B had a similar response to FL throughout the experimental period with inferior vegetative growth as well as later flowering and fruit production. After 70 days from transplantation and 84 days from the seedlings' exposure to the different light qualities, no differences were detected in fruit morphological and biochemical properties. It is concluded that a light recipe including red, blue and far-red wavelengths during the healing of grafted seedlings enhances the overall growth, flowering, and yield earliness of watermelon crops. Combined with results from nursery experiments which are presented in other publications, these findings highlight the effectiveness of 12B and 12B + FR wavelengths during healing to produce high-quality grafted watermelon seedlings with greater potential for vegetative growth and rapid flower blooming and fruiting in the field.

**Author Contributions:** Conceptualization, methodology, and data analysis: F.B. and A.K.; experimental measurements: F.B., A.G. and C.D.; writing—original draft preparation: F.B. and A.G.; writing—review and editing: F.B., A.G., C.D. and A.K.; supervision and project administration: A.K. All authors have read and agreed to the published version of the manuscript.

**Funding:** This research has been cofinanced by the European Union and Greek national funds through the Operational Program Competitiveness, Entrepreneurship and Innovation, under the call RESEARCH—CREATE—INNOVATE (project code: T1EDK-00960, LEDWAR.gr).

**Conflicts of Interest:** The authors declare no conflict of interest.

## References

1. Lee, J.-M. Cultivation of Grafted Vegetables I. Current Status, Grafting Methods, and Benefits. *HortScience* **1994**, *29*, 235–239. [CrossRef]
2. Olle, M.; Alsiņa, I. Influence of wavelength of light on growth, yield and nutritional quality of greenhouse vegetables. *Proc. Latv. Acad. Sci.* **2019**, *73*, 1–9. [CrossRef]
3. Yujin, P.; Runkle, E. Far-red radiation and photosynthetic photon flux density independently regulate seedling growth but interactively regulate flowering. *Environ. Exp. Bot.* **2018**, *155*, 206–216.

4. Zhen, S.; Bugbee, B. Substituting Far-Red for Traditionally Defined Photosynthetic Photons Results in Equal Canopy Quantum Yield for $CO_2$ Fixation and Increased Photon Capture during Long-Term Studies: Implications for Re-Defining PAR. *Front. Plant Sci.* **2020**, *11*, 581156. [CrossRef] [PubMed]
5. IPCC. *Climate Change 2021: The Physical Science Basis. Summary for Policymakers. Working Group. Contribution to the Sixth Assessment Report of the Intergovernmental Panel on Climate Change*; IPCC: Geneva, Switzerland, 2021; p. 151.
6. Bantis, F.; Panteris, E.; Dangitsis, C.; Carrera, E.; Koukounaras, A. Blue light promotes hormonal induced vascular reconnection, while red light boosts the physiological response and quality of grafted watermelon seedlings. *Sci. Rep.* **2021**, *11*, 21754. [CrossRef]
7. Bantis, F.; Koukounaras, A.; Siomos, A.S.; Fotelli, M.N.; Kintzonidis, D. Bichromatic red and blue LEDs during healing enhance the vegetative growth and quality of grafted watermelon seedlings. *Sci. Hortic.* **2020**, *261*, 109000. [CrossRef]
8. Brazaitytė, A.; Duchovskis, P.; Urbonaviciute, A.; Samuolienė, G.; Jankauskienė, J.; Kasiulevicìute-Bonakere, A.; Bliznikas, Z.; Novičkovas, A.; Breive, K.; Zukauskas, A. The effect of light-emitting diodes lighting on cucumber transplants and after-effect on yield. *Zemdir. Agric.* **2009**, *96*, 102–118.
9. Sager, J.C.; Smith, W.O.; Edwards, J.L.; Cyr, K.L. Photosynthetic Efficiency and Phytochrome Photoequilibria Determination Using Spectral Data. *Trans. ASAE* **1988**, *31*, 1882–1889. [CrossRef]
10. Singleton, V.L.; Rossi, J.A. Colorimetry of total phenolics with phosphomolybdic-phosphotungstic acid reagents. *Am. J. Enol. Vitic.* **1965**, *16*, 144–158.
11. Benzie, I.F.F.; Strain, J.J. The ferric reducing ability of plasma (FRAP) as a measure of 'antioxidant power': The FRAP assay. *Anal. Biochem.* **1996**, *239*, 70–76. [CrossRef]
12. Luterotti, A.; Marković, K.; Franko, M.; Bicanic, D.; Madžgalj, A.; Kljak, K. Comparison of spectrophotometric and HPLC methods for determination of carotenoids in foods. *Food Chem.* **2013**, *140*, 390–397. [CrossRef]
13. Scott, A.J.; Knott, M.A. Cluster analysis method for grouping means in the analysis of variance. *Biometrics* **1974**, *30*, 507–512. [CrossRef]
14. Bantis, F.; Dangitsis, C.; Siomos, A.S.; Koukounaras, A. Light Spectrum Variably Affects the Acclimatization of Grafted Watermelon Seedlings While Maintaining Fruit Quality. *Horticulturae* **2021**, *8*, 10. [CrossRef]
15. Bozokalfa, M.K. Short communication. Irrigation temperature effects on seedling growth and transplant quality of tomato, pepper and eggplant. *Span. J. Agric. Res.* **2008**, *6*, 120–124. [CrossRef]
16. Baoying, S.; Yinian, L.; Sanqin, Z.; Weimin, D.; Na, H.; Jie, L. Effect of dark period lighting regulation on cucumber seedling morphology and comprehensive evaluation analysis and comprehensive evaluation. *Trans. Chin. Soc. Agric. Eng.* **2014**, *30*, 201–208.
17. Bantis, F.; Koukounaras, A.; Siomos, A.; Menexes, G.; Dangitsis, C.; Kintzonidis, D. Assessing quantitative criteria for characterization of quality categories for grafted watermelon seedlings. *Horticulturae* **2019**, *5*, 16. [CrossRef]
18. Ouzounis, T.; Heuvelink, E.; Ji, H.; Schouten, H.J.; Visser, R.G.F.; Marcelis, L.F.M. Blue and red LED lighting effects on plant biomass, stomatal conductance, and metabolite content in nine tomato genotypes. *Acta Hortic.* **2016**, *1134*, 251–258. [CrossRef]
19. Novičkovas, A.; Brazaitytė, A.; Duchovskis, P.; Jankauskienė, J.; Samuolienė, G.; Virsilė, A.; Sirtautas, R.; Bliznikas, Z.; Zukauskas, A. Solid-state lamps (LEDs) for the short-wavelenght supplementary lighting in greenhouses: Experimental results with cucumber. *Acta Hortic.* **2012**, *927*, 723–730. [CrossRef]
20. Rabara, R.C.; Behrman, G.; Timbol, T.; Rushton, P.J. Effect of Spectral Quality of Monochromatic LED Lights on the Growth of Artichoke Seedlings. *Front. Plant Sci.* **2017**, *8*, 190. [CrossRef]
21. Hernandez, R.; Kubota, C. Physiological responses of cucumber seedlings under different blue and red photon flux ratios using LEDs. *Environ. Exp. Bot.* **2016**, *121*, 66–74. [CrossRef]
22. Javanmardi, J.; Emami, S. Response of Tomato and Pepper Transplants to Light Spectra Provided by Light Emitting Diodes. *Int. J. Veg. Sci.* **2013**, *19*, 138–149. [CrossRef]
23. Yan, Z.; He, D.; Niu, G.; Zhai, H. Evaluation of growth and quality of hydroponic lettuce at harvest as affected by the light intensity, photoperiod and light quality at seedling stage. *Sci. Hortic.* **2019**, *248*, 138–144. [CrossRef]
24. Demotes-Mainard, S.; Péron, T.; Corot, A.; Bertheloot, J.; Le Gourrierec, J.; Pelleschi- Travier, S.; Crespel, L.; Morel, P.; Huché-Thélier, L.; Boumaza, R.; et al. Plant responses to red and far-red lights, applications in horticulture. *Environ. Exp. Bot.* **2016**, *121*, 4–21. [CrossRef]
25. Huché-Thélier, L.; Crespel, L.; Gourrierec, J.; Le Morel, P.; Sakr, S.; Leduc, N. Light signaling and plant responses to blue and UV radiations-perspectives for applications in horticulture. *Environ. Exp. Bot.* **2016**, *121*, 22–38. [CrossRef]
26. Yanovsky, M.J.; Kay, S.A. Molecular basis of seasonal time measurement in Arabidopsis. *Nature* **2002**, *419*, 308–312. [CrossRef]
27. Suárez-López, P.; Wheatley, K.; Robson, F.; Onouchi, H.; Valverde, F.; Coupland, G. CONSTANS mediates between the circadian clock and the control of flowering in Arabidopsis. *Nature* **2001**, *410*, 1116–1120. [CrossRef]
28. Prisca, M.; Maarten, V.; Van Doorsselaere, J.; Bart, N.; Saeys, W.; Hytonen, T.; De Coninck, B.; Van de Poel, B. Blue and far-red light control flowering time of woodland strawberry (*Fragaria vesca*) distinctively via CONSTANS (CO) and FLOWERING LOCUS T1 (FT1) in the background of sunlight mimicking radiation. *Environ. Exp. Bot.* **2022**, *198*, 104865. [CrossRef]
29. Craig, D.S.; Runkle, E.S. An intermediate phytochrome photoequilibria from night interruption lighting optimally promotes flowering of several long-day plants. *Environ. Exp. Bot.* **2016**, *121*, 132–138. [CrossRef]
30. Riadi, M.F.; Esyanti, R.R.; Faizal, A. The Effect of Led Light on Production of Female Flowers in Cucumber (*Cucumis sativus* L.). *Int. J. Sci. Technol.* **2015**, *1*, 13–23.

31. Palmitessa, O.D.; Durante, M.; Leoni, B.; Montesano, F.; Renna, M.; Serio, F.; Somma, A.; Santamaria, P. Enhancement of a Landrace of Carosello (Unripe Melon) through the Use of Light-Emitting Diodes (LED) and Nutritional Characterization of the Fruit Placenta. *Sustainability* **2021**, *13*, 11464. [CrossRef]
32. Gajc-Wolska, J.; Kowalczyk, K.; Przybysz, A.; Mirgos, M.; Orliński, P. Photosynthetic Efficiency and Yield of Cucumber (*Cucumis sativus* L.) Grown under HPS and LED Lighting in Autumn–Winter Cultivation. *Plants* **2021**, *10*, 2042. [CrossRef] [PubMed]

agriculture MDPI

Article

# Proposed Light Wavelengths during Healing of Grafted Tomato Seedlings Enhance Their Adaptation to Transplant Shock

**Christos Melissas [1], Filippos Bantis [1], Christodoulos Dangitsis [2], Stefanos Kostas [1] and Athanasios Koukounaras [1,*]**

1 Department of Horticulture, Aristotle University, 54124 Thessaloniki, Greece; cmelissas@agro.auth.gr (C.M.); fbantis@agro.auth.gr (F.B.); skostas@agro.auth.gr (S.K.)
2 Agris S.A., Kleidi, 59300 Imathia, Greece; cdaggitsis@agris.gr
* Correspondence: thankou@agro.auth.gr

**Abstract:** Tomato, which is mainly established with grafted seedlings, is one of the most popular vegetables worldwide with a high nutritional value,. Market demand for grafted seedlings is high in specific seasons; thus, commercial nurseries face a problem of limited space availability during the healing stage. Light quality is an essential parameter during healing that can adjust seedling development towards desirable traits and lead to time and space saving during seedling production. Moreover, transplant shock constitutes another challenge that could limit crop yield. The objective of this study was to evaluate the overall quality of grafted tomato seedlings and their potential adjustment to transplant shock as affected by different light spectra during healing in a chamber. Evaluations were conducted immediately after exiting the healing chamber and after transplantation into pots. Light wavelengths were used from fluorescent lamps (FL) or light-emitting diodes with red (R), blue (B), red–blue combinations with 12 and 24% blue (12B and 24B), and white (W) emitting 11% blue. W enhanced the dry shoot biomass and the root architecture before and after transplantation. 24B led to an increased stem diameter, root development, and phenolic and antioxidant accumulation at both phases of the experiment. 12B enhanced the leaf area before transplantation and root development after transplantation. FL, R and B induced inferior seedling growth compared to the red–blue-containing LEDs, with B performing poorly in almost all tested parameters. Overall, red, including 11–24% blue, provides the optimum light conditions during the healing stage for the production of high-quality grafted tomato seedlings, with advanced capabilities of abiotic stress adaptation to transplant shock.

**Keywords:** *Solanum lycopersicon* L.; scion; rootstock; light-emitting diodes; light quality; nursery; growth chamber; photomorphogenesis; root system architecture; antioxidant activity

**Citation:** Melissas, C.; Bantis, F.; Dangitsis, C.; Kostas, S.; Koukounaras, A. Proposed Light Wavelengths during Healing of Grafted Tomato Seedlings Enhance Their Adaptation to Transplant Shock. *Agriculture* **2022**, *12*, 797. https://doi.org/10.3390/agriculture12060797

Academic Editor: Jie He

Received: 11 April 2022
Accepted: 28 May 2022
Published: 31 May 2022



## 1. Introduction

Tomato (*Solanum lycopersicum* L.) is a very popular vegetable worldwide, mostly known for its unique taste and high nutritional value. It constitutes a great source of health-promoting compounds such as minerals and antioxidants such as vitamin C, E, carotenoids, flavonoids and anthocyanins [1]. It is one of the most widely cultivated crops reaching 4.8 million ha globally, almost 500,000 of which were in the E.U., for the period 2010–2019 [2]. Tomato cultivation is mainly established with transplants due to their higher uniformity in size, and well-developed root systems and shoots, leading to constant and high-quality productions with reduced losses compared to seed planting [3].

Transplants might be grafted or nongrafted, although, in the last few decades, grafted transplants have become preferable in the market due to their increased capabilities such as tolerance to soilborne diseases, salinized cultivated lands, drought, heavy metals presence, etc. [3]. Grafting is the union of two intraspecific or interspecific plants or even intrafamilial plant parts, aiming for a successful connection between their vascular bundles to form one composite organism that functions as a single plant with combined genetic characteristics [4]. The grafting procedure includes four distinguishable stages: (a) the selection of the appropriate combination of a rootstock–scion, (b) the grafting union and connection of the rootstock–scion, (c) healing of the newly grafted plant and (d) hardening of the newly grafted plant [3]. Among the procedure steps, the healing stage of the grafted union is a very crucial process and, thus, requires experienced personnel and specific conditions of relative humidity (>90%), temperature (22–30 °C) and even lighting that favour tissue regeneration and a successful connection between the vascular bundles of the rootstock and scion [5,6]. The stage of healing can be accomplished in environmentally controlled spaces, including growth chambers, where the above-mentioned factors can be adjusted entirely. In addition, market demand for grafted seedlings is high in specific seasons; thus, commercial nurseries face a problem of limited space availability during the healing stage. Therefore, potential time saving through new techniques such as altering the light quality could be essential for saving space and reducing operational costs.

Light is an essential factor in the healing process as numerous cell divisions on the grafting union require a large amount of energy derived through respiration from the consumption of carbohydrates, which are produced in the photosynthesis process controlled by light [7]. Plants receive light radiation and efficiently absorb between wavelengths of 300–750 nm through photoreceptors, and they exhibit various responses depending on their genotype as well as the light intensity, quality, direction and duration [8]. Artificial lighting in horticulture is usually accomplished using fluorescent (FL) lamps, which are used in greenhouses and growth chambers especially due to their high performance and low cost while also having a balanced emission spectrum suitable for plant growing. In the last century, the expanding technology of light-emitting diodes (LEDs) has replaced conventional lamps in almost every artificial lighting application [9]. For example, LEDs are utilized during the healing of grafted watermelon seedlings, which account for over 90% of the total produced watermelon seedlings in some countries (e.g., Japan, Korea, Greece), and grafted tomato seedlings, which account for more than 25% of the total produced tomato seedlings in some countries (Japan, Taiwan, Korea 40%, USA 70%) [3,10]. Relatively narrow-band spectra for matching plants' photoreceptors, production of high light irradiations with low radiant heat and long-life cycles [11] are the major features of LEDs, along with their low energy consumption and small size [12]. According to McCree's study [13], a light environment including high portions of red (600–700 nm) and blue (400–500 nm) wavelengths is ideal for photosynthesis since they are the most photosynthetically efficient parts of the radiation spectra. A recent study involving LEDs for grafted tomato seedlings' production revealed that white comprised of a red/blue (R/B) ratio of 1.2 and a red/far-red (R/FR) ratio of 16 enhanced the transplant quality [14]. In another study with light quality during the healing and acclimatization of grafted tomato seedlings, blue light led to inferior growth compared to red and FL [15].

By the end of a successful grafting process, and the production and distribution of high-quality grafted seedlings, transplantation constitutes another challenge that could limit the crop yield. The transplantation process often diminishes root development of the newly planted seedlings through the destruction of the effective root area and root hairs, resulting in reduced water and nutrient uptake capacity, a phenomenon known as transplant shock [16]. This abiotic stress to plant metabolism is exacerbated when combined with unfavourable soil conditions [17].

The objective of this study was to evaluate the overall quality of grafted tomato seedlings by determining their important physiological and morphological characteristics after their exposure to different light spectra during healing in a chamber. Furthermore, the study aimed to assess the potential after-effect response of the seedlings after their transplantation as affected by different light wavelengths during healing.

## 2. Materials and Methods

### *2.1. Plant Material and Growing Conditions*

The experiment consisted of two separate phases. The first phase was executed in the facilities of a nursery company (Agris S.A. Kleidi, Imathia, Greece), and the second phase was performed at the greenhouse of the Laboratory of Floriculture of the Aristotle University of Thessaloniki, Greece. All measurements were conducted at the Laboratory of Vegetable Crops of the Aristotle University of Thessaloniki, Greece.

Two tomato (*Solanum lycopersicum* L.) hybrids, "Kabrera F1" and "Emperador F1", were used as scion and rootstock material, respectively. Kabrera F1 hybrid is a popularly grown tomato in Greece, which is usually grafted onto other Solanaceae plants (e.g., tomato and eggplant). Emperador F1 hybrid is a tomato rootstock that provides the scion with tolerance to low temperatures and nematodes. Kabrera × Emperador is a popular grafting combination for grafted tomato seedlings grown in greenhouses in Greece. Tomato seeds of both hybrids were sown in 128-cell plug trays (G.K. Rizakos S.A., Lamia, Greece) containing a 3:1 mixture of peat and vermiculite as substrate.

A schematic representation of the growth of grafted tomato seedlings along with the sampling times in the first and second phases is depicted in Figure 1. After sowing, trays were placed in a germination chamber of favourable conditions of 97% relative humidity and 24 °C temperature in darkness until germination (48–72 h). Upon seedlings' emergence, trays were placed in a Venlo-type greenhouse for 18 days until grafting. Scion seedlings were grown at 18 °C day temperature, while rootstock seedlings were grown at 21.5 °C, all at 60–75% relative humidity. The night temperature was common at 19 °C, and 18 h artificial lighting (100 ± 10 μmol $m^{-2}$ $s^{-1}$) was supplemented to both hybrids provided by high-pressure sodium lamps (MASTER GreenPower E40, Philips Lighting, Eindhoven, The Netherlands).

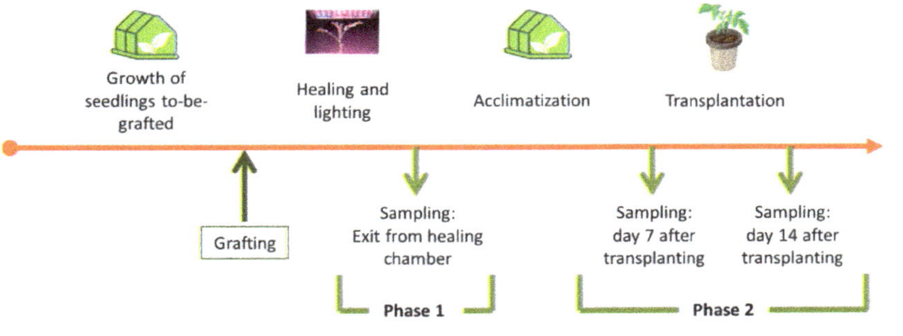


**Figure 1.** Schematic depiction of the growth of grafted tomato seedlings along with the sampling times in the first and second phases of our experiment.

### *2.2. Grafting and Healing Process*

Eighteen days later, at the stage of two true leaves, scion was grafted on rootstock hybrids through splice grafting, and the plug trays containing the newly grafted seedlings were immediately placed in a healing chamber for six days. Precise environmental conditions of high relative humidity at 90–95% and temperature of 22.5 °C were performed while the air was recirculating.

### *2.3. Light Treatments in the Healing Chamber (First Phase)*

Inside the healing chamber, sole artificial lighting was provided by 5 LEDs or fluorescent lamps mounted on vertically structured shelves (L:2 00 m × W:1.66 m × H:0.76 m). Plug trays were placed on every shelf where one lamp was installed at 30 cm above the plant top. The photoperiod was 18 h, and photosynthetic photon flux density (PPFD) at plant top was 85 μmol $m^{-2}$ $s^{-1}$, while the LEDs' spectra consisted of narrow-band red (R; peak wavelength at 661 nm), narrow-band blue (B; peak wavelength at 451 nm), two combinations of RB (12B and 24B) emitting 12% and 24% of blue, respectively, and white (W) emitting 11% of blue. The latter treatment is desirable due to the high colour rendering index (CRI > 50), which facilitates the activities inside the healing chamber. Light treatments' properties and spectral distributions were obtained with a HD 30.1 spectroradiometer (DeltaOhm Srl, Padova, Italy) and are presented in Table 1.

**Table 1.** Spectral distribution expressed as percentages of total photons reaching the seedling canopy. PPS: phytochrome photostationary state; FL: fluorescent lamps; W: white; B: blue; 24B: 24% blue; 12B: 12% blue; R: red.

| Waveband | Light Treatment | | | | | |
|---|---|---|---|---|---|---|
| | FL | W | B | 24B | 12B | R |
| Blue%; 400–499 nm | 35 | 11 | 100 | 24 | 12 | 0 |
| Green%; 500–599 nm | 24 | 18 | 0 | 0 | 0 | 0 |
| Red%; 600–699 nm | 37 | 70 | 0 | 76 | 88 | 100 |
| Far-red%; 700–780 nm | 4 | 1 | 0 | 0 | 0 | 0 |
| PPS | 0.82 | 0.89 | 0.51 | 0.89 | 0.89 | 0.89 |

### 2.4. Acclimatization Process

After the healing process, grafted seedlings were placed into a Venlo-type greenhouse for seven days with a mean temperature of 18 °C and relative humidity of 50–55% and 75–80% at day and night, respectively, for acclimatization. Supplemental artificial lighting was provided by HPS lamps for 18 h daily with a PPFD of 100 ± 10 μmol $m^{-2}$ $s^{-1}$ at the plant top.

### 2.5. Transplantation (Second Phase)

Seven days later, 25 grafted seedlings per light treatment (150 in total) were transplanted in larger pots (L:7.00 cm × W:7.00 cm × H:6.00 cm) containing a mixture of peat and perlite (2:1) and were placed in a greenhouse. Upon transplantation, seedlings were irrigated with 100 mL of Hoagland's solution [18] until runoff, followed by 20 mL every two days for a total of 14 days.

### 2.6. Sampling and Measurements

In the first phase of the experiment and upon exiting the healing chamber, ten grafted seedlings per light treatment were sampled randomly, while their quality parameters were evaluated. In the second phase of the experiment, at seven and fourteen days after transplantation, six randomized grafted seedlings per light treatment were sampled and assessed for their qualitative characteristics. These characteristics were suggested by Lee et al. [3] for the definition of grafted tomato seedling quality. Specifically, leaf area was measured using an AM350 area meter (ADC BioScientific Ltd., Hoddesdon, UK), while shoot length (i.e., the length between the apical bud and root collar) and stem diameter below the cotyledons were determined using a digital calliper. In addition, fresh and dry (after three days in an oven at 72 °C) shoot and root weights were measured. Shoot/root (S/R) ratio, shoot dry weight/shoot length (DW/L) ratio, root dry weight/surface area (R/SA) ratio and percentage of root dry weight (%DWR) were also calculated. Chlorophyll fluorescence was determined after 20 min dark adaptation on the first fully developed leaf with a pocket PEA Chlorophyll Fluorimeter (Hansatech Instruments Ltd., Norfolk, UK), and relative chlorophyll content was determined using a CCM-200 plus chlorophyll meter (Opti-Sciences, Hudson, NH, USA). Root growth parameters, including root diameter, root surface area and root length, were obtained through root scanning using a root scanner (EPSON Perfection V700, Nagano, Japan) after their flushing with clean water, and the results were acquired through an image analysis software (WinRHIZO Pro, Regent Instruments Inc., Quebec City, QC, Canada). Moreover, grafted seedlings' leaf lamellae were cooled with liquid nitrogen before their pulverization in a porcelain mortar and eventually stored at −30 °C for a week. Following this, 2.5 g was extracted into 25 mL 80% aqueous methanol, and total phenolic compounds, as well as total antioxidant capacity, were determined.

2.6.1. Total Phenolic Compounds

Phenolic compounds' concentration in the leaf lamellae extracts was determined according to the Folin–Ciocalteu method [19], where 2.5 mL of Folin–Ciocalteu and 2 mL of 7.5% sodium carbonate solution were added in 0.5 mL of methanolic plant extract and mixed under continuous stirring. The mixture was incubated at 50 °C for 5 min and cooled at room temperature for 3 min, and its absorbance was measured at 760 nm. The results were expressed as g of gallic acid (GAE)/g fresh weight.

2.6.2. Total Antioxidant Capacity (FRAP)

The same methanolic plant extract (from leaf lamellae) was used for the conductance of this method. The ferric reducing antioxidant power (FRAP) assay was produced according to [20], where 250 mL of $CH_3COONa$ buffer solution, 100 mL of TPTZ solution and 100 mL of $FeCl_2$ solution were mixed using a stirrer. Afterwards, 3 mL of this reagent was added to 0.1 mL methanolic plant extract and incubated at 37 °C for 4 min. The absorbance was measured at 593 nm, and the results were expressed as μg of plant extract's FRAP assay.

### 2.7. Statistical Analysis

Data were statistically analysed using IBM SPSS software (SPSS 23.0, IBM Corp., Armonk, NY, USA). After analysis of variance (ANOVA), post hoc test for comparisons between all the treatments was conducted using the LSD method (unprotected) at significance level $\alpha = 0.05$. Moreover, *t*-test was conducted for the comparison between FL and each LED treatment at significance level $\alpha = 0.05$. The experiment was performed two times, reaching similar conclusions. Herein, the results from the first repetition are presented.

## 3. Results

### 3.1. Exit from the Healing Chamber (First Phase)

Shoot length was significantly affected by the different light treatments, as seedlings exposed to W exhibited the highest values compared to the rest of the light treatments, while seedlings under FL were the shortest among all light treatments. Moreover, values in 24B were significantly higher than B and FL. Comparisons using the *t*-test showed that every LED treatment had significantly greater values compared to FL (Figure 2A). Stem diameter was also affected; seedlings under red–blue combinations (24B and 12B) developed the thickest stems, which were significantly greater compared to W and B LEDs. 24B was the only treatment with significantly greater stem diameter compared to FL (Figure 2B). Furthermore, red–blue combinations (24B and 12B) enhanced the seedlings' leaf area compared to R, B and FL (Figure 2C). In contradiction to the above, FL induced the development of a greater DW/L ratio compared to the rest of the light treatments, while B showed significantly lower values compared to 24B and 12B. According to the *t*-test, all LEDs had significantly lower DW/L compared to FL (Figure 2D). Shoot dry weight was greater under W and 24B compared to FL and B, while the latter also showed significantly lower values compared to 12B and R. Moreover, individual *t*-test comparisons showed significant differences between FL and W, 24B, 12B and R (Figure 2E). 24B also promoted root dry weight development compared to FL, B, 12B and R, while the latter showed lower values than W as well (Figure 2F).

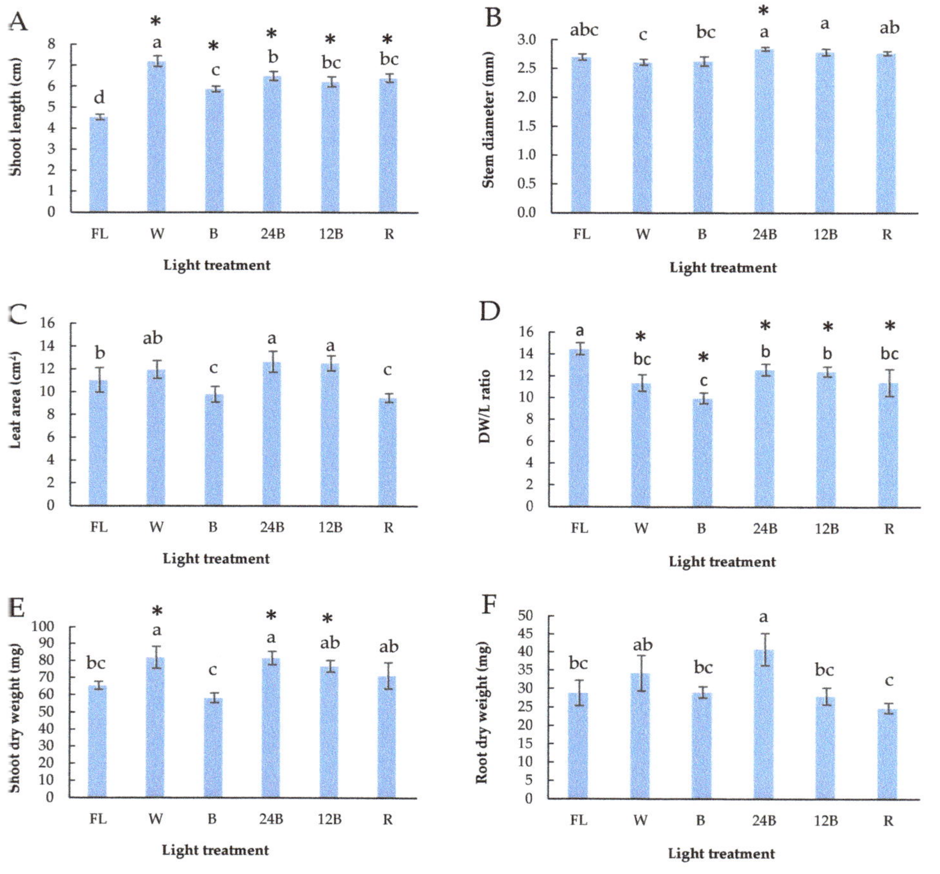


**Figure 2.** (**A**) Shoot length, (**B**) stem diameter, (**C**) leaf area, (**D**) dry weight/length (DW/L) ratio, (**E**) shoot dry weight and (**F**) root dry weight of 10 grafted tomato seedlings per light treatment after their exposure to six light treatments during healing. Bars (±SE) followed by different letters are significantly different ($p \leq 0.05$) according to the LSD method. Bars (±SE) followed by asterisks indicate significant differences between the LED treatments and the control (FL) at $p \leq 0.05$ according to *t*-test. FL: fluorescent lamps; W: white; B: blue; 24B: 24% blue; 12B: 12% blue; R: red.

### 3.2. *Transplantation (Second Phase)*

Seven days after transplantation, 24B enhanced stem diameter compared to FL and R, while 12B led to enhanced values compared to R (Figure 3A). Leaves were significantly expanded under W compared to B, 24B and R (Figure 3B). Shoot dry weight was enhanced under FL and W compared to B, while the latter light treatment also significantly inhibited the root dry weight production compared to FL, W, 24B and 12B (Figure 3C,D). Regarding root architecture analysis, W increased the root length compared to FL, B and R, while the root surface area was also enhanced under W compared to B. Moreover, W and 24B were also greater compared to FL according to the *t*-test (Figure 3E,F). The maximum quantum yield of the primary photochemistry (Fv/Fm) was significantly greater under B compared to 12B (Table 2). However, the total phenolic compounds and FRAP were not significantly affected by the different light treatments (Table 2). R/SA was significantly greater in FL and 24B compared to W, while %DWR at day 7 was not affected by the light treatments (Table 2).

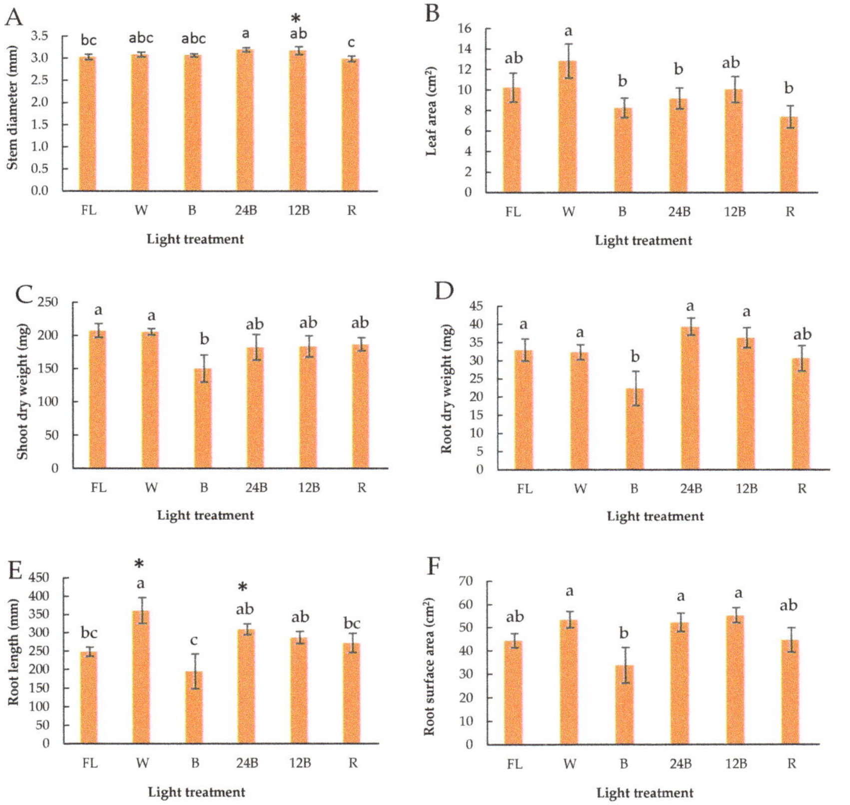


**Figure 3.** (**A**) Stem diameter, (**B**) leaf area, (**C**) shoot dry weight, (**D**) root dry weight, (**E**) root length and (**F**) root surface area of 6 grafted tomato seedlings per light treatment during healing, 14 days after acclimatization process plus seven days after transplantation. Bars (±SE) followed by different letters are significantly different ($p \leq 0.05$) according to the LSD method. Bars (±SE) followed by asterisks indicate significant differences between the LED treatments and the control (FL) at $p \leq 0.05$ according to *t*-test. FL: fluorescent lamps; W: white; B: blue; 24B: 24% blue; 12B: 12% blue; R: red.

Fourteen days after transplantation, the stem diameter was significantly promoted under R compared to the rest of the light treatments, except for FL, while B showed lower values compared to FL according to the *t*-test (Figure 4A). 12B enhanced leaf area development compared to FL and R (Figure 4B). Shoot dry weight was enhanced by W than B, while the *t*-test also showed greater values for W compared to FL (Figure 4C). Root dry weight was greater under 24B compared to FL, while the *t*-test also showed greater values for W, 24B and R compared to FL (Figure 4D). Roots were significantly longer in W and 24B compared to B (Figure 4E), while no significant differences were observed in the root surface area, except for the *t*-test, which showed greater values in 12B compared to FL (Figure 4F). Fv/Fm was not significantly different among the light treatments, but B and 24B had significantly lower values compared to FL according to the *t*-test (Table 2). The total phenolic compounds were greater under 24B and FL than B (Table 2), while the FRAP was enhanced under 24B compared to R and FL (Table 2). At 14 days, R/SA was not significantly affected, while the %DWR was significantly greater under B compared to FL (Table 2).

**Table 2.** Effect of light treatment on physiological parameters, biochemical compounds and calculated qualitative parameters of grafted tomato seedlings 7 and 14 days after transplantation. Fv/Fm: maximum quantum yield of primary photochemistry; TPC in mg/kg: total phenolic content; FRAP in μg/g; ferric reducing antioxidant power; R/SA in g/cm$^2$: root dry weight/surface area; %DWR: percentage of root dry weight; FL: fluorescent lamps; W: white; B: blue; 24B: 24% blue; 12B: 12% blue; R: red. Mean values followed by different letters are significantly different at $p \leq 0.05$ according to the LSD method. Asterisks indicate significant differences between the LED treatments and the control (FL) at $p \leq 0.05$ according to *t*-test.

| Parameter | Light Treatment | | | | | | |
|---|---|---|---|---|---|---|---|
| | FL | W | B | 24B | 12B | R | *p*-Value |
| Fv/Fm d7 | 0.84 ± 0.00 $^{ab}$ | 0.84 ± 0.00 $^{ab}$ | 0.84 ± 0.00 $^{a}$ | 0.84 ± 0.00 $^{ab}$ | 0.83 ± 0.00 $^{b}$ | 0.84 ± 0.00 $^{ab}$ | 0.136 |
| Fv/Fm d14 | 0.81 ± 0.00 $^{a}$ | 0.80 ± 0.01 $^{a}$ | 0.80 ± 0.00 $^{a,*}$ | 0.80 ± 0.00 $^{a,*}$ | 0.81 ± 0.00 $^{a}$ | 0.81 ± 0.00 $^{a}$ | 0.509 |
| TPC d7 | 0.33 ± 0.03 $^{a}$ | 0.29 ± 0.03 $^{a}$ | 0.30 ± 0.01 $^{a}$ | 0.34 ± 0.01 $^{a}$ | 0.29 ± 0.01 $^{a}$ | 0.32 ± 0.02 $^{a}$ | 0.435 |
| TPC d14 | 0.30 ± 0.01 $^{a}$ | 0.28 ± 0.01 $^{ab}$ | 0.25 ± 0.00 $^{b,*}$ | 0.29 ± 0.03 $^{a}$ | 0.26 ± 0.01 $^{ab}$ | 0.28 ± 0.00 $^{ab}$ | 0.060 |
| FRAP d7 | 155.6 ± 7.1 $^{a}$ | 168.1 ± 6.0 $^{a}$ | 162.3 ± 17.7 $^{a}$ | 165.6 ± 1.8 $^{a}$ | 166.9 ± 3.6 $^{a}$ | 168.1 ± 5.5 $^{a}$ | 0.866 |
| FRAP d14 | 146.5 ± 5.1 $^{c}$ | 166.4 ± 8.7 $^{abc}$ | 165.5 ± 8.3 $^{abc}$ | 179.1 ± 6.8 $^{a,*}$ | 171.4 ± 7.5 $^{ab}$ | 152.4 ± 2.4 $^{bc}$ | 0.048 |
| R/SA d7 | 0.74 ± 0.02 $^{a}$ | 0.61 ± 0.01 $^{b,*}$ | 0.68 ± 0.07 $^{ab}$ | 0.71 ± 0.02 $^{a}$ | 0.70 ± 0.01 $^{ab}$ | 0.69 ± 0.02 $^{ab}$ | 0.085 |
| R/SA d14 | 0.57 ± 0.01 $^{a}$ | 0.63 ± 0.03 $^{a}$ | 0.64 ± 0.03 $^{a}$ | 0.73 ± 0.09 $^{a}$ | 0.56 ± 0.11 $^{a}$ | 0.62 ± 0.03 $^{a}$ | 0.430 |
| %DWR d7 | 13.8 ± 1.4 $^{a}$ | 13.6 ± 0.9 $^{a}$ | 13.7 ± 4.2 $^{a}$ | 17.8 ± 0.8 $^{a}$ | 16.5 ± 0.2 $^{a}$ | 14.0 ± 0.8 $^{a}$ | 0.459 |
| %DWR d14 | 9.4 ± 0.1 $^{b}$ | 10.3 ± 0.5 $^{ab}$ | 13.0 ± 2.2 $^{a,*}$ | 12.3 ± 1.0 $^{ab}$ | 11.1 ± 1.2 $^{ab}$ | 10.6 ± 0.5 $^{ab}$ | 0.115 |

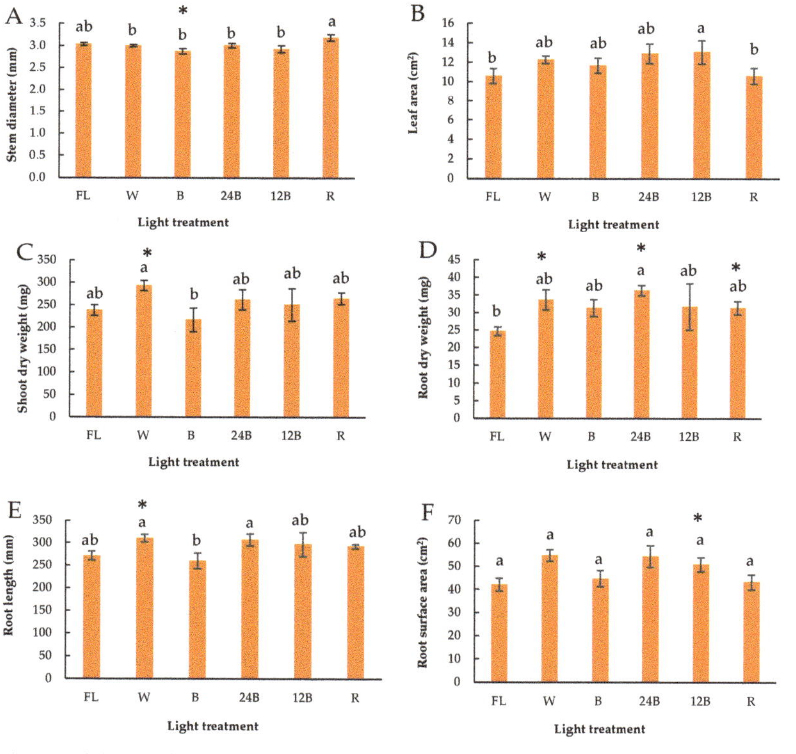


**Figure 4.** (**A**) Stem diameter, (**B**) leaf area, (**C**) shoot dry weight, (**D**) root dry weight, (**E**) root length and (**F**) root surface area of 6 grafted tomato seedlings per light treatment during healing, 14 days after acclimatization process plus 14 days after transplantation. Bars (±SE) followed by different letters are significantly different ($p \leq 0.05$) according to the LSD method. Bars (±SE) followed by asterisks indicate significant differences between the LED treatments and the control (FL) at $p \leq 0.05$ according to *t*-test. FL: fluorescent lamps; W: white; B: blue; 24B: 24% blue; 12B: 12% blue; R: red.

## 4. Discussion

The seedling production industry and market require the production of high-quality grafted seedlings. Seedlings of optimum morphological and physiological characteristics exhibit faster plant development and uniformity along with better transplantation success, thus leading to high, standard yields of excellent quality and marketable profit [3]. Subsequently, it is crucial to ensure the development of appropriate methods, including specific conditions during the healing process, along with selecting suitable light spectra for the proper development of high-quality grafted seedlings with the minimum possible cost and the lowest environmental impact. The introduction of LEDs in seedlings' production during the healing stage constitutes a highly efficient, nonchemical, sustainable solution for plant development regulation and quality enhancement [21].

LED is a significantly more expensive technology per photosynthetic photon compared to traditional light sources; thus, economic viability is based on decreased electric costs due to enhanced fixture efficiency. Among the important benefits of LED technology, the highly focused radiation can lead to the light reaching the plants with considerable efficiency and subsequently leading to reduced electricity costs [22].

### *4.1. Exit from the Healing Chamber (First Phase)*

Significant differences were recorded among light treatments in the majority of the evaluated parameters. It is obvious that treatments with an increased red light portion enhanced stem elongation through phytochromes, as reported by Li et al. [23]. In addition, blue light is known to decelerate stem elongation [24]. These findings agree with Javanmardi and Emami [25], who reported tomato and pepper seedlings' shoot elongation under LED light spectra compared to narrow-band blue and red and their combinations, as well as with Głowacka's [26] findings of an inhibiting effect of blue light on tomato transplant growth. Phytochromes (the red and far-red photoreceptors) and cryptochromes (the blue photoreceptors) are responsible for the so-called shade-avoidance responses, including stem elongation. A reduction in the R/FR ratio alters the level of phytochrome B leading to the activation of Phytochrome Interacting Factors (PIFs), and subsequently increasing the auxins' level [27].

Stem diameter was significantly enhanced under combinations of red and blue wavelengths (mainly 12B and 24B) compared to B, as reported by Hernandez et al. [28] in their study with grafted tomato seedlings exposed to different light wavelengths in a plant factory. Similar results were recorded on leaf area under 12B and 24B, which showed a beneficial effect compared to B and R spectra, in agreement with Ouzounis et al.'s [29] reports about the additive effect of red and blue light in leaf expansion. Both stem diameter and leaf area have been characterized as valuable quality indices for the determination of grafted watermelon seedling quality after the healing stage [30].

DW/L has been suggested as an efficient indicator of grafted tomato seedlings' quality [3], as also stated for grafted watermelon [30]. In our case, FL exhibited considerable DW/L, but both incorporated parameters, shoot dry weight and shoot length, were inferior compared to other light treatments. We concluded that seedlings treated with FL indeed reached a high quality but showed a much slower growth rate. Among the LEDs, B showed inferior DW/L, reaffirming the low quality displayed by other quality parameters such as stem diameter, leaf area and root dry weight.

Indeed, seedlings treated with B exhibited inferior shoot and root biomass formation compared to red–blue treatments and especially 24B. In green tissues, blue wavelengths are mainly absorbed by carotenoids and anthocyanins, which are not efficient energy transducers of the photosynthetic apparatus [31]. On the contrary, red light drives photosynthesis in a more efficient manner, while blue mediates photomorphogenic processes [8].

### *4.2. Transplantation (Second Phase)*

Seven days after transplantation, stem diameter was greater under a combination of red–blue (24B) compared to monochromatic red light, as also reported by Li et al. [23], who studied tomato seedlings' responses under six different light treatments for 30 days in an artificial climate chamber. Conversely, R significantly improved stem diameter 14 days after transplantation. It is possible that the long-term effects of light wavelengths could be contrasting compared to the effects during or immediately after plant exposure.

The leaf area of transplanted tomato seedlings was significantly enhanced by W after 7 days and 12B after 14 days from transplantation. Both light treatments emit a significant portion of red light (83–88%) supplemented with 11–12% blue, leading to the conclusion that red–blue wavelengths are beneficial for the growth and development of plant tissues, at least during the first stages of plant growth. These results are in agreement with a study involving cucumber supporting the beneficial effect of red and blue combinations on seedlings' development compared to monochromatic red light [32]. This conclusion is associated with chlorophyll pigments, which mainly absorb blue and red wavelengths; thus, red–blue combinations comprise a highly efficient, complete light source. Conversely, Wu et al. [33] reported that a sole red LED light imposed a beneficial effect on leaf area expansion compared to a white LED light.

The B wavelength decelerated the shoot and root biomass production both at 7 and 14 days after transplantation, except for the root dry weight at 14 days. Regarding the analysis of the root architecture after transplantation, the image is similar to the previously reported results. As a general rule, B induced the development of the least expanded root system as displayed by the shorter total root length and the lowest root surface values (the latter only at seven days after transplantation). Conversely, LEDs containing red and blue wavelengths enhanced the root system development. These observations highlight the long-term effect of sole blue light spectra on the inhibition of cell expansion and division [34]. Similarly, a study with cucumber showed that the dry shoot mass was enhanced under red light supplemented with blue [35].

All plants were healthy at both time intervals after transplantation, as shown by the high Fv/Fm values (0.80–0.84). Björkman and Demmig [36] stated that values of 0.78–0.86 are indicative of healthy plants with efficient photosynthetic activity.

Plants produce and accumulate secondary metabolites such as phenolic compounds as a response to several biotic and abiotic stressful situations. In some cases, light quality can impose significant stress on plant tissues, thus leading to the increased biosynthesis of such compounds [37]. Phenolics are antioxidant compounds associated with the scavenging of reactive oxygen species. It is assumed that an increased antioxidant capacity may be related to sturdier and better acclimated plants, which may show enhanced vegetative growth in the long run. In our study, transplanted tomato seedlings showed increased phenolic compounds and antioxidant capacity under 24B, indicating the necessity of both red and blue wavelengths in specific portions for enhanced plant development and quality. Similarly to our results, a 70% red/30% blue treatment led to the increased phenolic content of lamb's lettuce compared with monochromatic red and blue [38], while spinach did not show significant differences in total phenolics and antioxidants when grown under broad-spectra light treatments [39].

Overall, individual comparisons showed that FL performed similar to B leading to the production of seedlings with poor capacity to intercept light (i.e., leaf area), and thus lower overground and underground biomass production. This effect was continued after transplantation in the greenhouse when both light treatments, and especially B, showed inferior root development. Conversely, LED treatments emitting relatively high amounts of red light favoured the production of seedlings with greater potential to overcome the transplanting shock and develop into plants with a vast root system. This study revealed potential research gaps that require further research and attention. For example, the experimental procedure can also be applied for different scion–rootstock hybrid combinations, or even for different species with the ability to be grafted. Moreover, light quality is known to affect the flowering of plants; thus, field or greenhouse cultivation would enhance our understanding of the potential after-effect of light quality on flowering and even crop yield and quality.

## 5. Conclusions

B performed poorly in almost all tested parameters in both experiment phases, reaffirming the wavelength's inhibitory effect on plant growth when used alone. R also induced inferior seedling growth compared to red–blue-containing LEDs, but not as much as B, indicating the increased importance of the red wavelength for plant growth compared to blue. Seedling growth was also decelerated under FL compared to red–blue-containing LEDs proving once more the superiority of the latter light source for the production of high-quality seedlings due to better spectral distribution. Overall, LED treatments emitting at least a portion of red and blue wavelengths (i.e., W, 12B and 24B) enhanced several developmental characteristics of grafted tomato seedlings after healing and up to 14 days from transplantation. It is concluded that red, including 11–24% blue provides the optimum light conditions during the healing stage for the production of high-quality grafted tomato seedlings, including higher antioxidant activity and abiotic stress adaptation to transplant shock. The addition of a small portion of green light in the latter wavelength is optional but beneficial for the visualization of white light, which facilitates scouting in the healing chamber.

**Author Contributions:** Conceptualization, methodology and data analysis: C.M., F.B. and A.K.; experimental measurements: C.M., F.B., C.D. and S.K.; writing—original draft preparation: C.M. and F.B.; writing—review and editing: C.M., F.B., C.D., S.K. and A.K.; supervision and project administration: A.K. All authors have read and agreed to the published version of the manuscript.

**Funding:** This research received no external funding.

**Conflicts of Interest:** The authors declare no conflict of interest.

## References

1. Dorais, M.; Ehret, D.L.; Papadopoulos, A.P. Tomato (*Solanum lycopersicum*) health components: From the seed to the consumer. *Phytochem. Rev.* **2008**, *7*, 231. [CrossRef]
2. Food and Agriculture Organization of the United Nations. *FAOSTAT Database*; FAO: Rome, Italy, 2012.
3. Lee, J.M.; Kubota, C.; Tsao, S.J.; Bie, Z.; Echevarria, P.H.; Morra, L.; Oda, M. Current status of vegetable grafting: Diffusion, grafting techniques, automation. *Sci. Hortic.* **2010**, *127*, 93–105. [CrossRef]
4. Mudge, K.; Janick, J.; Scofield, S.; Goldschmidt, E.E. A history of grafting. *Hortic. Rev.* **2009**, *35*, 437–493.
5. Nieves, F.G.; Carvajal, M.; Olmos, E. Graft union formation in tomato plants: Peroxidase and catalase involvement. *Ann. Bot.* **2004**, *93*, 53–60.
6. Yang, X.; Hu, X.; Zhang, M.; Xu, J.; Ren, R.; Liu, G.; Chen, X. Effect of low night temperature on graft union formation in watermelon grafted onto bottle gourd rootstock. *Sci. Hortic.* **2016**, *212*, 29–34. [CrossRef]
7. Hunter, D.C.; Burritt, D.J. Light quality influences adventitious shoot production from cotyledon explants of lettuce (*Lactuca sativa* L.). *In Vitro Cell. Dev. Biol. Plant* **2004**, *40*, 215–220. [CrossRef]
8. Bantis, F.; Smirnakou, S.; Ouzounis, T.; Koukounaras, A.; Ntagkas, N.; Radoglou, K. Current status and recent achievements in the field of horticulture with the use of light-emitting diodes (LEDs). *Sci. Hortic.* **2018**, *235*, 437–451. [CrossRef]
9. Olle, M.; Viršile, A. The effects of light-emitting diode lighting on greenhouse plant growth and quality. *Agric. Food Sci.* **2013**, *22*, 223–234. [CrossRef]
10. Bantis, F.; Panteris, E.; Dangitsis, C.; Carrera, E.; Koukounaras, A. Blue light promotes vascular reconnection, while red light boosts the physiological response and quality of grafted watermelon seedlings. *Sci. Rep.* **2021**, *11*, 21754. [CrossRef]
11. Morrow, R.C. LED lighting in horticulture. *Hortic. Sci.* **2008**, *43*, 1947–1950. [CrossRef]
12. D'Souza, C.; Yuk, H.G.; Khoo, G.H.; Zhou, W. Application of light-emitting diodes in food production, postharvest preservation, and microbiological food safety. *Compr. Rev. Food Sci. Food Saf.* **2015**, *14*, 719–740. [CrossRef]
13. McCree, K.J. The action spectrum, absorptance and quantum yield of photosynthesis in crop plants. *Agric. Metereol.* **1972**, *9*, 191–216. [CrossRef]
14. Zheng, J.; Gan, P.; Ji, F.; He, D.; Yang, P. Growth and Energy Use Efficiency of Grafted Tomato Transplants as Affected by LED Light Quality and Photon Flux Density. *Agriculture* **2021**, *11*, 816. [CrossRef]
15. Vu, N.-T.; Kim, Y.-S.; Kang, H.-M.; Kim, I.-S. Influence of short-term irradiation during pre- and post-grafting period on the graft-take ratio and quality of tomato seedlings. *Hortic. Environ. Biotechnol.* **2014**, *55*, 27–35. [CrossRef]
16. Sharma, N.; Abrams, S.R.; Waterer, D.R. Uptake, movement, activity, and persistence of an abscisic acid analog (8′ acetylene ABA methyl ester) in marigold and tomato. *J. Plant Growth Regul.* **2005**, *24*, 28–35. [CrossRef]
17. Passioura, J.B. Soil conditions and plant growth. *Plant Cell Environ.* **2002**, *25*, 311–318. [CrossRef]

18. Waheed, H.; Javaid, M.M.; Shahid, A.; Ali, H.H.; Nargis, J.; Mehmood, A. Impact of foliar-applied Hoagland's nutrient solution on growth and yield of mash bean (*Vigna mungo* L.) under different growth stages. *J. Plant Nutr.* **2019**, *42*, 1133–1141. [CrossRef]
19. Al-Shwaiyat, M.; Denisenko, T.; Miekh, Y.; Vishnikin, A. Spectrophotometric determination of polyphenols in green teas with 18-molybdodiphosphate. *Chem. Chem. Technol.* **2018**, *12*, 135–142. [CrossRef]
20. Fidrianny, I.; Sari, P.I.; Wirasutisna, K.R. Antioxidant activities in various peel extracts of four varieties rambutan (*Nephelium lappaceum*) using DPPH, FRAP assays. *Int. J. Pharmac. Phytoc. Res.* **2015**, *7*, 280–285.
21. Bantis, F.; Koukounaras, A.; Siomos, A.S.; Fotelli, M.N.; Kintzonidis, D. Bichromatic red and blue LEDs during healing enhance the vegetative growth and quality of grafted watermelon seedlings. *Sci. Hortic.* **2020**, *261*, 109000. [CrossRef]
22. Nelson, J.; Bugbee, B. Economic analysis of greenhouse lighting: Light emitting diodes vs. high intensity discharge fixtures. *PLoS ONE* **2014**, *9*, e99010. [CrossRef]
23. Li, Y.; Xin, G.; Wei, M.; Shi, Q.; Yang, F.; Wang, X. Carbohydrate accumulation and sucrose metabolism responses in tomato seedling leaves when subjected to different light qualities. *Sci. Hortic.* **2017**, *225*, 490–497. [CrossRef]
24. Dougher, T.A.; Bugbee, B. Long-term blue light effects on the histology of lettuce and soybean leaves and stems. *J. Am. Soc. Hortic. Sci.* **2004**, *129*, 467–472. [CrossRef]
25. Javanmardi, J.; Emami, S. Response of tomato and pepper transplants to light spectra provided by light emitting diodes. *Int. J. Veget. Sci.* **2013**, *19*, 138–149. [CrossRef]
26. Głowacka, B. The effect of blue light on the height and habit of the tomato (*Lycopersicon esculentum* Mill.) transplant. *Folia Hortic.* **2004**, *16*, 3–10.
27. Casal, J.J. Photoreceptor signaling networks in plant responses to shade. *Annu. Rev. Plant Biol.* **2013**, *64*, 403–427. [CrossRef]
28. Hernández, R.; Eguchi, T.; Deveci, M.; Kubota, C. Tomato seedling physiological responses under different percentages of blue and red photon flux ratios using LEDs and cool white fluorescent lamps. *Sci. Hortic.* **2016**, *213*, 270–280. [CrossRef]
29. Ouzounis, T.; Heuvelink, E.; Ji, Y.; Schouten, H.J.; Visser, R.G.F.; Marcelis, L.F.M. Blue and red LED lighting effects on plant biomass, stomatal conductance, and metabolite content in nine tomato genotypes. *Acta Hortic.* **2016**, *1134*, 251–258. [CrossRef]
30. Bantis, F.; Koukounaras, A.; Siomos, A.; Menexes, G.; Dangitsis, C.; Kintzonidis, D. Assessing quantitative criteria for characterization of quality categories of grafted watermelon seedlings. *Horticulturae* **2019**, *5*, 16. [CrossRef]
31. Cope, K.R.; Snowden, M.C.; Bugbee, B. Photobiological interactions of blue light and photosynthetic photon flux: Effects of monochromatic and broad-spectrum light sources. *Photochem. Photobiol.* **2014**, *90*, 574–584. [CrossRef]
32. Hogewoning, S.W.; Trouwborst, G.; Maljaars, H.; Poorter, H.; van Ieperen, W.; Harbinson, J. Blue light dose–responses of leaf photosynthesis, morphology, and chemical composition of *Cucumis sativus* grown under different combinations of red and blue light. *J. Exp. Bot.* **2010**, *61*, 3107–3117. [CrossRef] [PubMed]
33. Wu, M.C.; Hou, C.Y.; Jiang, C.M.; Wang, Y.T.; Wang, C.Y.; Chen, H.H.; Chang, H.M. A novel approach of LED light radiation improves the antioxidant activity of pea seedlings. *Food Chem.* **2007**, *101*, 1753–1758. [CrossRef]
34. Folta, K.M.; Lieg, E.J.; Durham, T.; Spalding, E.P. Primary inhibition of hypocotyl growth and phototropism depend differently on phototropin-mediated increases in cytoplasmic calcium induced by blue light. *Plant Physiol.* **2003**, *133*, 1464–1470. [CrossRef] [PubMed]
35. Hernández, R.; Kubota, C. Physiological responses of cucumber seedlings under different blue and red photon flux ratios using LEDs. *Environ. Exp. Bot.* **2016**, *121*, 66–74. [CrossRef]
36. Björkman, O.; Demmig, B. Photon yield of $O_2$ evolution and chlorophyll fluorescence characteristics at 77 K among vascular plants of diverse origin. *Planta* **1987**, *170*, 489–504. [CrossRef]
37. Lattanzio, V.; Lattanzio, V.M.; Cardinali, A. Role of phenolics in the resistance mechanisms of plants against fungal pathogens and insects. *Phytochem. Adv. Res.* **2006**, *661*, 23–67.
38. Długosz-Grochowska, O.; Kołton, A.; Wojciechowska, R. Modifying folate and polyphenol concentrations in Lamb's lettuce by the use of LED supplemental lighting during cultivation in greenhouses. *J. Funct. Foods* **2016**, *26*, 228–237. [CrossRef]
39. Bantis, F.; Fotelli, M.; Ilic, Z.S.; Koukounaras, A. Physiological and Phytochemical Responses of Spinach Baby Leaves Grown in a PFAL System with LEDs and Saline Nutrient Solution. *Agriculture* **2020**, *10*, 574. [CrossRef]

*agriculture*

MDPI

*Article*

# Enabling Year-round Cultivation in the Nordics-Agrivoltaics and Adaptive LED Lighting Control of Daily Light Integral

**Marco Hernandez Velasco** [1,2]

[1] Department of Energy and Built Environments, Dalarna University, 791 88 Falun, Sweden; mhv@du.se
[2] Department of Materials Science and Engineering, Solar Cell Technology, Uppsala University, 751 21 Uppsala, Sweden

**Abstract:** High efficacy LED lamps combined with adaptive lighting control and greenhouse integrated photovoltaics (PV) could enable the concept of year-round cultivation. This concept can be especially useful for increasing the production in the Nordic countries of crops like herbaceous perennials, forest seedlings, and other potted plants not native of the region, which are grown more than one season in this harsh climate. Meteorological satellite data of this region was analyzed in a parametric study to evaluate the potential of these technologies. The generated maps showed monthly average temperatures fluctuating from −20 °C to 20 °C throughout the year. The natural photoperiod and light intensity also changed drastically, resulting in monthly average daily light integral (DLI) levels ranging from 45–50 mol·m$^{-2}$·d$^{-1}$ in summer and contrasting with 0–5 mol·m$^{-2}$·d$^{-1}$ during winter. To compensate, growth room cultivation that is independent of outdoor conditions could be used in winter. Depending on the efficacy of the lamps, the electricity required for sole-source lighting at an intensity of 300 µmol·m$^{-2}$·s$^{-1}$ for 16 h would be between 1.4 and 2.4 kWh·m$^{-2}$·d$^{-1}$. Greenhouses with supplementary lighting could help start the cultivation earlier in spring and extend it further into autumn. The energy required for lighting highly depends on several factors such as the natural light transmittance, the light threshold settings, and the lighting control protocol, resulting in electric demands between 0.6 and 2.4 kWh·m$^{-2}$·d$^{-1}$. Integrating PV on the roof or wall structures of the greenhouse could offset some of this electricity, with specific energy yields ranging from 400 to 1120 kWh·kW$^{-1}$·yr$^{-1}$ depending on the region and system design.

**Keywords:** daily light integral (DLI) maps; LED grow lights; greenhouse integrated PV; adaptive lighting control; year-round cultivation; agrivoltaics

**Citation:** Hernandez Velasco, M. Enabling Year-round Cultivation in the Nordics-Agrivoltaics and Adaptive LED Lighting Control of Daily Light Integral. *Agriculture* **2021**, *11*, 1255. https://doi.org/10.3390/agriculture11121255

Academic Editors: Athanasios Koukounaras and Filippos Bantis

Received: 20 September 2021
Accepted: 8 December 2021
Published: 11 December 2021

**Publisher's Note:** MDPI stays neutral with regard to jurisdictional claims in published maps and institutional affiliations.



## 1. Introduction

The climate in the Nordics (Denmark, Sweden, Finland, Norway, and Iceland) as well as in the Baltic countries (Estonia, Latvia, and Lithuania) is characterized by strong seasonal variations with short, moderately warm, and moist summers contrasted by long, very cold winters. These extreme variations considerably restrict the vegetation period, which is the time of the year when the ambient temperature stays above a certain threshold, usually of +5 °C, and allows for outdoor plant cultivation [1]. For the Nordic and Baltic countries, depending on the location, the start of the vegetation period can be delayed to halfway through spring and last just until early autumn.

Controlled environment agriculture consists in using structures that allow manipulating the indoor climate conditions and provide shelter to the crops [2]. Greenhouse cultivation allows optimizing the use of space and resources, giving protection and maintaining optimal conditions during the growth of the plants. In Northern Europe, greenhouses enable growers to cultivate their crops for longer periods and grow products locally that would have normally been imported [3]. However, this usually comes with high energy demands for heating to achieve a better indoor environment [4–6].

Besides the cold temperatures of the winter months, the low amounts of sunlight during this period further restrict plant cultivation in northern latitudes. While artificial lighting can be used as a remedy, using high-intensity discharge (HID) lamps to provide all the light that the plants need can result in prohibitive electricity costs and for most species, it is rarely done in practice [7,8]. Instead, the strategies adopted by growers consist of extending the day length by providing photoperiodic lighting or increasing the photosynthesis by delivering supplementary light additional to sunlight [9–11].

Plants use to light both as an energy source to drive photosynthesis as well as an information source with signals that trigger different processes [12–14]. Photosynthesis is mainly driven by photons of wavelengths between 400 nm and 700 nm. This spectral range is known as photosynthetically active radiation (PAR) [15,16]. Although the light quality, which is determined by the spectral distribution, has a strong influence on the plant's morphology [17–20]; it is the PAR quantity that mainly affects growth and biomass production [21–25]. The total amount of radiation reaching the canopy is determined by the light intensity at the surface and the duration of exposure. When the photosynthetic photon flux density (PPFD, $\mu mol \cdot m^{-2} \cdot s^{-1}$) is integrated over the course of a day (photoperiod, $h \cdot d^{-1}$), the resulting daily light integral (DLI, $mol \cdot m^{-2} \cdot d^{-1}$) represents the accumulated PAR photons delivered on the given area during that day [26].

The DLI has proven to be a very useful and reliable tool for greenhouse cultivation, allowing growers to assess their light requirements with a simple quantity, similar to a "rain gauge" that accumulates all the PAR photons received in an area each day [27,28]. For northern latitudes where the natural light varies considerably throughout the year, the DLI can help determine the need for supplementary lighting and the strategy to use, either to reach an intensity threshold or to extend the photoperiod [29–32].

Improvements in light-emitting diodes (LED) have made them a feasible option for greenhouse lighting [33–35] that is economically viable [7,36] and can be adjusted to provide additional benefits for plant growth [37,38]. These new technologies have opened the possibility for new lighting strategies such as adaptive control. This consists in regulating in real-time the intensity and even the spectral output of the lamps based on the outdoor conditions to supplement only the necessary light [39–42]. This type of flexibility was not previously possible with HID lamps since they have a fixed spectral output, are usually not dimmable, and could be damaged if they are switched and cycled too often [43].

Modern artificial lights with higher efficiencies and lower heat production have also enabled the development of year-round cultivation concepts in areas where the outdoor conditions would not allow it [44–47]. LED lighting can now be used as sole-source lighting for indoor cultivation in single or multi-layered growth rooms also called vertical farms or plant factories [48–54]. While one of the main drawbacks of using artificial lights for plant growth is the high electricity consumption [55], renewable energy technologies such as solar photovoltaics (PV) could provide an alternative to offset some of this energy demand and produce the electricity at the place where it is needed.

Although PV essentially competes with plants for sunlight, recent years have shown an increased interest in finding ways to combine agricultural production and PV energy generation in the form of agrivoltaics [56,57]. Besides mounting PV on agricultural open fields [58,59], integrating PV directly onto the structure of greenhouses was presented as a realistic option about a decade ago [60–62] and since then this idea has been widely implemented in several countries [63]. This solution requires an optimization to balance the PV energy output and the crop yield [64–66] as well as the economic aspects [67,68], but studies have shown that it is possible to find an adequate PV roof coverage depending on the location that still allows sufficient light into the greenhouse for the plants [59,69,70].

The main goal of this work is to make a broad assessment through a parametric analysis of the potential for a year-round cultivation concept in Northern Europe using:

- sole-source LED lighting for indoor cultivation during the winter months;
- adaptive LED lighting control of DLI to supplement the light changes and extend the greenhouse growing season through spring and autumn;
- an outdoor cultivation phase in the summer months;

- greenhouses or other buildings with integrated PV to produce electricity.

## 2. Materials and Methods

### 2.1. Data Sources

To analyze these different cultivation concepts, calculations were done based on meteorological and geographical data publicly available from the European Commission Joint Research Centre in Ispra, Italy; through their online service Photovoltaic Geographical Information System-PVGIS version 5.1 [71,72]. The weather and solar radiation data source used in this study consist of a reanalysis made by the European Centre for Medium-range Weather Forecast (ECMWF-ERA-5) which includes hourly values extended between 2005 and 2016 at a spatial resolution of 0.25° for latitude and longitude forming roughly a 30 km global grid [73,74]. Despite having higher uncertainties, the ECMWF-ERA-5 dataset is the default source for the Nordic countries in PVGIS because it provides data for these regions where geostationary satellites normally have no cover.

The values were retrieved using the non-interactive service of PVGIS, using a grid cell of $0.1° \times 0.1°$ in both latitude and longitude for the Nordic countries (except for Iceland) and the Baltics. The coordinates for this region went from 54.5° N to 70.0° N in latitude and from 4.5° E to 31.5° E in longitude: forming cells of approximately 11 km in latitude by 6 km in longitude in the south part of the region. Due to the Earth's curvature, as the latitude increases, the distance along the longitude is reduced to 4 km per cell. When the coordinates of the requested points did not match the data points available, PVGIS' algorithm resolved this by interpolation [75]. In total, 33,051 coordinate pairs were retrieved without counting the points over the ocean. For comparability, the cells of the geographical grid were selected with the same angular distance ($0.1° \times 0.1°$) as those reported in similar maps for the United States [28].

A different grid was queried for Iceland since it is separated from the rest of the countries and is relatively small in comparison. The data retrieved went from 63° N to 67.5° N in latitude and from 13° W to 25° W in longitude using a cell size of $0.05° \times 0.05°$ in both latitude and longitude. Since the values were very similar to those at the same latitudes, it was decided to present the corresponding figures for Iceland in Appendix A.

The values extracted were hourly averages of a representative day for each month. This means that for every hour of the day, the average was calculated from all the days in that month and from all years when data was available. This resulted in 24 values for each month at each point in the geographic grid (one per hour of the representative day of the month; 288 values in total for the year). The variables for which data were extracted were:

- Global irradiance on the horizontal plane ($G_h$, $W \cdot m^{-2}$);
- Direct irradiance component on the horizontal plane ($G_b$, $W \cdot m^{-2}$);
- Diffuse irradiance component on the horizontal plane ($G_d$, $W \cdot m^{-2}$);
- Ambient temperature calculated at 2 m above the ground ($T_a$, °C).

### 2.2. Ambient Temperature

The role of temperature in the development of plants is so important, that it allows methods like the Growing Degree Days to accurately estimate different growth stages using only the accumulation of heat units [76–78]. In northern climates, temperature determines the vegetation period, contributes to the cold acclimation of plants during autumn, and is one of the main cues for de-acclimation in spring together with the photoperiod [79]. Based on the monthly average ambient temperature presented in Figure 1 for the Nordic and Baltic countries (for Iceland see Appendix A, Figure A1), it was estimated that outdoor plant cultivation would be limited to the periods from [1]:

- April to October in southern latitudes (at 55° N, about 210 days);
- May to September in the central regions (at 63° N, about 180 days);
- June to August in Iceland and the northernmost regions (at 70° N, about 100 days).

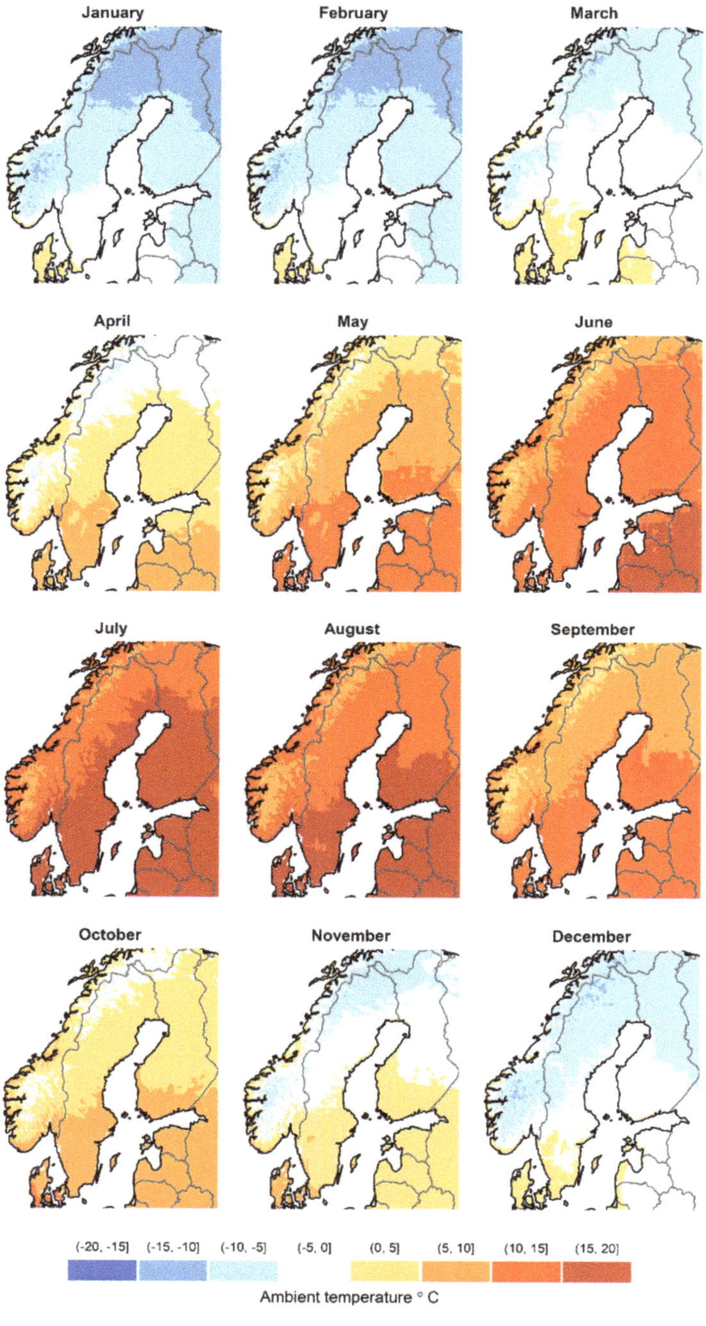


**Figure 1.** Monthly average daily ambient temperature ($T_a$, °C) maps for the Nordic and Baltic countries. Values from the ECMWF-ERA-5 dataset with a coverage period of 2005–2016 retrieved from PVGIS as a 0.1° × 0.1° grid both in latitude and longitude, extending from 54.5° N to 70° N in latitude and 4.5° E to 31.5° E in longitude.

### 2.3. *Outdoor Light Conditions*

To develop a useful year-round cultivation concept, the following properties of light relevant to plant growth were analyzed: the duration of the photoperiod, the spectral composition (PAR), the intensity (PPFD), and the daily availability (DLI) [12]. Other characteristics such as light uniformity, ray direction, and intermittency might become more relevant in the future as the use of sole-source artificial lighting spreads [80].

#### 2.3.1. Natural Daylength and Photoperiod

The seasonal differences in the duration of a day occur because of the orbital and translational movements of the Earth. These variations become more evident as the latitude increases and they are independent of changes in the global climate because they depend only on the latitude and time of the year. Plants have evolved to detect these differences in the natural photoperiod and use them as signals to adjust their development phases to the seasons and the local weather conditions [81].

Successful cultivation practices require considering the local photoperiod at the cultivation site together with the conditions at the provenance of the seeds used. Modern greenhouses and growth rooms allow to adapt and regulate the light duration by either extending the photoperiod [82,83], providing night interruption cycles [84,85], or creating short days with reduced hours of light by restricting the light [86–91].

Although most photosynthesis happens while the sun is above the horizon, plants' photoreceptors are very sensitive and able to detect considerably small amounts of radiation (PPFD < 1 $\mu mol \cdot m^{-2} \cdot s^{-1}$). This allows them to also use the faint light before sunrise and after sunset as cues for the daily photoperiod [92]. The twilight is this illumination in the atmosphere when the sun disk is below the horizon, with civil twilight defined when the sun is 6° below the horizon and nautical twilight until 12° below [93]. Using the geometric relationships from Earth's movements [94,95], Figure 2 shows the natural daylength together with the civil and nautical twilights for different latitudes throughout the year.

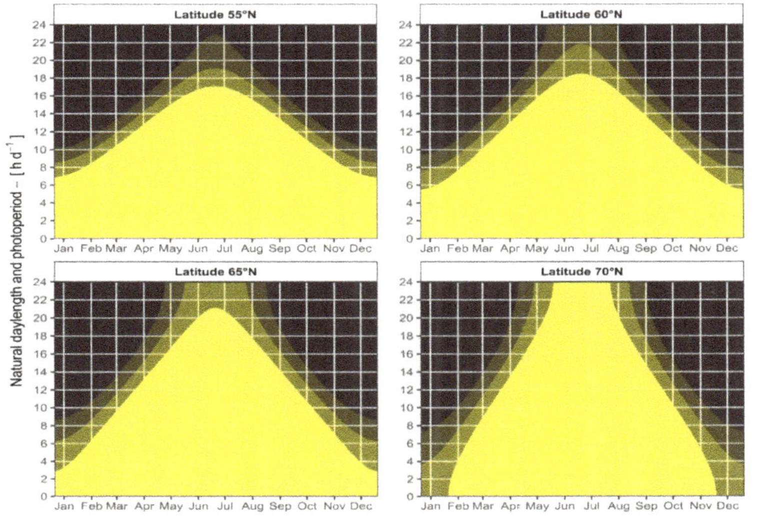


**Figure 2.** Natural daylength and photoperiod ($h \cdot d^{-1}$) throughout the year in four northern latitudes (55°, 60°, 65°, and 70°). The solid yellow region represents the daytime or period when the sun is above the horizon. The two increasing levels of transparency correspond to the civil twilight (0° to 6° below the horizon) and the nautical twilight (6° to 12° below the horizon) respectively.

2.3.2. Estimating PPFD from Global Irradiance

All PPFD quantities presented in this work were calculated using the traditional definition of the PAR range (400–700 nm) [16]. However, research in recent years has shown that far-red photons (up to 750 nm) also contribute and influence photosynthetic activity. This suggests that updating the PAR concept towards an extended range (ePAR; 400–750 nm) might be needed as this understanding improves [96,97].

The amount of photosynthetic radiation at a place is not among the most commonly used meteorological parameters and is therefore not usually contained in standard satellite datasets. In addition, despite being relatively low-priced and easily available, quantum sensors are not normally installed in weather stations. Instead, sensors for measuring solar shortwave radiation ($SW_i$) such as pyranometers that have a broad spectral response (normally between 280–3000 nm) are included. To use the measurements of these instruments for plant growth, numerous studies have been made in different locations to understand the relationship between PAR and $SW_i$ as well as the dependence on the atmospheric conditions and seasonal changes [98–107].

To estimate the outdoors $PPFD_{sun}$ from commonly reported $G_h$, the proportion of PAR in the $SW_i$ is needed ($PAR_{(400-700\ nm)}/SW_{i,(280-3000\ nm)}$, unitless). Then, the amount of energy of the photons within the PAR range ($E_{PAR}$, $J \cdot \mu mol^{-1}$) has to be measured or calculated based on the photons' wavelengths. These two amounts can be then substituted in Equation (1) to obtain the PPFD [107]:

$$PPFD_{sun} = G_h \left( \frac{PAR}{SW_i} \right) \left( \frac{1}{E_{PAR}} \right) \quad (1)$$

A recent study on PAR proportion showed that if a constant of $PAR/SW_i = 0.45$ and an $E_{PAR} = 0.223\ J \cdot \mu mol^{-1}$ were assumed (corresponding to the mean of the different sky conditions observed), the calculated $PPFD_{sun}$ deviated less than 5% from the measured value [107]. These constants can be conveniently combined with a conversion factor (0.0036 $J \cdot mol \cdot Wh^{-1} \cdot \mu mol^{-1}$) to modify Equation (1) into a summation over 24 h to obtain the DLI (see Equation (2)) [28]. Since the average $G_h$ values are normally reported in hourly steps, it is common to assume a constant irradiance throughout each interval.

$$DLI_{sun} = \sum_{t=1}^{24} G_{h,t} \cdot 0.0072646 \left( mol \cdot Wh^{-1} \right) \quad (2)$$

Equation (2) implies that 45% of the solar spectrum is within the PAR region and there is 0.0072646 mol of PAR photons for every Wh of $SW_i$. Figure 3 was calculated using the $G_h$ data extracted from PVGIS and applying Equation (2) (see Figure A2 for Iceland). The hourly radiometric data ($W \cdot m^{-2}$) was accumulated into daily quantum units ($mol \cdot m^{-2} \cdot d^{-1}$). This generated monthly average DLI maps with data bins of 5 $mol \cdot m^{-2} \cdot d^{-1}$ and a range between 0–50 $mol \cdot m^{-2} \cdot d^{-1}$, similar to those existing for the United States [28].

In reality, both the ratio $PAR/SW_i$ and the $E_{PAR}$ are not constant and depend heavily on the solar spectrum. This is in turn affected by the atmospheric conditions and the distance that light has to travel. When the sun is low in the sky e.g., at sunset or during winter, the path traveled by sunlight is longer and more photons in the PAR region are absorbed compared to those with longer wavelengths [107]. Air pollution and aerosols can also significantly scatter and attenuate the PAR irradiance [104,108]. Conversely, cloudy days with an overcast sky present a higher fraction of PAR because water vapor mainly absorbs longer wavelengths of near-infrared radiation (NIR, 760–4000 nm) [106,109].

Diffuse radiation is the light that gets scattered in the atmosphere before reaching the ground [71,94]. Figure 4 provides the monthly average diffuse fraction of global horizontal radiation ($G_d/G_h$) calculated from the extracted data (see Figure A3 for Iceland). Due to the influence of the atmospheric conditions on the solar spectrum, knowing the proportion of diffuse radiation in a place can be important as an indicator of the cloudiness [107] and help estimate the overall light transmittance into the greenhouse [110].

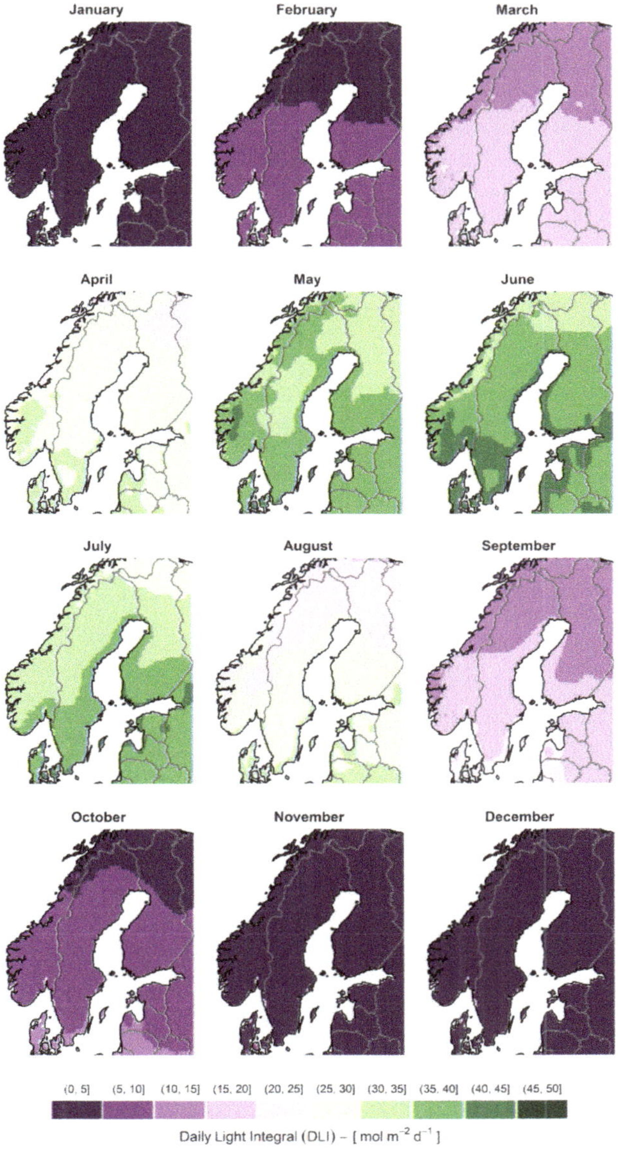


**Figure 3.** Monthly average photosynthetic daily light integral ($DLI_{sun}$, $mol{\cdot}m^{-2}{\cdot}d^{-1}$) maps for the Nordic and Baltic countries. Values from the ECMWF-ERA-5 dataset with a coverage period of 2005–2016 retrieved from PVGIS as a 0.1° × 0.1° grid both in latitude and longitude, extending from 54.5° N to 70° N in latitude and 4.5° E to 31.5° E in longitude.

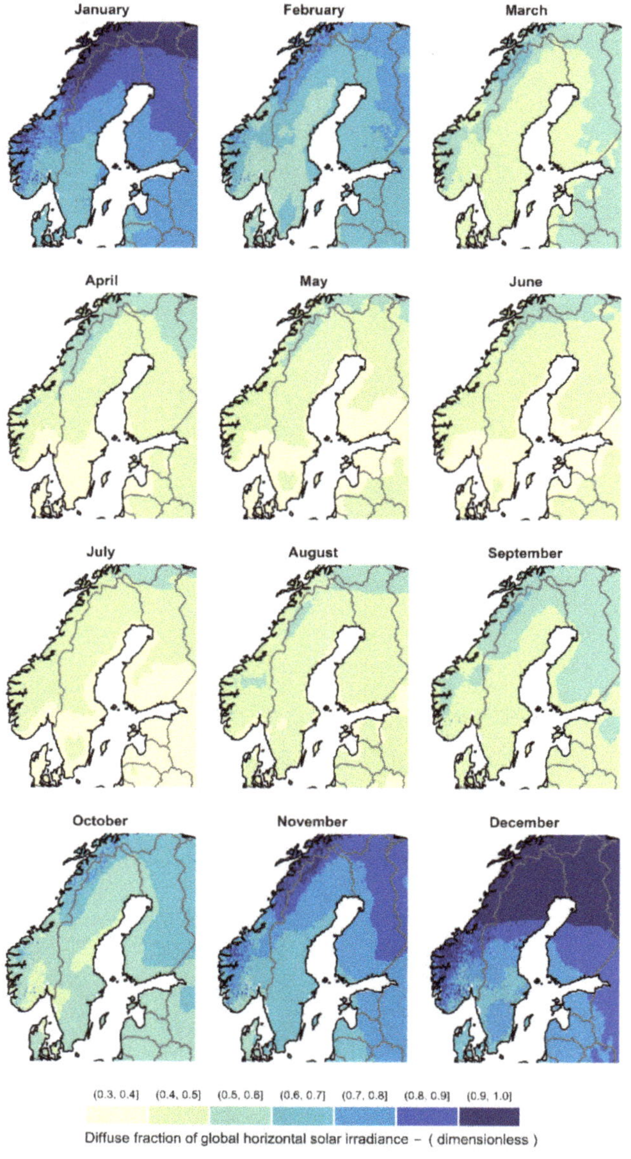


**Figure 4.** Monthly average diffuse fraction of global horizontal irradiance ($G_d/G_h$) for the Nordic and Baltic countries. Values from the ECMWF-ERA-5 dataset with a coverage period of 2005–2016 retrieved from PVGIS as a 0.1° × 0.1° grid both in latitude and longitude, extending from 54.5° N to 70° N in latitude and 4.5° E to 31.5° E in longitude.

### 2.4. *Indoor Light Conditions*

During the indoor cultivation phases, either in a growth room or a greenhouse, at least part of the light would come from artificial sources. It is therefore important to define the characteristics of the lamps and the lighting control protocols planned.

2.4.1. Sole-Source Lighting

The daily output of the lamps ($DLI_{lamps}$, $mol \cdot m^{-2} \cdot d^{-1}$) can be calculated by adding the photosynthetic light intensity at the cultivation surface ($PPFD_{lamps}$, $\mu mol \cdot m^{-2} \cdot s^{-1}$) for each timestep during the photoperiod ($h \cdot d^{-1}$). Equation (3) assumes hourly timesteps and adjusts the units with a conversion factor:

$$DLI_{lamps} = \sum_{t=1}^{h} PPFD_{lamps,t} \cdot 0.0036 \left(mol \cdot s \cdot \mu mol^{-1} \cdot h^{-1}\right) \tag{3}$$

In a closed growth room where the light intensity is constant all day, the $DLI_{lamps}$ can be simply calculated by multiplying the $PPFD_{lamps}$ at the cultivation area by the duration of the photoperiod ($h \cdot d^{-1}$) and adjusting the units (see Equation (4)):

$$DLI_{lamps} = PPFD_{lamps} \cdot photoperiod \cdot 0.0036 \left(mol \cdot s \cdot \mu mol^{-1} \cdot h^{-1}\right) \tag{4}$$

The efficiency of a growth lamp is measured as the ratio of the luminous power output to the electrical power input ($W_{light}/W_{el}$). However, for cultivation purposes, it is more useful to compare the efficacy which is a measure of photon output rate (measured in $\mu mol \cdot s^{-1}$) against the electrical power input ($W_{el}$). If only the PAR photons are considered, then the parameter is called photosynthetic photon efficacy (PPE, $\mu mol \cdot J^{-1}$) [7,54].

When the lamp efficacy measurements are done using a flat-plane integration method and assuming no other energy losses [7]; the PPE ($\mu mol \cdot J^{-1}$) can be used to calculate the daily electrical energy consumption ($E_{el,\ lamps}$, $kWh \cdot m^{-2} \cdot d^{-1}$) per cultivation surface. Equation (5) accounts also for the conversion from quantum to energy units:

$$E_{el,\ lamps} = \frac{DLI_{lamps}}{PPE \cdot 3.6 \left(mol \cdot J \cdot \mu mol^{-1} \cdot kWh^{-1}\right)} \tag{5}$$

Equations (4) and (5) can then be combined into Equation (6) to calculate $E_{el,\ lamps}$ directly from the $PPFD_{lamps}$, the photoperiod, and the lamp efficacy. This can be useful to directly compare and select the best lamp for a specific growth light protocol.

$$E_{el,\ lamps} = \frac{PPFD_{lamps} \cdot photoperiod}{PPE \cdot 1000 \left(W \cdot kW^{-1}\right)} \tag{6}$$

2.4.2. Greenhouse Light Transmission

The amount of outdoor light that is transmitted into a greenhouse depends in great measure on the covering material properties together with the angle of incidence at which the irradiance arrives [110]. When the light reaches the surface, it is either reflected, absorbed, or transmitted through the material [111,112]. The highest transmittance of PAR occurs when the incoming light is at 90° from the roof surface; this is called the perpendicular PAR transmittance ($\tau_{p,\ PAR}$, %) [112]. However, since the position of the sun in the sky varies with time and latitude, so does the proportion of light reaching inside. A better indicator of the natural light available is the global or hemispherical PAR transmittance ($\tau_{h,\ PAR}$, %) which considers the total transmission inside the greenhouse [111,112].

Due to the broad number of locations in this study, it was decided to use the assumption of a uniformly bright sky [113]. This implies that all sunlight reaching the greenhouse surface is diffuse and has the same spectrum. Thus, $\tau_{h,PAR}$ is a constant value regardless of season or roof orientation. The light inside the greenhouse ($PPFD_{greenhouse}$) was then calculated as a percentage of the outdoor $PPFD_{sun}$ using Equation (7) [112]:

$$PPFD_{greenhouse} = \tau_{h,\ PAR} \cdot PPFD_{sun} \tag{7}$$

To compensate for the shortcomings of this simplification, the calculations were done for different values of $\tau_{h,PAR}$ (40%, 55% or 70%) selected to be within the range of those reported in different studies of various greenhouse materials and at various locations, obtained through simulations or from measurements [65,110,111,114–118].

2.4.3. Greenhouse Lighting Control Protocols

Two previously established lighting control protocols were compared in this study: an on-off control vs. an adaptive control [42]. In both cases, dimmable LED fixtures were assumed to be used to maintain a predefined minimum light level ($PPFD_{threshold}$) throughout the entire photoperiod. In the two protocols, when the natural light transmitted inside the greenhouse reached or surpassed the threshold ($PPFD_{greenhouse} \geq PPFD_{threshold}$), the lamps would turn off and emit no light ($PPFD_{lamps} = 0$).

With the on-off control protocol (Equation (8)), if the threshold was not reached with natural light ($PPFD_{greenhouse} < PPFD_{threshold}$), the lamps would then produce exactly the amount of light at the setting level ($PPFD_{lamps} = PPFD_{threshold}$). This would be equivalent to having lamps set exactly at the $PPFD_{threshold}$ and just switching them on and off. In contrast, the adaptive control protocol (Equation (9)) considers the dimmability of the LED fixtures. The output of the lamps is assumed to be continuously adjusting and regulated to provide only the necessary output to reach the desired threshold while considering the incoming sunlight ($PPFD_{lamps} = PPFD_{threshold} - PPFD_{greenhouse}$).

$$PPFD_{on-off} = \begin{cases} 0, & PPFD_{greenhouse} \geq PPFD_{threshold} \\ PPFD_{threshold}, & PPFD_{greenhouse} < PPFD_{threshold} \end{cases} \quad (8)$$

$$PPFD_{adaptive} = \begin{cases} 0, & PPFD_{greenhouse} \geq PPFD_{threshold} \\ PPFD_{threshold} - PPFD_{greenhouse}, & PPFD_{greenhouse} < PPFD_{threshold} \end{cases} \quad (9)$$

*2.5. Year-Round Cultivation Concept*

In the proposed year-round concept, indoor cultivation in northern Europe would need to be done for at least 6 months and up to 9 months depending on the location. After evaluating Figures 1–3, it was decided to consider an extreme scenario with about 3 months outdoor cultivation complemented by 9 months of indoor cultivation as follows:

- late-May to late-August: transfer outside for outdoor cultivation;
- late-August to October: greenhouse cultivation with supplementary lighting;
- November to January: indoor growth room cultivation with sole-source lighting;
- February to late May: greenhouse cultivation with supplementary lighting.

2.5.1. Outdoor Cultivation

The outdoor cultivation phase was assumed to last 92 days, from 21 May to 20 August. This is of course a generalization for the whole region during a fictitious average year. A detailed plan should be done when introducing a similar concept to a particular location adjusting to the actual climatic conditions.

2.5.2. Indoor Cultivation in Growth Room with Sole-Source Lighting

The indoor growth room cultivation phase under sole-source lighting was assumed to be of 92 days, from 1 November to 31 January using a photoperiod of 16 $h{\cdot}d^{-1}$ at a constant $PPFD_{lamps}$ of 300 $\mu mol{\cdot}m^{-2}{\cdot}s^{-1}$. These conditions are similar to those that have already been tested in different studies in the region [18,25,82,90,119,120]. Using Equation (4), the resulting $DLI_{lamps}$ is 17.3 $mol{\cdot}m^{-2}{\cdot}d^{-1}$ which is considered enough for most cut flowers, greenhouse vegetables, and forest seedlings [8,9,28,31]. The lamps were assumed to be adjustable LED fixtures with efficacies equal to either:

- PPE = 2.0 $\mu mol{\cdot}J^{-1}$, considering a standard LED luminaire;
- PPE = 2.75 $\mu mol{\cdot}J^{-1}$, for a state-of-the-art standard, LED luminaire;

- PPE = 3.5 $\mu mol \cdot J^{-1}$, to account for upcoming developments.

2.5.3. Greenhouse Cultivation with Supplementary Lighting

The assumed greenhouse cultivation period consisted of 181 days, from 1 February to 20 May and from 20 August to 31 October. The objective was to maintain at least the set $PPFD_{threshold}$ (100, 200, or 300 $\mu mol \cdot m^{-2} \cdot s^{-1}$) inside the greenhouse during a 16 $h \cdot d^{-1}$ photoperiod (04:00–20:00) to reach the minimum equivalent $DLI_{threshold}$ value.

The hourly average natural light levels inside the greenhouse ($PPFD_{greenhouse}$) were calculated using Equation (7) for the different values of $\tau_{h,\,PAR}$ (40%, 55%, or 70%). Then, the two lighting control protocols were compared using Equations (8) and (9) to calculate the hourly supplementary light provided by the lamps ($PPFD_{lamps}$). The average daily supplementary lighting requirement ($DLI_{lamps}$) was determined by adding the hourly $PPFD_{lamps}$ values using Equation (3) and dividing it by the number of cultivation days.

The same three PPE values as in the sole-source lighting phase were assumed. The energy consumption corresponding to the $DLI_{lamps}$ was estimated using Equation (5) for the different $PPFD_{threshold}$ settings and added for the complete season at each location in the geographic grid. Table 1 shows a summary of the parameters used for the calculations during the greenhouse cultivation period.

**Table 1.** Values used in the parametric study for the greenhouse cultivation period.

| Lighting Control | $PPFD_{threshold}$ | $DLI_{threshold}$ [1] | $\tau_{h,\,PAR}$ | PPE |
|---|---|---|---|---|
| On-off control | 100 $\mu mol \cdot m^{-2} \cdot s^{-1}$ | 5.8 $mol \cdot m^{-2} \cdot d^{-1}$ | 40% | 2.0 $\mu mol \cdot J^{-1}$ |
| Adaptive control | 200 $\mu mol \cdot m^{-2} \cdot s^{-1}$ | 11.5 $mol \cdot m^{-2} \cdot d^{-1}$ | 55% | 2.75 $\mu mol \cdot J^{-1}$ |
| | 300 $\mu mol \cdot m^{-2} \cdot s^{-1}$ | 17.3 $mol \cdot m^{-2} \cdot d^{-1}$ | 70% | 3.5 $\mu mol \cdot J^{-1}$ |

[1] Equivalent $DLI_{threshold}$ when maintaining the $PPFD_{threshold}$ for 16 $h \cdot d^{-1}$ photoperiod.

*2.6. Photovoltaics on Greenhouses*

The nominal peak power of a PV installation ($P_{PV,\,STC}$, W) is a quantity used in industry to describe the output measured at standard test conditions (STC). The STC are defined for the device temperature ($T_{PV}$ = 25 °C), input irradiance ($G_{STC}$ = 1000 $W \cdot m^{-2}$), and light spectrum (Air mass 1.5) [121]. The efficiency at STC ($\eta_{STC}$, %) relates the $P_{PV,\,STC}$ to the input irradiance ($G_{STC}$) over the PV area ($A_{PV}$, $m^2$) as shown in Equation (10).

Commercially available PV modules have efficiencies ranging from 12–22% depending on their material [122]. Different PV technologies have differences in their spectral responsivity and thermal behaviors. In practice, however, it is usually assumed that systems with equal $P_{PV,STC}$ will have a similar yearly energy output [123]. When the $\eta_{STC}$ is known, Equation (10) can be used to calculate the surface needed for a specific $P_{PV,nom}$.

$$\eta_{STC,\%} = \frac{P_{PV,\,STC}}{G_{STC} \cdot A_{PV}} \cdot 100 \tag{10}$$

The ratio between the $A_{PV}$ and the available area on the ground or roof ($A_{ground}$ or $A_{roof}$, $m^2$) is called the PV coverage ratio ($PV_R$) (Equation (11)). Selecting the ideal $PV_R$ value is an optimization task that should balance the electricity output goals with the light requirements of the species to cultivate, also considering the irradiance available at the location and the transmittance of the greenhouse [59,64,65,70].

$$PV_{R,\%} = \frac{A_{PV}}{A_{roof}} \cdot 100 \tag{11}$$

Once the target $PV_R$ is decided either by calculations or measurements, Equation (12) can be used to estimate the maximum recommended $P_{PV,STC}$ that should be installed as a function of the efficiency of the PV modules and the roof area of the greenhouse.

$$P_{PV,STC} \leq (PV_R \cdot A_{roof})(G_{STC} \cdot \eta_{STC}) \tag{12}$$

Neither a specific $P_{PV,STC}$ or $PV_R$ were defined for this study; instead, relative energy output values were estimated which can then be used later to choose the system size.

PV System Parameters

The energy output of a PV system depends on many aspects besides the installed peak power and the irradiance at the location. Among the main factors influencing the yield are the temperature, the tilt and orientation of the surface (also known as slope and azimuth angle), and the material of the solar cells [123]. Considering these parameters [71], PVGIS was used to estimate the monthly energy output of different PV system configurations in multiple locations. The same geographic grid and data source (ECMWF-ERA-5) were used as previously described.

The PV modules were assumed to be mounted either on the roof with a slope of 25° or the wall (slope of 90°) of a wide-span greenhouse with a double-pitched roof, typical of northern Europe [116,124]. A symmetrical yearly irradiance around the North-South line was assumed; implying that an inclined surface with an azimuth towards the East receives the same amount of sunshine as one with the same angle mirrored towards the West. Four different orientations were compared using the direction perpendicular to the roof ridge as reference. The corresponding azimuth angle was measured clockwise using the navigation convention with North = 0° [93]. For some of the orientations, the planned $P_{PV,STC}$ was divided between two surfaces in opposing directions (see Table 2).

**Table 2.** Orientation and nominal peak power of PV systems analyzed.

| | | | | | | | | |
|---|---|---|---|---|---|---|---|---|
| **Roof-mounted systems** slope: 25° | 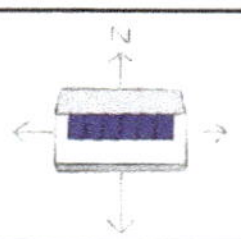 | | 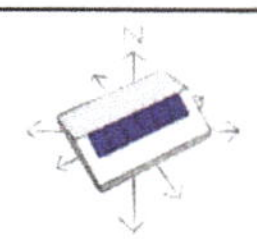 | | 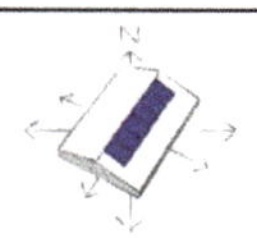 | | 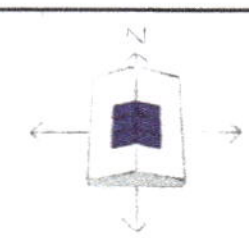 | |
| orientation | **N—S** | | **NW—SE** | | **NW—SE** | | **W—E** | |
| azimuth | 0° | 180° | 330° | 150° | 300° | 120° | 270° | 90° |
| $P_{PV,STC}$ | | 6 kW | | 6 kW | | 6 kW | 3 kW | 3 kW |
| $P_{PV,STC}$ [1] | | | 1 kW | 5 kW | 2 $kW_p$ | 4 kW | | |
| **Wall-mounted systems** slope: 90° | 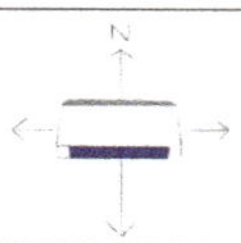 | | 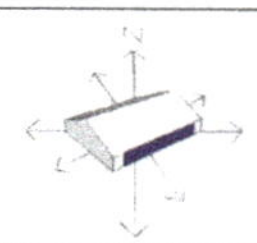 | | 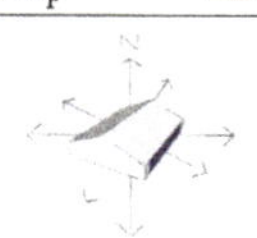 | | 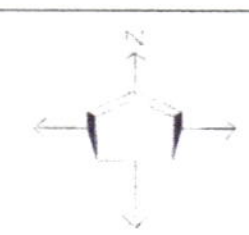 | |
| orientation | **N—S** | | **NW—SE** | | **NW—SE** | | **W—E** | |
| azimuth | 0° | 180° | 330° | 150° | 300° | 120° | 270° | 90° |
| $P_{PV,STC}$ | | 6 kW | | 6 kW | | 6 kW | 3 kW | 3 kW |

[1] Additional energy yield estimation dividing the $P_{PV,STC}$ between the two roofs.

The following system parameters were used in common for all systems and locations; a detailed explanation of the parameters can be found in the user manual of PVGIS [70]:

- $P_{PV,STC}$: 6 kW;
- PV technology: crystalline silicon (c-Si);
- mounting position: fixed and building-integrated;
- horizon: yes;
- estimated system losses: 14% *(PVGIS default)*.

A system size of 6 kW was chosen for all the configurations because this amount allows to easily divide the planned $P_{PV,STC}$ among the two sides of the roof using whole number ratios (6:0, 5:1, 4:2, 3:3) that can be later scaled to larger systems. Finally, the yearly energy output ($E_{PV}$, $kWh \cdot yr^{-1}$) extracted from PVGIS was divided by the $P_{PV,STC}$ to obtain the PV specific energy yield ($E_{rel,PV}$, $kWh \cdot kW^{-1} \cdot yr^{-1}$) for all locations.

### 2.7. Data Analysis

The analysis of the different parameters was performed using the statistical software *R* version 4.1.1 [125] with the *tidyverse* package collection for the various calculations [126,127] including the division in bins and interpolations for the contour plots. The vector data for the maps were plotted using *ggplot2* [128,129] and *sf* packages [130,131]. The geographical data of the different regions were retrieved from the public dataset *Natural Earth* [132] using the *R* package *rnaturalearth* [133].

## 3. Results

### 3.1. Outdoor Ambient Conditions

The environmental data extracted confirmed the strong seasonal variations in the north European region. The monthly average daily ambient temperature (Figures 1 and A1 for Iceland) ranged during the winter between −20 °C and −15 °C in the northern and mountainous areas. For most of the regions, the temperature remained below 0 °C for several months. Only in Denmark and the southernmost part of Sweden, the monthly average temperature stays above freezing for the complete year. In contrast, the summer months exhibited average temperatures close to 20 °C for the Baltic countries, Denmark, and half of Sweden and Finland. Norway and Iceland had lower summer temperatures averaging closer to 15 °C. The coldest temperatures were registered during January and February while the warmest period was in July and August. Since these are averaged values, the actual temperatures are expected to reach beyond these levels, accounting for differences of more than 60 °C between summer and winter.

The available light also varied drastically between the seasons and along the different latitudes analyzed; changing the duration of the photoperiod (Figure 2) as well as the daily amount of light suitable for photosynthesis (Figures 3 and A2). The daylength in the south (latitude 55° N) changed from 6.5 h in the winter to almost 18 h in summer. In the northernmost latitudes, the day length shifted between days of constant sunshine in summer to periods in the winter when the sun remained below the horizon.

The DLI levels (Figures 3 and A2) also exhibited a large variation, reaching a maximum range of 45–50 $mol{\cdot}m^{-2}{\cdot}d^{-1}$ during summer; contrasting to the minimum range in winter between 0–5 $mol{\cdot}m^{-2}{\cdot}d^{-1}$. Figures 4 and A3 show how the proportion of scattered light ($G_d/G_h$) increases after the summer months, being more than 40% from August to March in most locations. Although the climate is a very complex system with many factors intertwined, the regions with lower ratios of diffuse light (which is related to the cloudiness) match well to the regions with higher DLI and warmer summers.

### 3.2. Sole-Source Lighting Requirements

The DLI maps and the photoperiod lengths (Figures 2, 3 and A2) can be used as a reference for the light levels required during indoor cultivation. The chosen photoperiod of 16 $h{\cdot}d^{-1}$ roughly corresponds to the daylengths during early spring (March–April) or early autumn (late August–September) depending on the latitude. Setting the intensity of the lamps to $PPFD_{lamps}$ = 300 $\mu mol{\cdot}m^{-2}{\cdot}s^{-1}$ during the entire photoperiod creates a $DLI_{lamps}$ of 17.3 $mol{\cdot}m^{-2}{\cdot}d^{-1}$ which also matches well the outdoor $DLI_{sun}$ levels during those seasons.

Having these settings and the resulting $DLI_{lamps}$, Figure 5 combines Equations (3)–(6) to calculate the daily electric energy consumption for the cultivation area based on the efficacy of the lamps. A standard LED fixture with a PPE = 2 $\mu mol{\cdot}J^{-1}$ would require daily 2.4 $kWh{\cdot}m^{-2}$ while a state-of-the-art luminaire with PPE = 2.75 $\mu mol{\cdot}J^{-1}$ could reduce the energy consumption by 27% to 1.75 $kWh{\cdot}m^{-2}{\cdot}d^{-1}$. Expected technological improvements in LEDs (PPE $\approx$ 3.5 $\mu mol{\cdot}J^{-1}$) could bring the savings further down for consumption of 1.4 $kWh{\cdot}m^{-2}{\cdot}d^{-1}$ under the same settings. During the assumed 92 days of indoor cultivation with sole-source lighting, the growth room would require 221 $kWh{\cdot}m^{-2}$, 161 $kWh{\cdot}m^{-2}$, or 129 $kWh{\cdot}m^{-2}$ of electricity for lighting considering these respective lamp efficacies.

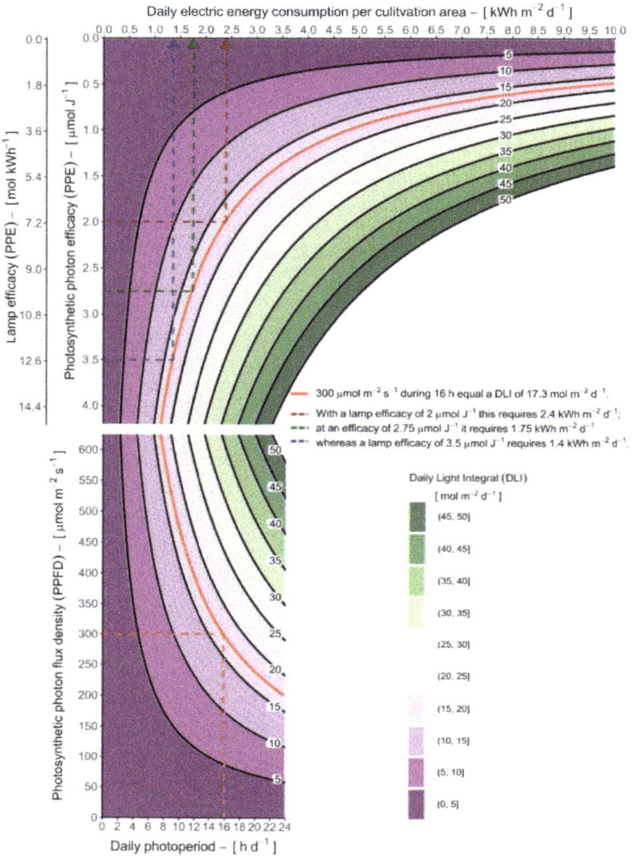


**Figure 5.** Daily PAR output of sole-source lighting ($DLI_{lamps}$, $mol \cdot m^{-2} \cdot d^{-1}$) and the corresponding electric energy consumption per cultivation area ($E_{el,\ lamps}$, $kWh \cdot m^{-2} \cdot d^{-1}$) based on the photoperiod ($h \cdot d^{-1}$) and PPFD ($\mu mol \cdot m^{-2} \cdot s^{-1}$) settings; as well as the efficacy of the lamps (PPE, $\mu mol \cdot J^{-1}$) used, expressed also in the equivalent units typically found commercially ($mol \cdot kWh^{-1}$) for convenience. The red dashed lines indicate the growth-room settings selected in this study while the contiguous red line is the resulting $DLI_{lamps}$. The three dashed lines of different colors in represent the lamp efficacies by matching their corresponding energy consumptions on the right.

### *3.3. Greenhouse Supplementary Lighting Requirements*

The parametric analysis of the greenhouse supplementary lighting showed a large variation across the different settings. When comparing only the control protocols for all locations while maintaining the other variables equal, the average supplementary light delivered was always lower for the adaptive control (Figures 6 and A4 for Iceland; panels a vs. b). The differences ranged between 0.3 and 1.5 $mol \cdot m^{-2} \cdot d^{-1}$ for the lowest $PPFD_{threshold}$ of 100 $\mu mol \cdot m^{-2} \cdot s^{-1}$, from 1.1 to 3.6 $mol \cdot m^{-2} \cdot d^{-1}$ at 200 $\mu mol \cdot m^{-2} \cdot s^{-1}$, and for 300 $\mu mol \cdot m^{-2} \cdot s^{-1}$ the difference range went from 2.4 to 5.7 $mol \cdot m^{-2} \cdot d^{-1}$.

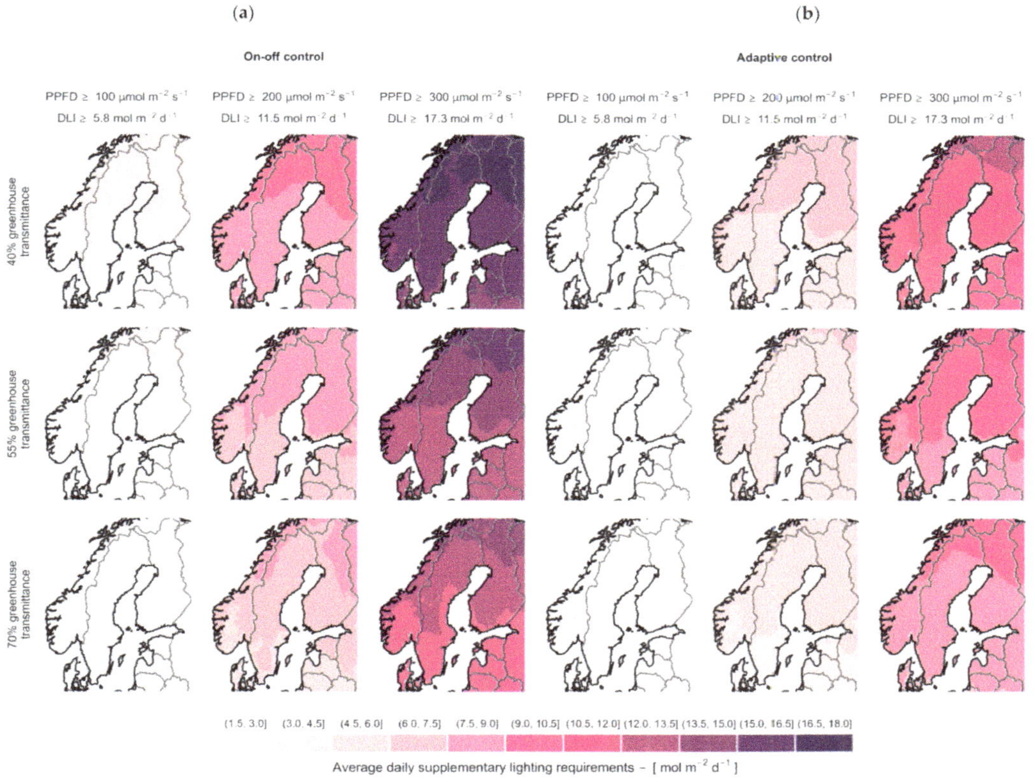


**Figure 6.** Average daily supplementary lighting requirements ($DLI_{lamps}$, $mol \cdot m^{-2} \cdot d^{-1}$) maps of the Nordic and Baltic counTable 181 days: from 1 February to 20 May and from 20 August to 31 October. A photoperiod of 16 $h \cdot d^{-1}$ was assessed considering three minimum light settings ($PPFD_{threshold}$: 100, 200 or 300 $\mu mol \cdot m^{-2} \cdot s^{-1}$) and three possible greenhouse hemispherical transmittances ($\tau_{th,\,PAR}$: 40%, 55% or 70%). The two panels compare the different control protocols: (**a**) On-off control vs. (**b**) Adaptive control. Values presented as a 0.1° × 0.1° grid both in latitude and longitude, ex-tending from 54.5° N to 70° N in latitude and 4.5° E to 31.5° E in longitude.

At the lowest $PPFD_{threshold}$ of 100 $\mu mol \cdot m^{-2} \cdot s^{-1}$, regardless of the control protocol, the light transmitted into the greenhouse was often enough to reach the threshold resulting in a lower need for supplementary light. Only when the transmittance was below 55% combined with the on-off control protocol did the supplementary lighting requirements increase to 4.5 $mol \cdot m^{-2} \cdot d^{-1}$ in some places (see Figure 6, upper left corner). The transmission levels caused larger differences when using the on-off protocol (observed when comparing by rows in Figures 6 and A4). The reason is that with the on-off protocol the lamps are assumed to be turned on at full power when the threshold is not reached; with lower transmittance and higher thresholds settings, this happens more often.

As the $PPFD_{threshold}$ increased, these combined effects were more evident. In the "worst-case scenario", with the lowest transmittance and the highest $PPFD_{threshold}$, the differences between control protocols were higher (Figures 6 and A4, upper right corner in each panel). In this setting, the available natural light inside the greenhouse did not reach the minimum threshold during most of the day. While the adaptive control was able to make use of the low available light and provide only the difference to reach the threshold; with the on-off protocol, the lamps were active at full power throughout the entire photoperiod ($DLI_{threshold} = DLI_{lamps}$).

The electrical energy needed for lighting was calculated considering the three different lamp efficacies. Figures 7 and A5 show the estimated yearly consumption during the 181 days of greenhouse cultivation using the highest $PPFD_{threshold}$ of 300 μmol·m$^{-2}$·s$^{-1}$.

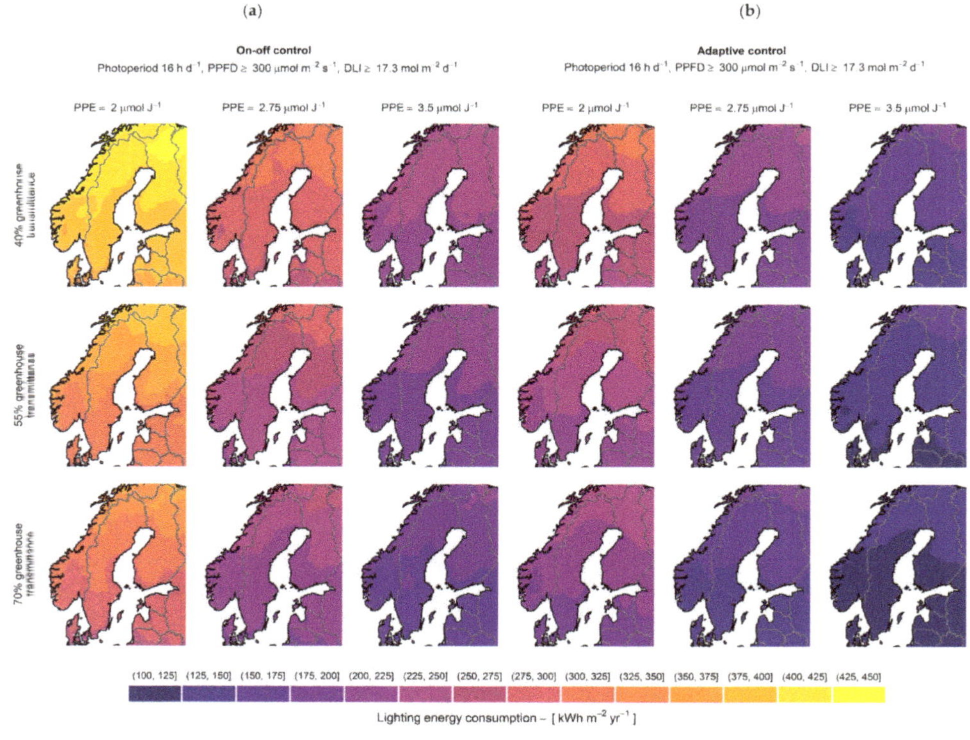


**Figure 7.** Yearly lighting energy consumption ($E_{el,\ lamps}$, kWh·m$^{-2}$·yr$^{-1}$) maps of the Nordic and Baltic countries for the chosen greenhouse cultivation period of 181 days: from 1 February to 20 May and from 20 August to 31 October. The electric energy demand of the lamps was calculated for a photoperiod of 16 h·d$^{-1}$ with a $PPFD_{threshold}$ of 300 μmol·m$^{-2}$·s$^{-1}$, considering three different photosynthetic photon efficacies (PPE: 2, 2.75 or 3.5 μmol·J$^{-1}$) and three possible greenhouse hemispherical transmittances ($\tau_{h,\ PAR}$: 40%, 55% or 70%). The two panels compare different lighting control protocols: (**a**) On-off control vs. (**b**) Adaptive control. Values presented as a 0.1° × 0.1° grid both in latitude and longitude, extending from 54.5° N to 70° N in latitude and 4.5° E to 31.5° E in longitude.

The variations in the $E_{el,\ lamps}$ follow similar trends as the supplementary light requirements regarding greenhouse transmittance and lighting control protocols. However, the differences were more pronounced between the $PPFD_{threshold}$ levels due to the interaction effects with the PPE of the lamps, where lower efficacy values resulted in higher energy consumptions (comparing by column in Figures 7 and A5).

The energy consumption approaches infinity as the PPE decreases (Figure 5). This means that at some point the $DLI_{threshold}$ can become too high and it is not possible for lamps with very low efficacy to supply it. When the $PPFD_{threshold}$ was set to 100 μmol·m$^{-2}$·s$^{-1}$, the maximum $E_{el,\ lamps}$ range was 110–120 kWh·m$^{-2}$·yr$^{-1}$; doubling the value raised the highest range to 280–300 kWh·m$^{-2}$·yr$^{-1}$ (data not shown). Finally, the $PPFD_{threshold}$ of 300 μmol·m$^{-2}$·s$^{-1}$ (Figure 7 and A5) resulted in a maximum range of 425–450 kWh·m$^{-2}$·yr$^{-1}$.

### 3.4. PV Systems Specific Energy Yield

The specific energy yield of the roof-mounted PV systems varied between the different regions and system designs, ranging from around 400 to 1120 kWh·kW$^{-1}$·yr$^{-1}$. This almost threefold increase is clearly shown in Figure 8 (data for Iceland not shown); where the output decreases noticeably as the latitude increases. The output also gradually decreases as the azimuth turns away from the optimal south-facing orientation (Figure 8a) towards an east-west layout (Figure 8f). Shifting some of the installed power from the south-east-facing roof onto the north-west roof side reduced the relative output (Figure 8, panel b vs. c and panel d vs. e). The coastal regions, particularly along the Baltic Sea presented higher PV outputs compared to areas inland at the same latitude.

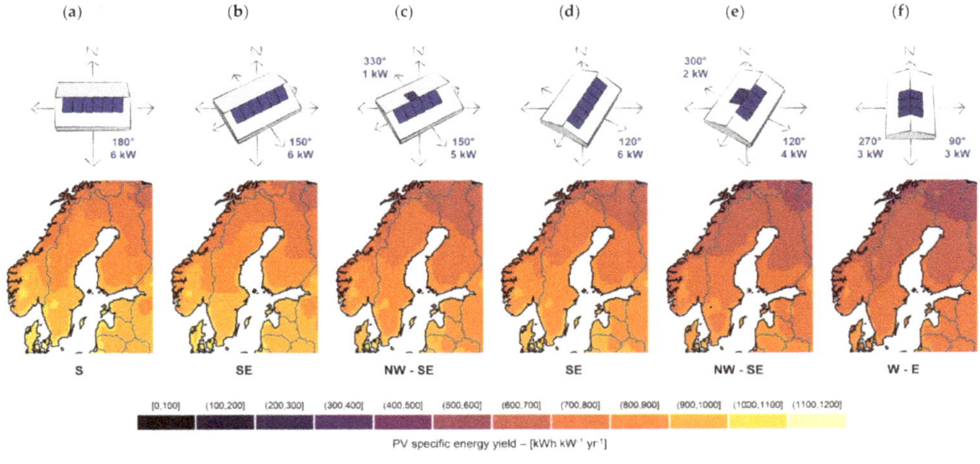


**Figure 8.** Yearly PV specific energy yield ($E_{rel,\,PV}$, kWh·kW$^{-1}$·yr$^{-1}$) maps of the Nordic and Baltic countries for roof-mounted PV systems at a 25° slope with an installed peak power ($P_{PV,STC}$) of 6 kW. The values were estimated using PVGIS version 5.1 assuming the solar cell technology was c-Si and the estimated system losses were 14%. The irradiance values used were from the ECMWF-ERA-5 dataset with a coverage period of 2005–2016 retrieved as a 0.1° × 0.1° grid both in latitude and longitude, extending from 54.5° N to 70° N in latitude and 4.5° E to 31.5° E in longitude. Depending on the orientation of the building and the distribution of the PV modules among the two roof directions, the studied system designs were: (**a**) 6 kW at azimuth 180°; (**b**) 6 kW at azimuth 150°; (**c**) 5 kW at azimuth 150° combined with 1 kW at azimuth 330°; (**d**) 6 kW with azimuth 120°; (**e**) 4 kW at azimuth 120° combined with 2 kW at azimuth 300°; (**f**) 3 kW at azimuth 90° combined with 3 kW at azimuth 270°.

The yearly specific yields of wall-mounted PV systems also presented a wide variation from around 400 to 925 kWh·kW$^{-1}$·yr$^{-1}$ for the same installed power (Figure 9). However, with vertically mounted PV modules, the effect of the latitude became less important when the azimuth rotated away from 180° facing south. Systems distributed at a combined azimuth of 90° and 270° towards the east-west (Figure 9z), had a much lower variation along the latitude with all the locations presenting specific system yields in the range of 400–600 kWh·kW$^{-1}$·yr$^{-1}$. Interestingly, PV modules on the south and south-east facade (azimuths 180° and 150°; in Figure 9w,x), presented similar specific energy yield as the combined roof-mounted systems with azimuths 120° and 300° (Figure 8e) and azimuths 90° and 270° (Figure 8f) respectively.

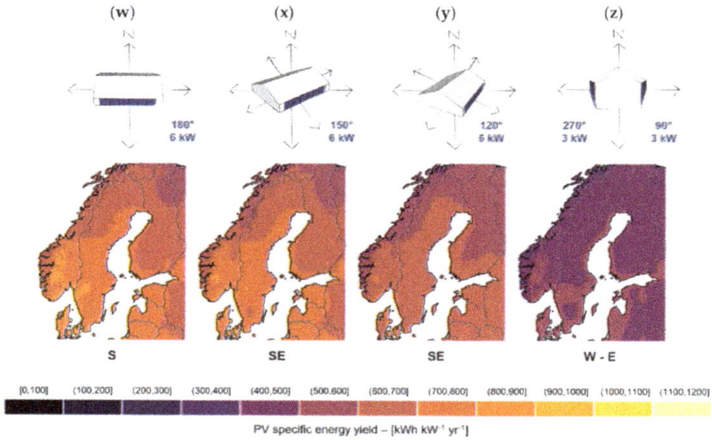


**Figure 9.** Yearly PV specific energy yield ($E_{rel,PV}$, $kWh\cdot kW^{-1}\cdot yr^{-1}$) maps of the Nordic and Baltic countries for wall-mounted PV systems at a 90° slope with an installed peak power ($P_{PV,STC}$) of 6 kW. The values were estimated using PVGIS version 5.1 assuming the solar cell technology was c-Si and the estimated system losses were 14%. The irradiance values used were from the ECMWF-ERA-5 dataset with a coverage period of 2005–2016 retrieved as a 0.1° × 0.1° grid both in latitude and longitude, extending from 54.5° N to 70° N in latitude and 4.5° E to 31.5° E in longitude. Depending on the orientation of the building and the distribution of the PV modules, the studied system designs were: **(w)** 6 kW at azimuth 180°; **(x)** 6 kW at azimuth 150°; **(y)** 6 kW with azimuth 120°; **(z)** 3 kW at azimuth 90° combined with 3 kW at azimuth 270°.

### *3.5. Year-Round Concept Implementation for One Location*

Perhaps the best way to describe the year-round cultivation concept is to calculate the values for one location. Table 3 shows the average monthly values for the environmental variables extracted for the city of Borlänge, Sweden (lat. 60.5° N, lon. 15.4° E); showing the different cultivation phases as well. The day length can be found in Figure 2.

**Table 3.** Summary of monthly values for the year-round cultivation concept in Borlänge, Sweden.

| **Monthly Average Values** | **Jan** | **Feb** | **Mar** | **Apr** | **May** | **Jun** | **Jul** | **Aug** | **Oct** | **Nov** | **Dec** |
|---|---|---|---|---|---|---|---|---|---|---|---|
| Ambient temperature-$T_a$ (°C) | −4.8 | −4.4 | −0.9 | 4.7 | 10.0 | 14.1 | 16.7 | 15.1 | 11.3 | 5.5 | 1.4 |
| Daily global horizontal irradiation—$H_h$ ($kWh\cdot m^{-2}\cdot d^{-1}$) | 0.3 | 0.9 | 2.3 | 3.9 | 4.8 | 5.3 | 4.8 | 3.7 | 2.4 | 1.1 | 0.4 |
| Daily Light Integral-$DLI_{sun}$ ($mol\cdot m^{-2}\cdot d^{-1}$) | 2.2 | 6.6 | 17.1 | 28.3 | 34.8 | 38.8 | 35.2 | 26.9 | 17.5 | 8.1 | 2.5 |
| Diffuse to global horizontal irradiance ratio ($G_d/G_h$) | 67% | 59% | 44% | 41% | 41% | 40% | 41% | 43% | 46% | 51% | 64% |
| Indoor cultivation with sole-source LED lighting [1] | → | | | | | | | | | → | → |
| Greenhouse cultivation with supp. LED lighting | | → | → | → | | | | → | → | | |
| Outdoor cultivation with natural light | | | | | → | → | → | | | | |

[1] Arrows and shaded regions indicate the months when each cultivation type could be used.

Using the following parameters for indoor cultivation at $PPFD_{lamps}$ = 300 $\mu mol\cdot m^{-2}\cdot s^{-1}$ during a photoperiod of 16 $h\cdot d^{-1}$ would result in a $DLI_{lamps}$ of 17.3 $mol\cdot m^{-2}\cdot d^{-1}$. These settings agree well with the average $DLI_{sun}$ values for March and October in this location (Table 3). Depending on the efficacy of the lamps (using Figure 5b), 92 days of indoor cultivation would require 221 $kWh\cdot m^{-2}$, 161 $kWh\cdot m^{-2}$, or 129 $kWh\cdot m^{-2}$.

Table 4 presents the estimated daily supplementary lighting requirements and corresponding electric energy consumption for lighting during the greenhouse cultivation phase (181 days), according to the different parameter values for the two control protocols.

**Table 4.** Parametric comparison of the estimated daily supplementary lighting and corresponding electric energy consumption for a 181 days greenhouse cultivation from February to May and from August to October in Borlänge, Sweden.

| **Greenhouse parameters** | **On−off lighting control protocol** | | | | | | | | |
|---|---|---|---|---|---|---|---|---|---|
| Minimum PAR level ($PPFD_{threshold}$) | **100 μmol·m$^{-2}$·s$^{-1}$** | | | **200 μmol·m$^{-2}$·s$^{-1}$** | | | **300 μmol·m$^{-2}$·s$^{-1}$** | | |
| Transmittance ($\tau_{h,\ PAR}$) | **40%** | **55%** | **70%** | **40%** | **55%** | **70%** | **40%** | **55%** | **70%** |
| Daily supplementary lighting requirements ($DLI_{lamps}$, mol·m$^{-2}$·d$^{-1}$) | 2.9 | 2.5 | 2.3 | 8.4 | 7.3 | 6.3 | 15.6 | 12.9 | 12.0 |
| **Lamps' photosynthetic photon efficacy (PPE)** | **Yearly [1] lighting energy consumption ($E_{el,\ lamps}$, kWh·m$^{-2}$·yr$^{-1}$)** | | | | | | | | |
| PPE = 2 μmol·J$^{-1}$ | 73 | 63 | 58 | 210 | 183 | 159 | 393 | 325 | 302 |
| PPE = 2.75 μmol·J$^{-1}$ | 53 | 46 | 43 | 153 | 133 | 116 | 286 | 236 | 219 |
| PPE = 3.5 μmol·J$^{-1}$ | 42 | 36 | 33 | 120 | 104 | 91 | 225 | 185 | 172 |
| **Greenhouse parameters** | **Adaptive lighting control protocol** | | | | | | | | |
| Minimum PAR light level ($PPFD_{threshold}$) | **100 μmol·m$^{-2}$·s$^{-1}$** | | | **200 μmol·m$^{-2}$·s$^{-1}$** | | | **300 μmol·m$^{-2}$·s$^{-1}$** | | |
| Transmittance ($\tau_{h,\ PAR}$) | **40%** | **55%** | **70%** | **40%** | **55%** | **70%** | **40%** | **55%** | **70%** |
| Daily supplementary lighting requirements ($DLI_{lamps}$, mol·m$^{-2}$·d$^{-1}$) | 2.2 | 2.0 | 1.8 | 5.8 | 5.0 | 4.5 | 10.5 | 9.0 | 8.1 |
| **Lamps' photosynthetic photon efficacy (PPE)** | **Yearly [1] lighting energy consumption ($E_{el,\ lamps}$, kWh·m$^{-2}$·yr$^{-1}$)** | | | | | | | | |
| PPE = 2 μmol·J$^{-1}$ | 54 | 49 | 46 | 146 | 126 | 114 | 264 | 227 | 204 |
| PPE = 2.75 μmol·J$^{-1}$ | 39 | 36 | 34 | 106 | 91 | 83 | 192 | 165 | 148 |
| PPE = 3.5 μmol·J$^{-1}$ | 31 | 28 | 27 | 83 | 72 | 65 | 151 | 130 | 116 |

[1] Calculation made for 180 greenhouse cultivation days from 1 February to 20 May and from 20 August to 31 October.

Assuming a moderate greenhouse transmittance of 55% and using an on-off control protocol, the lamps would provide on average 12.9 mol·m$^{-2}$·d$^{-1}$ of supplementary lighting. Depending on the efficacy, the corresponding energy consumption for the whole greenhouse cultivation period of 181 days would be 325 kWh·m$^{-2}$yr$^{-1}$ for standard LED lamps, 236 kWh·m$^{-2}$yr$^{-1}$ if state-of-the-art standard LED luminaires are used and future developments could potentially reduce it to 185 kWh·m$^{-2}$yr$^{-1}$ (see Table 4, orange shades).

For comparison, the adaptive control protocol (Table 4, purple shades) would provide instead 9.0 mol·m$^{-2}$·d$^{-1}$ of supplementary lighting. Which, depending on the efficacy of the lamps, would translate into 227 kWh·m$^{-2}$yr$^{-1}$, 165 kWh·m$^{-2}$yr$^{-1}$ or 130 kWh·m$^{-2}$yr$^{-1}$ for the same period. Finally, selecting state-of-the-art standard LED luminaires for both indoor and greenhouse cultivation as well as adaptive control would require per year an estimate of 326 kWh·m$^{-2}$yr$^{-1}$ (161 kWh·m$^{-2}$ for 92 days of indoor cultivation and 165 kWh·m$^{-2}$ for 181 days of greenhouse cultivation).

Knowing the energy requirements makes it possible to estimate the needed $P_{PVSTC}$ and as well as how many m$^2$ of cultivation with supplementary lighting can be compensated with the installed PV. The yearly specific energy yields for the different greenhouse integrated PV systems at this location are shown in Table 5. The different system types correspond to those presented in Table 2 and Figures 8 and 9.

**Table 5.** Yearly specific energy yields for the different greenhouse integrated PV systems in Borlänge, Sweden.

| | **Roof-Mounted (25°)** | | | | | | **Wall-Mounted (90°)** | | | |
|---|---|---|---|---|---|---|---|---|---|---|
| **Greenhouse Integrated PV Systems** | a | b | c | d | e | f | w | x | y | z |
| Yearly PV specific energy yield ($E_{rel,PV}$, kWh·kW$^{-1}$·yr$^{-1}$) | 904 | 888 | 823 | 825 | 743 | 708 | 745 | 732 | 648 | 484 |

In a commercial greenhouse (assuming a roof area, $A_{roof}$ = 2000 m$^2$) cultivating of a shade-intolerant species that would be negatively affected if $PV_{R,\%} \geq 15\%$; the maximum $P_{PV,STC}$ installed could be 54 kW with PV modules of 18% efficiency (using Equation (12)). If all PV modules are installed on a south-facing roof (Table 5, system a), the expected yearly electrical yield would be 48,816 kWh·yr$^{-1}$. Considering an example of a forest nursery producing Scots pine (*Pinus sylvestris* L.) seedlings on a year-round basis using an LED growth room during germination and early growth [18,25,86,134], the electricity generated by the PV of the greenhouse would be enough to compensate for the lighting of roughly 300 $m^2$ of indoor cultivation for 92 days during the winter months.

## 4. Discussion

Although in optimal conditions, many species could grow better and produce higher quality plants when growing outdoors due to the higher light levels, better air movement, lower relative humidity, and less overheating [28,31]; damages to the crops due to climate phenomena such as droughts and floods as well as biological menaces such as pests and diseases considerably affect and reduce productivity. For species that can be cultivated entirely in greenhouses, this option presents a more sustainable and resource-efficient form of cultivation [6,10].

In the Nordic countries, the outdoor vegetation period is very short and limited to only a few months, restricting the outdoor cultivation of crops like herbaceous perennials, forest seedlings, and other potted plants that are grown for more than one season. During the summer, there are long days with abundant sunlight, the $DLI_{sun}$ levels peak in June, and the warmest temperatures come in July and August (Figures 1–3). In contrast, during the winter the available PAR light is almost negligible, and the temperatures can go well below the freezing point during several weeks in most parts of the region.

The year-round cultivation concept presented here aims to work with the harsh northern climate and adapt to the natural conditions instead of trying to go against them. When the outside environment is not suitable anymore for cultivation, the plants can be transported inside modern greenhouses with transparent covers which present the best option to provide additional heating and supplementary lighting to the plants [10,110].

The amount of natural light reaching inside a greenhouse depends heavily on multiple factors including the building design [114,124], roof shape and inclination [110,115], location and orientation [135], the position of the supporting structures as well as hanging objects inside like pots or lamps [111,136], and the covering material of the greenhouse [116,117,137,138]. Weathering and aging of the glazing as well as other local factors like condensation, dust, snow accumulation, and buildings nearby can obstruct the light that enters a greenhouse [112].

From all these factors, the covering material is probably one of the most important for light transmission and therefore has been broadly studied [110]. Distinct wavelengths behave differently depending on the material composition and the angle of incidence, affecting the spectral distribution at the cultivation area [137,139]. Determining the natural light inside a greenhouse involves complex calculations once the main specifications such as location, structure design, and covering material have been decided [4,64,113,140]; or requires extensive simultaneous measurements of the outdoors and indoors illumination levels [110,111,116,117]. Instead, the parametric study here described aimed to provide a broad perspective and a range of plausible hemispherical PAR transmittance values within reasonable limits ranging from 40% to 70%.

Compared to other environmental variables, the amount of PAR inside modern greenhouses is usually less accurately regulated and depends heavily on outdoor conditions [9,112]. Different greenhouse lighting control protocols have been proposed in the past including strategies for extending the day length [9,10,141], ensuring a consistent DLI [39], and using adaptive control to reach a minimum PPFD threshold [40,42]. Usually, the implementation of these concepts has been restricted to the technical limitations of the available HID lamps. However, as LED grow lights become commonly adopted, more advanced and dynamic control protocols can benefit from the LEDs flexible spectra, dimmability, rapid response, and tolerance to numerous switching cycles [142].

Two supplementary lighting control protocols were compared in this study: on-off and adaptive control. Regardless of parameter combinations and locations, the adaptive LED lighting control had the best outcomes with the lowest supplementary lighting requirements. The adaptive control allowed to reach and maintain the $PPFD_{threshold}$ throughout the entire photoperiod while considering the natural light inside the greenhouse. This resulted in lower energy consumption for lighting. The potential energy savings that followed could justify doing a lamp retrofit in relatively new greenhouses that are in good conditions but still use HID lamps [7,51,54].

Although greenhouses can extend the cultivation period during spring and autumn in the Nordics, at some moment in the year, there is very low radiation gain from the sun and too high heat loss through the glazing. In the summer, the opposite happens as it becomes difficult to naturally dissipate the excess heat from the abundant sunlight [4,116].

Closed plant production systems with sole-source lighting are designed to maximize the yield using multi-level cultivation and have the advantage of a standardized and controlled environment independent from outdoor conditions. Nevertheless, they also have the evident disadvantage of excluding sunlight which is a valuable and free resource. This results in the high energy consumption for lighting and climate control [55,112]. It has been estimated that between 70% and 80% of the electricity consumption in indoor growth rooms is due to lighting while the rest is mainly cooling and dehumidification [53,142].

Increasing the planting density assures a better use of the space and selecting efficient lamps prevents unnecessary energy consumption. The efficacy of growth lights has developed significantly in the past years, especially for LEDs. In 2014 the best-in-class HID and LED lamps had similar efficacies of around 1.70 $\mu mol \cdot J^{-1}$ [7,143]. While HID lamps were improved to 2.1 $\mu mol \cdot J^{-1}$ by 2018 [7,144,145]; LED fixtures in 2020 reached efficacies between 2.5 and 2.8 $\mu mol \cdot J^{-1}$ and are expected to reach even higher values between 3.4 and 4.1 $\mu mol \cdot J^{-1}$ during this decade [54]. Irrespectively of how advanced the lighting technologies are, there will always exist a certain investment and energy cost for artificial lighting compared to sunlight [51]. Under some circumstances and for certain species, the advantages of a fully controlled growth space and the possibility of year-round cultivation might outweigh some of those costs [52,53,55].

To partly offset the energy used for supplementary lighting during the darker months, some of the solar energy abundant during the summer can be transformed into electricity integrating PV on the roof of the greenhouses. Agrivoltaics can reduce the competition for space between energy and plants production but if not planned carefully it can increase competition for sunlight [57]. Hence, proper optimization for the particular location and targeted plant species is required, depending on if the main goal is to maximize crop yield or increase the electrical output [58,59].

Using Equation (10), it was possible to calculate that 1 kW of PV with 18% efficiency takes up approximately 5.6 $m^2$ and if it is mounted on a south-facing roof of a greenhouse or growth chamber (Table 5, system a), each $m^2$ of PV will have a yearly output of 162.5 $kWh \cdot m^{-2} \cdot yr^{-1}$. This corresponds more or less with the amount of electricity for lighting needed per $m^2$ in a location in mid-Sweden, either for indoor cultivation for 92 days or greenhouse cultivation for 181 days. Nevertheless, concluding that 1 $m^2$ of PV would compensate the lighting for 1 $m^2$ of cultivation would not be correct. The scenarios presented in this study assumed only 92 days of indoor cultivation with sole-source lighting, however, one of the main purposes of a growth room is to save space by using multi-layer cultivation on a year-round basis. This means that the growth area and cultivation time indoors would be by several factors larger for the available roof area directly on the growth room [52,55].

An adequate $PV_R$ requires considering the shading effect that the integrated PV will have inside the greenhouse and its impact on the light levels for the crop. Plant species with a low (5–10 $mol \cdot m^{-2} \cdot d^{-1}$) and moderate light needs (10–20 $mol \cdot m^{-2} \cdot d^{-1}$) are more tolerant to shading compared to plants with high (20–30 $mol \cdot m^{-2} \cdot d^{-1}$) or very high (DLI > 30 $mol \cdot m^{-2} \cdot d^{-1}$) requirements [28–32]. Greenhouses in southern Europe with a $PV_{R,\%}$ between 10% and 15% showed a very small decrease in the crop yields, even when cultivating shade-intolerant plants like tomatoes [66,69]. Roof coverage ratios between 20–25% showed minimal negative effects on the yield of low and moderate light-demanding species [59,70]. However, at higher coverage ratios ($PV_{R,\%} \geq 60\%$) even species requiring moderate light were affected [70].

Naturally, parametric studies like the one here presented come with clear limitations due to the great number of variables assumed. For the implementation of greenhouses with integrated PV in the Nordics, additional studies with detailed simulations for specific locations are necessary to determine the ideal PV coverage ratios for different species when cultivated in the region and how the greenhouse environment can influence the PV energy yield. Practical implementation of the concept will increase the understanding and validate the results while providing valuable data for the improvement of the concept.

Finally, the electrical equipment used inside the growth rooms and greenhouses will have an important effect on the ambient temperature in the cultivation area. The lamps' efficiency, as well as the amount of time they are used, can impact the amount of heating and cooling load necessary. Although not included in the scope of this study, these types of calculations are necessary for the year-round cultivation concept using LED lamps and some examples regarding the thermal energy requirements for controlled-environment cultivation exist already in the literature [4,55,146–148].

## 5. Conclusions

New technological developments such as LED grow lights, adaptive lighting control protocols, as well as greenhouse integrated PV, have the potential of making the concept of year-round cultivation a feasible option, even in the harsh conditions of the Nordic countries. After analyzing satellite meteorological data for the region, it can be concluded that:

- Ambient temperature and natural DLI levels are suitable for outdoor cultivation during at least three months in the summer for most of the region.
- Greenhouses can be used to start the cultivation earlier in spring and extend the vegetation period until later in autumn.
- Transmittance levels of natural light inside the greenhouse can significantly influence the supplementary lighting needed for the plants.
- Among the options compared, LED lamps with adaptive lighting control have the highest energy-saving potential. They can benefit from the available sunlight inside the greenhouse, avoiding unnecessary energy waste and supplementing only enough light to reach the DLI target for the cultivated species.
- During the winter months, indoor cultivation in closed growth chambers offers a standardized and controlled environment independent from outdoor conditions.
- Light intensity and duration of the photoperiod together with the efficacy of the lamps determine the amount of electricity needed for lighting in the growth room.
- Greenhouses with integrated PV provide an alternative for using the abundant sunshine in the summer and offsetting some of the electricity used for lighting during the darker months.
- To avoid negative effects on the plants caused by excessive shading from the solar panels, careful planning is required based on the design, location, and orientation of the greenhouse.
- Additional studies that consider the investment and running costs for heating and cooling of the growth rooms and greenhouses with regard to the lighting control protocols and LED lamps are necessary. Only when both electrical and thermal energy requirements are evaluated together can the feasibility of the year-round cultivation concept be truly evaluated.

**Funding:** This research project was supported by FP7 Environment [Grant Number ZEPHYR-Project ID: 308313].

**Data Availability Statement:** Publicly available datasets were analyzed in this study. This data can be found here: PVGIS © European Communities, 2001–2021.

**Acknowledgments:** The author would like to thank Anders Mattsson (Dalarna University), Frank Fiedler (Dalarna University), Uwe Zimmermann (Uppsala University), and Sofia Lindblad (Uppsala University) for their constructive comments on the content and general improvements. The irradiance and meteorological data as well as the PV simulations were extracted from PVGIS © European Communities, 2001–2021.

**Conflicts of Interest:** The authors declare no conflict of interest. The funders had no role in the design of the study; in the collection, analyses, or interpretation of data; in the writing of the manuscript, or in the decision to publish the results.

## Appendix A

The figures presented in this appendix are similar to those in the main text but for the region covering Iceland. The data was retrieved from the same source, from 63° N to 67.5° N in latitude and from 13° E to 25° E in longitude using a cell size of 0.05° × 0.05° in both latitude and longitude.

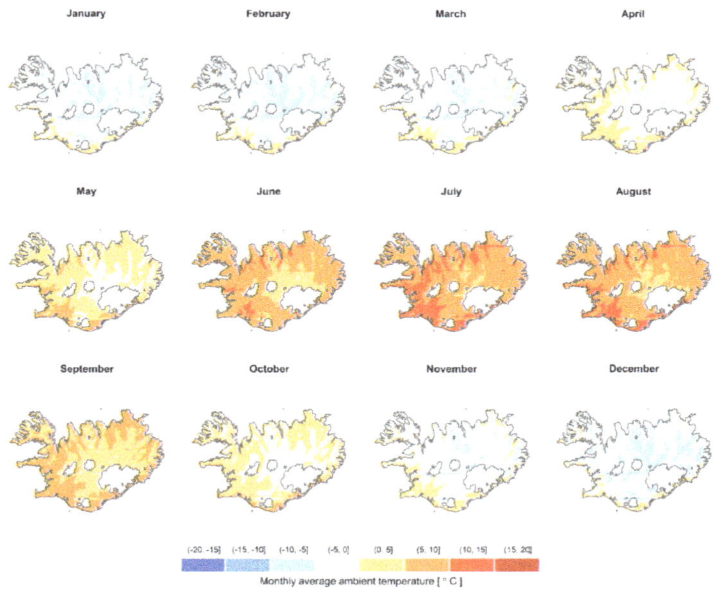


**Figure A1.** Monthly average daily ambient temperature maps for Iceland. Values from the ECMWF-ERA-5 dataset with a coverage period of 2005–2016 retrieved from PVGIS as a 0.05° × 0.05° grid both in latitude and longitude, extending from 63° N to 67.5° N in latitude and 13° W to 25° W in longitude.

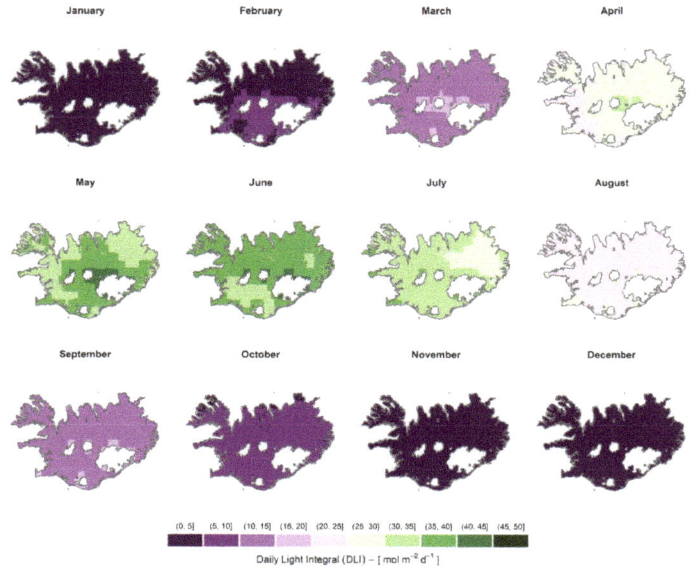


**Figure A2.** Monthly average photosynthetic daily light integral ($DLI_{sun}$, mol·m$^{-2}$·d$^{-1}$) maps for Iceland. Values from the ECMWF-ERA-5 dataset with a coverage period of 2005–2016 retrieved from PVGIS as a 0.05° × 0.05° grid both in latitude and longitude, extending from 63° N to 67.5° N in latitude and 13° W to 25° W in longitude.

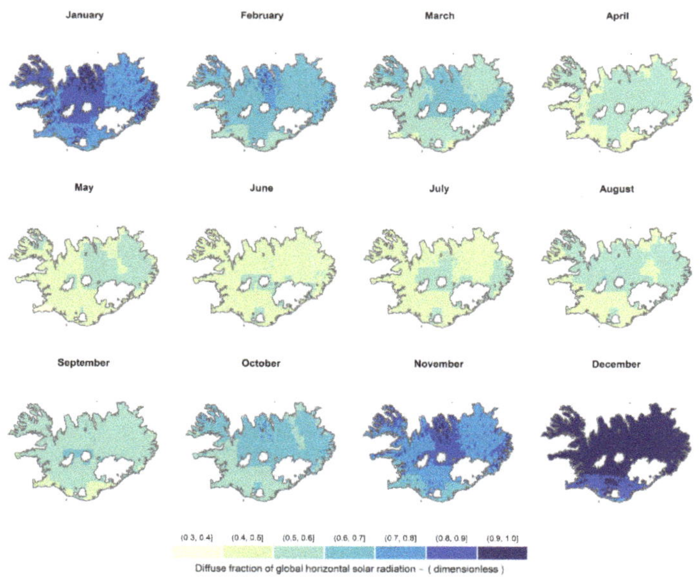


**Figure A3.** Monthly average diffuse fraction of global horizontal irradiance ($G_d/G_h$) for Iceland. Values from the ECMWF-ERA-5 dataset with a coverage period of 2005–2016 retrieved from PVGIS as a 0.05° × 0.05° grid both in latitude and longitude, extending from 63° N to 67.5° N in latitude and 13° W to 25° W in longitude.

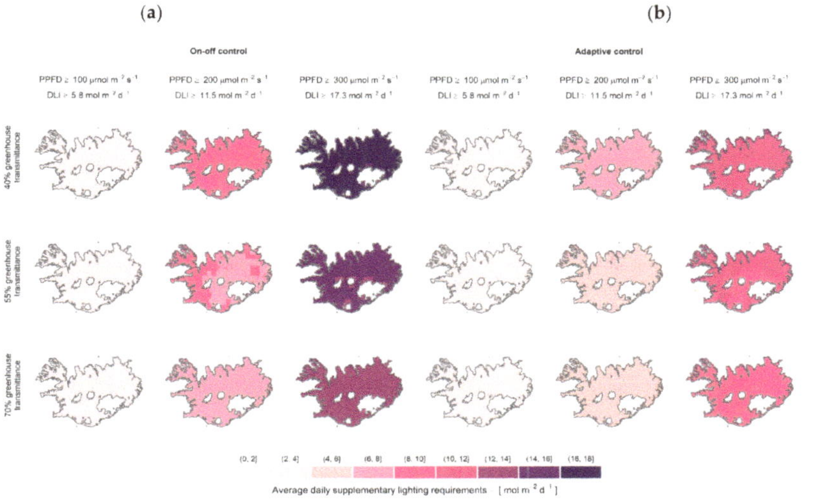


**Figure A4.** Average daily supplementary lighting requirements ($DLI_{lamps}$, $mol{\cdot}m^{-2}{\cdot}d^{-1}$) maps of Iceland for the chosen greenhouse cultivation period of 181 days: from 1 February to 20 May and from 20 August to 31 October. A photoperiod of 16 $h{\cdot}d^{-1}$ was assessed considering three minimum light settings ($PPFD_{threshold}$: 100, 200 or 300 $\mu mol{\cdot}m^{-2}{\cdot}s^{-1}$) and three possible greenhouse hemispherical transmittances ($\tau_{h,\,PAR}$: 40%, 55% or 70%). The two panels compare different control protocols: (**a**) On-off control vs. (**b**) Adaptive control. Values presented as a 0.05° × 0.05° grid both in latitude and longitude, extending from 63° N to 67.5° N in latitude and 13° W to 25° W in longitude.

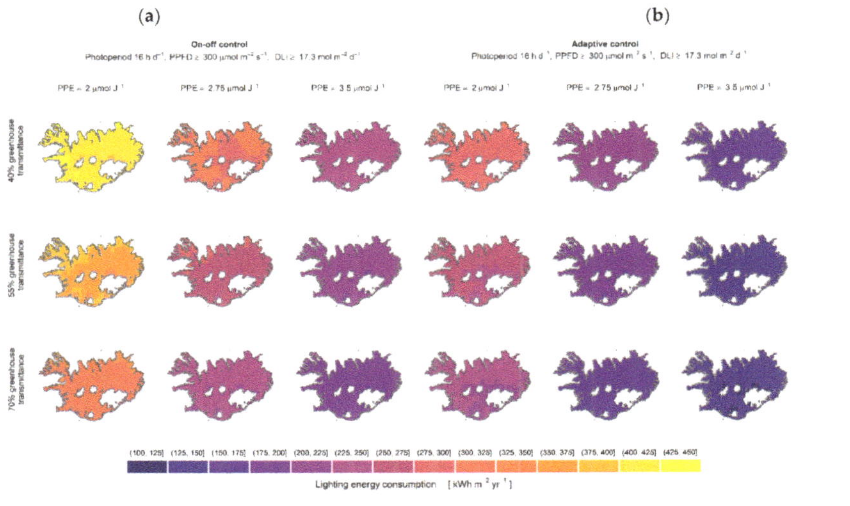


**Figure A5.** Yearly lighting energy consumption ($E_{el,\,lamps}$, $kWh{\cdot}m^{-2}{\cdot}yr^{-1}$) maps of Iceland for the chosen greenhouse cultivation period of 181 days: from 1 February to 20 May and from 20 August to 31 October. The electric energy demand of the lamps was calculated for a photoperiod of 16 $h{\cdot}d^{-1}$ with a $PPFD_{threshold}$ of 300 $\mu mol{\cdot}m^{-2}{\cdot}s^{-1}$, considering three different photosynthetic photon efficacies (PPE: 2, 2.75 or 3.5 $\mu mol{\cdot}J^{-1}$) and three possible greenhouse hemispherical transmittances ($\tau_{h,\,PAR}$: 40%, 55% or 70%). The two panels compare different lighting control protocols: (**a**) On-off control; (**b**) Adaptive control. Values presented as a 0.05° × 0.05° grid both in latitude and longitude, extending from 63° N to 67.5° N in latitude and 13° W to 25° W in longitude.

## References


1. Swedish Meteorological and Hydrological Institute (SMHI) Climate Indicators—Length of Vegetation Period. Available online: https://www.smhi.se/en/climate/climate-indicators/climate-indicators-length-of-vegetation-period-1.91482 (accessed on 23 August 2020).
2. Albright, L.D.; Langhans, R.W. *Controlled Environment Agriculture Scoping Study*; Electric Power Research Institute: Ithaca, NY, USA, 1996.
3. Despommier, D. *The Vertical Farm: Feeding the World in the 21st Century*; Macmillan: New York, NY, USA, 2010.
4. Vadiee, A.; Martin, V. Energy Analysis and Thermoeconomic Assessment of the Closed Greenhouse—The Largest Commercial Solar Building. *Appl. Energy* **2013**, *102*, 1256–1266. [CrossRef]
5. Nimmermark, S.; Nielsen, J.M. *Klimatisering, Belysning, Bevattning Och Mekanisering I Växthus*; Alnarp, Sweden, 2014; p. 30.
6. Stanghellini, C.; Van't Ooster, B.; Heuvelink, E. Greenhouses: Why? In *Greenhouse Horticulture*; Wageningen Academic Publishers: Wageningen, The Netherlands, 2018; pp. 15–25. ISBN 978-90-8686-329-7.
7. Nelson, J.A.; Bugbee, B. Economic Analysis of Greenhouse Lighting: Light Emitting Diodes vs. High Intensity Discharge Fixtures. *PLoS ONE* **2014**, *9*, e99010. [CrossRef] [PubMed]
8. Fisher, P.; Both, A.J.; Bugbee, B. Supplemental Lighting—Technology, Costs, and Efficiency. In *Light Management in Controlled Environments*; Lopez, R., Runkle, E.S., Eds.; CreateSpace Independent Publishing Platform: Willoughby, OH, USA, 2017; pp. 74–81. ISBN 978-1-5442-5449-4.
9. Landis, T.D.; Tinus, R.W.; McDonald, S.E.; Barnett, J.P. Chapter 3: Light. In *Atmospheric Environment—The Container Tree Nursery Manual*; Agriculture Handbook 674; USDA Forest Service: Washington, DC, USA, 1992; Volume 3, p. 145.
10. Nelson, P. *Greenhouse Operation and Management*, 7th ed.; Pearson: Boston, MA, USA, 2011; ISBN 978-0-13-243936-7.
11. Runkle, E. *Strategies for Supplemental Lighting*; Greenhouse Product News: Sparta, MI, USA, 2009; p. 50.
12. Bugbee, B. *Effects of Radiation Quality, Intensity, and Duration on Photosynthesis and Growth*; Plants, Soils, and Biometeorology Department Utah State University: Logan, UT, USA, 1994; N96-18128; p. 12.
13. Aphalo, P.J.; Ballaré, C.L.; Scopel, A.L. Plant-Plant Signalling, the Shade-Avoidance Response and Competition. *J. Exp. Bot.* **1999**, *50*, 1629–1634. [CrossRef]
14. Aphalo, P.J. *Light Signals and the Growth and Development of Plants—A Gentle Introduction*; Department of Biosciences Plant Biology University of Helsinki: Helsinki, Finland, 2010.
15. McCree, K.J. The Action Spectrum, Absorptance and Quantum Yield of Photosynthesis in Crop Plants. *Agric. Meteorol.* **1971**, *9*, 191–216. [CrossRef]
16. McCree, K.J. Test of Current Definitions of Photosynthetically Active Radiation against Leaf Photosynthesis Data. *Agric. Meteorol.* **1972**, *10*, 443–453. [CrossRef]
17. Craver, J.K.; Lopez, R.G. Control of Morphology by Manipulating Light Quality and Daily Light Integral Using LEDs. In *LED Lighting for Urban Agriculture*; Springer: Berlin/Heidelberg, Germany, 2016; pp. 203–217.
18. Riikonen, J.; Kettunen, N.; Gritsevich, M.; Hakala, T.; Särkkä, L.; Tahvonen, R. Growth and Development of Norway Spruce and Scots Pine Seedlings under Different Light Spectra. *Environ. Exp. Bot.* **2016**, *121*, 112–120. [CrossRef]
19. Bantis, F.; Radoglou, K. Morphology, Development, and Transplant Potential of *Prunus avium* and *Cornus sanguinea* Seedlings Growing under Different LED Lights. *Turk. J. Biol.* **2017**, *41*, 314–321. [CrossRef]
20. Hernandez, R.; Kubota, C. Light Quality and Photomorphogenensis. In *Light Management in Controlled Environments*; Lopez, R., Runkle, E.S., Eds.; CreateSpace Independent Publishing Platform: Willoughby, OH, USA, 2017; pp. 29–37. ISBN 978-1-5442-5449-4.
21. Stuefer, J.F.; Huber, H. Differential Effects of Light Quantity and Spectral Light Quality on Growth, Morphology and Development of Two Stoloniferous Potentilla Species. *Oecologia* **1998**, *117*, 1–8. [CrossRef] [PubMed]
22. Marcelis, L.F.M.; Broekhuijsen, A.G.M.; Meinen, E.; Nijs, E.; Raaphorst, M.G.M. Quantification of the Growth Response to Light Quantity of Greenhouse Grown Crops. In Proceedings of the V International Symposium on Artificial Lighting in Horticulture 711, Lillehammer, Norway, 21–24 June 2005; pp. 97–104.
23. Gerovac, J.R.; Craver, J.K.; Boldt, J.K.; Lopez, R.G. Light Intensity and Quality from Sole-Source Light-Emitting Diodes Impact Growth, Morphology, and Nutrient Content of *Brassica* Microgreens. *HortScience* **2016**, *51*, 497–503. [CrossRef]
24. Poorter, H.; Niinemets, Ü.; Ntagkas, N.; Siebenkäs, A.; Mäenpää, M.; Matsubara, S.; Pons, T. A Meta-Analysis of Plant Responses to Light Intensity for 70 Traits Ranging from Molecules to Whole Plant Performance. *New Phytol.* **2019**, *223*, 1073–1105. [CrossRef]
25. Hernandez Velasco, M.; Mattsson, A. Light Quality and Intensity of Light-Emitting Diodes during Pre-Cultivation of *Picea abies* (L.) Karst. and *Pinus sylvestris* L. Seedlings—Impact on Growth Performance, Seedling Quality and Energy Consumption. *Scand. J. For. Res.* **2019**, *34*, 159–177. [CrossRef]
26. Canham, A.E.; Canham, A.F. Plants and Light. *Sci. Hortic.* **1964**, *17*, 155–160.
27. Hutchinson, V.A.; Currey, C.J.; Lopez, R.G. Photosynthetic Daily Light Integral during Root Development Influences Subsequent Growth and Development of Several Herbaceous Annual Bedding Plants. *HortScience* **2012**, *47*, 856–860. [CrossRef]
28. Faust, J.E.; Logan, J. Daily Light Integral: A Research Review and High-Resolution Maps of the United States. *HortScience* **2018**, *53*, 1250–1257. [CrossRef]
29. Spaargaren, J.J. *Supplemental Lighting for Greenhouse Crops*; Hortilux Schreder: Monster, The Netherlands, 2001.
30. Korczynski, P.C.; Logan, J.; Faust, J.E. Mapping Monthly Distribution of Daily Light Integrals across the Contiguous United States. *HortTechnology* **2002**, *12*, 12–16. [CrossRef]

31. Faust, J.E. Light. In *Ball RedBook: Crop Production*; Hamrick, D., Ed.; Ball Publishing: Batavia, IL, USA, 2003; Volume 2, pp. 71–84. ISBN 978-1-883052-35-5.
32. Torres, A.P.; Lopez, R.G. *Measuring Daily Light Integral in a Greenhouse*; Purdue University Extension; Purdue Department of Horticulture and Landscape Architecture: West Lafayette, IN, USA, 2010; p. 7.
33. Morrow, R.C. LED Lighting in Horticulture. *HortScience* **2008**, *43*, 1947–1950. [CrossRef]
34. Mitchell, C.A.; Dzakovich, M.P.; Gomez, C.; Lopez, R.; Burr, J.F.; Hernández, R.; Kubota, C.; Currey, C.J.; Meng, Q.; Runkle, E.S.; et al. Light-Emitting Diodes in Horticulture. In *Horticultural Reviews*; John Wiley & Sons: Hoboken, NJ, USA, 2015; Volume 43, pp. 1–88.
35. Bantis, F.; Smirnakou, S.; Ouzounis, T.; Koukounaras, A.; Ntagkas, N.; Radoglou, K. Current Status and Recent Achievements in the Field of Horticulture with the Use of Light-Emitting Diodes (LEDs). *Sci. Hortic.* **2018**, *235*, 437–451. [CrossRef]
36. Bugbee, B. Economics of LED Lighting. In *Light Emitting Diodes for Agriculture: Smart Lighting*; Dutta Gupta, S., Ed.; Springer: Singapore, 2017; pp. 81–99. ISBN 978-981-10-5807-3.
37. Massa, G.D.; Kim, H.-H.; Wheeler, R.M.; Mitchell, C.A. Plant Productivity in Response to LED Lighting. *HortScience* **2008**, *43*, 1951–1956. [CrossRef]
38. Craver, J.K.; Boldt, J.K.; Lopez, R.G. Comparison of Supplemental Lighting Provided by High-Pressure Sodium Lamps or Light-Emitting Diodes for the Propagation and Finishing of Bedding Plants in a Commercial Greenhouse. *HortScience* **2019**, *54*, 52–59. [CrossRef]
39. Albright, L.D.; Both, A.-J.; Chiu, A.J. Controlling Greenhouse Light to a Consistent Daily Integral. *Trans. ASAE* **2000**, *43*, 421. [CrossRef]
40. Pinho, P.; Hytönen, T.; Rantanen, M.; Elomaa, P.; Halonen, L. Dynamic Control of Supplemental Lighting Intensity in a Greenhouse Environment. *Light Res. Technol.* **2013**, *45*, 295–304. [CrossRef]
41. Clausen, A.; Maersk-Moeller, H.M.; Soerensen, J.C.; Joergensen, B.N.; Kjaer, K.H.; Ottosen, C.O. *Integrating Commercial Greenhouses in the Smart Grid with Demand Response Based Control of Supplemental Lighting*; Atlantis Press: Dordrecht, The Netherlands, 2015; pp. 199–213.
42. van Iersel, M.W.; Gianino, D. An Adaptive Control Approach for Light-Emitting Diode Lights Can Reduce the Energy Costs of Supplemental Lighting in Greenhouses. *HortScience* **2017**, *52*, 72–77. [CrossRef]
43. Ji, Y.; Davis, R.; O'Rourke, C.; Chui, E.W.M. Compatibility Testing of Fluorescent Lamp and Ballast Systems. *IEEE Trans. Ind. Appl.* **1999**, *35*, 1271–1276. [CrossRef]
44. Moe, R.; Grimstad, S.O.; Gislerod, H.R. The Use of Artificial Light in Year Round Production of Greenhouse Crops in Norway. *Acta Hortic.* **2006**, *711*, 35–42. [CrossRef]
45. Panova, G.G.; Chernousov, I.N.; Udalova, O.R.; Alexandrov, A.V.; Karmanov, I.V.; Anikina, L.M.; Sudakov, V.L.; Yakushev, V.P. Scientific Basis for Large Year-Round Yields of High-Quality Crop Products under Artificial Lighting. *Russ. Agric. Sci.* **2015**, *41*, 335–339. [CrossRef]
46. Särkkä, L.E.; Jokinen, K.; Ottosen, C.-O.; Kaukoranta, T. Effects of HPS and LED Lighting on Cucumber Leaf Photosynthesis, Light Quality Penetration and Temperature in the Canopy, Plant Morphology and Yield. *Agric. Food Sci.* **2017**, *26*, 102–110. [CrossRef]
47. Van Delm, T.; Stoffels, K.; Melis, P.; Vervoort, M.; Vanderbruggen, R. Overcoming Climatic Limitations: Cultivation Systems and Winter Production under Assimilation Lighting Resulting in Year-Round Strawberries. *Acta Hortic.* **2017**, *1156*, 517–528. [CrossRef]
48. Despommier, D. The Vertical Farm: Controlled Environment Agriculture Carried out in Tall Buildings Would Create Greater Food Safety and Security for Large Urban Populations. *J. Verbraucherschutz. Leb.* **2010**, *6*, 233–236. [CrossRef]
49. Goto, E. Plant Production in a Closed Plant Factory with Artificial Lighting. In Proceedings of the VII International Symposium on Light in Horticultural Systems 956, Wageningen, The Netherlands, 14–18 October 2012.
50. Randall, W.C.; Lopez, R.G. Comparison of Bedding Plant Seedlings Grown Under Sole-Source Light-Emitting Diodes (LEDs) and Greenhouse Supplemental Lighting from LEDs and High-Pressure Sodium Lamps. *HortScience* **2015**, *50*, 705–713. [CrossRef]
51. Pattison, P.M.; Tsao, J.Y.; Brainard, G.C.; Bugbee, B. LEDs for Photons, Physiology and Food. *Nature* **2018**, *563*, 493–500. [CrossRef]
52. Stanghellini, C.; Van't Ooster, B.; Heuvelink, E. Vertical Farms. In *Greenhouse Horticulture*; Wageningen Academic Publishers: Wageningen, The Netherlands, 2018; pp. 279–285. ISBN 978-90-8686-329-7.
53. Kozai, T.; Niu, G. Role of the Plant Factory with Artificial Lighting (PFAL) in Urban Areas. In *Plant Factory*, 2nd ed.; Kozai, T., Niu, G., Takagaki, M., Eds.; Academic Press: Cambridge, MA, USA, 2020; pp. 7–34. ISBN 978-0-12-816691-8.
54. Kusuma, P.; Pattison, P.M.; Bugbee, B. From Physics to Fixtures to Food: Current and Potential LED Efficacy. *Hortic. Res.* **2020**, *7*, 56. [CrossRef]
55. Graamans, L.; Baeza, E.; van den Dobbelsteen, A.; Tsafaras, I.; Stanghellini, C. Plant Factories versus Greenhouses: Comparison of Resource Use Efficiency. *Agric. Syst.* **2018**, *160*, 31–43. [CrossRef]
56. Dinesh, H.; Pearce, J.M. The Potential of Agrivoltaic Systems. *Renew. Sustain. Energy Rev.* **2016**, *54*, 299–308. [CrossRef]
57. Trommsdorff, M.; Kang, J.; Reise, C.; Schindele, S.; Bopp, G.; Ehmann, A.; Weselek, A.; Högy, P.; Obergfell, T. Combining Food and Energy Production: Design of an Agrivoltaic System Applied in Arable and Vegetable Farming in Germany. *Renew. Sustain. Energy Rev.* **2021**, *140*, 110694. [CrossRef]

58. Campana, P.E.; Stridh, B.; Amaducci, S.; Colauzzi, M. Optimization of Vertically Mounted Agrivoltaic Systems. *arXiv* **2021**, arXiv:2104.02124.
59. Toledo, C.; Scognamiglio, A. Agrivoltaic Systems Design and Assessment: A Critical Review, and a Descriptive Model towards a Sustainable Landscape Vision (Three-Dimensional Agrivoltaic Patterns). *Sustainability* **2021**, *13*, 6871. [CrossRef]
60. Yano, A.; Furue, A.; Kadowaki, M.; Tanaka, T.; Hiraki, E.; Miyamoto, M.; Ishizu, F.; Noda, S. Electrical Energy Generated by Photovoltaic Modules Mounted inside the Roof of a North–South Oriented Greenhouse. *Biosyst. Eng.* **2009**, *103*, 228–238. [CrossRef]
61. Sonneveld, P.J.; Swinkels, G.L.A.M.; Campen, J.; van Tuijl, B.A.J.; Janssen, H.J.J.; Bot, G.P.A. Performance Results of a Solar Greenhouse Combining Electrical and Thermal Energy Production. *Biosyst. Eng* **2010**, *106*, 48–57. [CrossRef]
62. Poncet, C.; Muller, M.M.; Brun, R.; Fatnassi, H. Photovoltaic Greenhouses, Non-Sense or a Real Opportunity for the Greenhouse Systems? *Acta Hortic.* **2012**, *927*, 75–79. [CrossRef]
63. Yano, A.; Cossu, M. Energy Sustainable Greenhouse Crop Cultivation Using Photovoltaic Technologies. *Renew. Sustain. Energy Rev.* **2019**, *109*, 116–137. [CrossRef]
64. Fatnassi, H.; Poncet, C.; Bazzano, M.M.; Brun, R.; Bertin, N. A Numerical Simulation of the Photovoltaic Greenhouse Microclimate. *Sol. Energy* **2015**, *120*, 575–584. [CrossRef]
65. Cossu, M.; Cossu, A.; Deligios, P.A.; Ledda, L.; Li, Z.; Fatnassi, H.; Poncet, C.; Yano, A. Assessment and Comparison of the Solar Radiation Distribution inside the Main Commercial Photovoltaic Greenhouse Types in Europe. *Renew. Sustain. Energy Rev.* **2018**, *94*, 822–834. [CrossRef]
66. López-Díaz, G.; Carreño-Ortega, A.; Fatnassi, H.; Poncet, C.; Díaz-Pérez, M. The Effect of Different Levels of Shading in a Photovoltaic Greenhouse with a North–South Orientation. *Appl. Sci.* **2020**, *10*, 882. [CrossRef]
67. Barbera, E.; Sforza, E.; Vecchiato, L.; Bertucco, A. Energy and Economic Analysis of Microalgae Cultivation in a Photovoltaic-Assisted Greenhouse: Scenedesmus Obliquus as a Case Study. *Energy* **2017**, *140*, 116–124. [CrossRef]
68. Bambara, J.; Athienitis, A.K. Energy and Economic Analysis for the Design of Greenhouses with Semi-Transparent Photovoltaic Cladding. *Renew. Energy* **2019**, *131*, 1274–1287. [CrossRef]
69. Ureña-Sánchez, R.; Callejón-Ferre, Á.J.; Pérez-Alonso, J.; Carreño-Ortega, Á. Greenhouse Tomato Production with Electricity Generation by Roof-Mounted Flexible Solar Panels. *Sci. Agric.* **2012**, *69*, 233–239. [CrossRef]
70. Cossu, M.; Yano, A.; Solinas, S.; Deligios, P.A.; Tiloca, M.T.; Cossu, A.; Ledda, L. Agricultural Sustainability Estimation of the European Photovoltaic Greenhouses. *Eur. J. Agron.* **2020**, *118*, 126074. [CrossRef]
71. Huld, T.; Müller, R.; Gambardella, A. A New Solar Radiation Database for Estimating PV Performance in Europe and Africa. *Sol. Energy* **2012**, *86*, 1803–1815. [CrossRef]
72. PVGIS© European Communities JRC Photovoltaic Geographical Information System (PVGIS 5.1). Available online: https://ec.europa.eu/jrc/en/pvgis (accessed on 10 September 2021).
73. Hersbach, H.; Bell, B.; Berrisford, P.; Biavati, G.; Horányi, A.; Muñoz Sabater, J.; Nicolas, J.; Peubey, C.; Radu, R.; Rozum, I. ERA5 Hourly Data on Single Levels from 1979 to Present. In *Copernicus Climate Change Service (C3S) Climate Data Store (CDS)*; European Union: Luxembourg, 2018; Volume 10. [CrossRef]
74. Hersbach, H.; Bell, B.; Berrisford, P.; Hirahara, S.; Horányi, A.; Muñoz-Sabater, J.; Nicolas, J.; Peubey, C.; Radu, R.; Schepers, D.; et al. The ERA5 Global Reanalysis. *Q. J. R. Meteorol. Soc.* **2020**, *146*, 1999–2049. [CrossRef]
75. Suri, M.; Huld, T.; Dunlop, E.; Albuisson, M.; Wald, L. Online Data and Tools for Estimation of Solar Electricity in Africa: The PVGIS Approach. In Proceedings of the International Conference; WIP-Renewable Energies, Dresden, Germany, 4 October 2006; pp. 2623–2626.
76. Baskerville, G.L.; Emin, P. Rapid Estimation of Heat Accumulation from Maximum and Minimum Temperatures. *Ecology* **1969**, *50*, 514–517. [CrossRef]
77. McMaster, G.S.; Wilhelm, W.W. Growing Degree-Days: One Equation, Two Interpretations. *Agric. For. Meteorol.* **1997**, *87*, 291–300. [CrossRef]
78. Miller, P.; Lanier, W.; Brandt, S. Using Growing Degree Days to Predict Plant Stages. *AgExtension Commun. Coord. Commun. Serv. Mont. State Univ.-Bozeman Bozeman MO* **2001**, *59717*, 994–2721.
79. Bigras, F.J.; Ryyppö, A.; Lindström, A.; Stattin, E. Cold Acclimation and Deacclimation of Shoots and Roots of Conifer Seedlings. In *Conifer Cold Hardiness*; Bigras, F.J., Colombo, S.J., Eds.; Tree Physiology; Springer: Dordrecht, The Netherlands, 2001; pp. 57–88. ISBN 978-94-015-9650-3.
80. Xu, Y. Seven Dimensions of Light in Regulating Plant Growth. In Proceedings of the VIII International Symposium on Light in Horticulture 1134, East Lansing, MI, USA, 22–26 May 2016; pp. 445–452.
81. Serrano-Bueno, G.; Romero-Campero, F.J.; Lucas-Reina, E.; Romero, J.M.; Valverde, F. Evolution of Photoperiod Sensing in Plants and Algae. *Curr. Opin. Plant Biol.* **2017**, *37*, 10–17. [CrossRef]
82. Clapham, D.H.; Dormling, I.; Ekberg, L.; Eriksson, G.; Qamaruddin, M.; Vince-Prue, D. Latitudinal Cline of Requirement for Far-Red Light for the Photoperiodic Control of Budset and Extension Growth in *Picea abies* (Norway Spruce). *Physiol. Plant.* **1998**, *102*, 71–78. [CrossRef]
83. Chiang, C.; Aas, O.T.; Jetmundsen, M.R.; Lee, Y.; Torre, S.; Fløistad, I.S.; Olsen, J.E. Day Extension with Far-Red Light Enhances Growth of Subalpine Fir (*Abies lasiocarpa* (Hooker) Nuttall) Seedlings. *Forests* **2018**, *9*, 175. [CrossRef]

84. Riikonen, J.; Lappi, J. Responses of Norway Spruce Seedlings to Different Night Interruption Treatments in Autumn. *Can. J. Res.* **2016**, *46*, 478–484. [CrossRef]
85. Riikonen, J. Efficiency of Night Interruption Treatments with Red and Far-Red Light-Emitting Diodes (LEDs) in Preventing Bud Set in Norway Spruce Seedlings. *Can. J. For. Res.* **2018**, *48*, 1001–1006. [CrossRef]
86. Mattsson, A. Regeneration Practices in Scandinavia: State-of-the-Art and New Trends. *Thin Green Line* **2005**, 25–30.
87. Fløistad, I.S.; Granhus, A. Timing and Duration of Short-Day Treatment Influence Morphology and Second Bud Flush in *Picea Abies* Seedlings. *Silva Fenn.* **2013**, *47*, 1–10. [CrossRef]
88. Mattsson, A. Reforestation Challenges in Scandinavia. *Reforesta* **2016**, *1*, 67–85. [CrossRef]
89. Wallin, E.; Gräns, D.; Jacobs, D.F.; Lindström, A.; Verhoef, N. Short-Day Photoperiods Affect Expression of Genes Related to Dormancy and Freezing Tolerance in Norway Spruce Seedlings. *Ann. For. Sci.* **2017**, *74*, 59. [CrossRef]
90. Riikonen, J.; Luoranen, J. Use of Short-Day Treatment in the Production of Norway Spruce Mini-Plug Seedlings under Plant Factory Conditions. *Scand. J. For. Res.* **2018**, *33*, 625–632. [CrossRef]
91. Riikonen, J.; Luoranen, J. An Assessment of Storability of Norway Spruce Container Seedlings in Freezer Storage as Affected by Short-Day Treatment. *Forests* **2020**, *11*, 692. [CrossRef]
92. Bennie, J.; Davies, T.W.; Cruse, D.; Gaston, K.J. Ecological Effects of Artificial Light at Night on Wild Plants. *J. Ecol.* **2016**, *104*, 611–620. [CrossRef]
93. Bowditch, N. *2019 American Practical Navigator Bowditch Vol 1 & 2 Combined Edition*; Paradise Cay Publications: Blue Lake, CA, USA, 2020; ISBN 978-1-951116-05-7.
94. Duffie, J.; Beckman, W. *Solar Engineering of Thermal Processes*, 3nd ed.; John Wiley & Sons: Hoboken, NJ, USA, 2006; ISBN 978-0-471-69867-8.
95. Herbert Glarner Length of Day and Twilight. Available online: http://www.gandraxa.com/length_of_day.xml (accessed on 31 August 2021).
96. Zhen, S.; Bugbee, B. Substituting Far-Red for Traditionally Defined Photosynthetic Photons Results in Equal Canopy Quantum Yield for CO2 Fixation and Increased Photon Capture During Long-Term Studies: Implications for Re-Defining PAR. *Front. Plant Sci.* **2020**, *11*, 1433. [CrossRef] [PubMed]
97. Zhen, S.; van Iersel, M.; Bugbee, B. Why Far-Red Photons Should Be Included in the Definition of Photosynthetic Photons and the Measurement of Horticultural Fixture Efficacy. *Front. Plant Sci.* **2021**, *12*, 1158. [CrossRef]
98. Britton, C.M.; Dodd, J.D. Relationships of Photosynthetically Active Radiation and Shortwave Irradiance. *Agric. Meteorol.* **1976**, *17*, 1–7. [CrossRef]
99. Thimijan, R.W.; Heins, R.D. Photometric, Radiometric, and Quantum Light Units of Measure: A Review of Procedures for Interconversion. *HortScience* **1983**, *18*, 818–822.
100. Rodskjer, N. Spectral Daily Insolation at Uppsala, Sweden. *Arch. Meteorol. Geophys. Bioclimatol. Ser. B* **1983**, *33*, 89–98. [CrossRef]
101. Howell, T.A.; Meek, D.W.; Hatfield, J.L. Relationship of Photosynthetically Active Radiation to Shortwave Radiation in the San Joaquin Valley. *Agric. Meteorol.* **1983**, *28*, 157–175. [CrossRef]
102. Papaioannou, G.; Papanikolaou, N.; Retalis, D. Relationships of Photosynthetically Active Radiation and Shortwave Irradiance. *Theor. Appl. Climatol.* **1993**, *48*, 23–27. [CrossRef]
103. Alados, I.; Foyo-Moreno, I.; Alados-Arboledas, L. Photosynthetically Active Radiation: Measurements and Modelling. *Agric. For. Meteorol.* **1996**, *78*, 121–131. [CrossRef]
104. Jacovides, C.P.; Timbios, F.; Asimakopoulos, D.N.; Steven, M.D. Urban Aerosol and Clear Skies Spectra for Global and Diffuse Photosynthetically Active Radiation. *Agric. For. Meteorol.* **1997**, *87*, 91–104. [CrossRef]
105. Hassika, P.; Berbigier, P. Annual Cycle of Photosynthetically Active Radiation in Maritime Pine Forest. *Agric. For. Meteorol.* **1998**, *90*, 157–171. [CrossRef]
106. Akitsu, T.; Kume, A.; Hirose, Y.; Ijima, O.; Nasahara, K.N. On the Stability of Radiometric Ratios of Photosynthetically Active Radiation to Global Solar Radiation in Tsukuba, Japan. *Agric. For. Meteorol.* **2015**, *209–210*, 59–68. [CrossRef]
107. Blonquist, J.M.; Bugbee, B. Solar, Net, and Photosynthetic Radiation. In *Agroclimatology*; John Wiley & Sons: Hoboken, NJ, USA, 2018; pp. 1–49. ISBN 978-0-89118-358-7.
108. Jacovides, C.P.; Steven, M.D.; Asimakopoulos, D.N. Spectral Solar Irradiance and Some Optical Properties For Various Polluted Atmospheres. *Sol. Energy* **2000**, *69*, 215–227. [CrossRef]
109. Escobedo, J.F.; Gomes, E.N.; Oliveira, A.P.; Soares, J. Modeling Hourly and Daily Fractions of UV, PAR and NIR to Global Solar Radiation under Various Sky Conditions at Botucatu, Brazil. *Appl. Energy* **2009**, *86*, 299–309. [CrossRef]
110. Von Zabeltitz, C. Light Transmittance of Greenhouses. In *Integrated Greenhouse Systems for Mild Climates: Climate Conditions, Design, Construction, Maintenance, Climate Control*; Von Zabeltitz, C., Ed.; Springer: Berlin, Heidelberg, 2011; pp. 137–143. ISBN 978-3-642-14582-7.
111. Both, A.J.; Faust, J.E. Light Transmission: The Impact of Glazing Material and Greenhouse Design. In *Light Management in Controlled Environments*; Lopez, R., Runkle, E.S., Eds.; CreateSpace Independent Publishing Platform: Willoughby, OH, USA, 2017; pp. 74–81. ISBN 978-1-5442-5449-4.
112. Stanghellini, C.; Van't Ooster, B.; Heuvelink, E. Radiation, Greenhouse Cover and Temperature. In *Greenhouse Horticulture*; Wageningen Academic Publishers: Wageningen, The Netherlands, 2018; pp. 15–25. ISBN 978-90-8686-329-7.

113. Kozai, T.; Goudriaan, J.; Kimura, T. *Light Transmission and Photosynthesis in Greenhouses*; Pudoc: Wageningen, The Netherlands, 1978; ISBN 90-220-0646-8.
114. Roberts, W.J. Glazing Materials, Structural Design, and Other Factors Affecting Light Transmission in Greenhouses. In *GreenHouse Glazing and Solar Radiation Transmission Workshop*; Rutgers University: New Brunswick, NJ, USA, 1998; Volume 8.
115. Von Elsner, B.; Briassoulis, D.; Waaijenberg, D.; Mistriotis, A.; von Zabeltitz, C.; Gratraud, J.; Russo, G.; Suay-Cortes, R. Review of Structural and Functional Characteristics of Greenhouses in European Union Countries: Part I, Design Requirements. *J. Agric. Eng. Res.* **2000**, *75*, 1–16. [CrossRef]
116. Mattsson, A. *Odlingsmiljö: En analys av 80-talets Växthus för Produktion av Skogsplantor*; Institutionen För Skogsproduktion; Sveriges Lantbruksuniversitet: Garpenberg, Sweden, 1982; Volume 7, ISBN 91-576-1214-5.
117. Tantau, H.-J.; Hinken, J.; von Elsner, B.; Max, J.; Ulbrich, A.; Schurr, U.; Hofmann, T.; Reisinger, G. Solar Transmittance of Greenhouse Covering Materials. *Acta Hortic.* **2012**, *956*, 441–448. [CrossRef]
118. Hemming, S.; Mohammadkhani, V.; Dueck, T. Diffuse Greenhouse Covering Materials-Material Technology, Measurements and Evaluation of Optical Properties. In Proceedings of the International Workshop on Greenhouse Environmental Control and Crop Production in Semi-Arid Regions 797, Tucson, AZ, USA, 20–24 October 2008; pp. 469–475.
119. Bergstrand, K.-J.; Mortensen, L.M.; Suthaparan, A.; Gislerød, H.R. Acclimatisation of Greenhouse Crops to Differing Light Quality. *Sci. Hortic.* **2016**, *204*, 1–7. [CrossRef]
120. Navidad, H.; Fløistad, I.S.; Olsen, J.E.; Torre, S. Subalpine Fir (*Abies Laciocarpa*) and Norway Spruce (*Picea Abies*) Seedlings Show Different Growth Responses to Blue Light. *Agronomy* **2020**, *10*, 712. [CrossRef]
121. IEC 60904-1:2020. *Photovoltaic Devices—Part 1: Measurement of Photovoltaic Current-Voltage Characteristics*; International Electrotechnical Commission: Geneva, Switzerland, 2020; Volume 67.
122. Kurtz, S.; Repins, I.; Metzger, W.K.; Verlinden, P.J.; Huang, S.; Bowden, S.; Tappan, I.; Emery, K.; Kazmerski, L.L.; Levi, D. Historical Analysis of Champion Photovoltaic Module Efficiencies. *IEEE J. Photovolt.* **2018**, *8*, 363–372. [CrossRef]
123. Deutsche Gesellschaft für Sonnenenergie (Ed.) *Planning and Installing Photovoltaic Systems: A Guide for Installers, Architects and Engineers*; Routledge: Oxfordshire, UK, 2013.
124. Von Elsner, B.; Briassoulis, D.; Waaijenberg, D.; Mistriotis, A.; von Zabeltitz, C.; Gratraud, J.; Russo, G.; Suay-Cortes, R. Review of Structural and Functional Characteristics of Greenhouses in European Union Countries, Part II: Typical Designs. *J. Agric. Eng. Res.* **2000**, *75*, 111–126. [CrossRef]
125. R Core Team. *R: A Language and Environment for Statistical Computing*; R Foundation for Statistical Computing: Vienna, Austria, 2021.
126. Wickham, H. Tidy Data. *J. Stat. Softw.* **2014**, *59*, 1–23. [CrossRef]
127. Wickham, H.; Averick, M.; Bryan, J.; Chang, W.; McGowan, L.D.; François, R.; Grolemund, G.; Hayes, A.; Henry, L.; Hester, J.; et al. Welcome to the Tidyverse. *J. Open Source Softw.* **2019**, *4*, 1686. [CrossRef]
128. Wickham, H. *Ggplot2: Elegant Graphics for Data Analysis*; Springer: New York, NY, USA, 2009; ISBN 978-0-387-98140-6.
129. Wickham, H.; Chang, W.; Henry, L.; Pedersen, T.L.; Takahashi, K.; Wilke, C.; Woo, K.; Yutani, H.; Dunnington, D. RStudio Ggplot2: Create Elegant Data Visualisations Using the Grammar of Graphics; CRAN.R Project 2021. Available online: https://ggplot2.tidyverse.org (accessed on 29 August 2021).
130. Pebesma, E. Simple Features for R: Standardized Support for Spatial Vector Data. *R J.* **2018**, *10*, 439–446. [CrossRef]
131. Pebesma, E.; Bivand, R.; Racine, E.; Sumner, M.; Cook, I.; Keitt, T.; Lovelace, R.; Wickham, H.; Ooms, J.; Müller, K.; et al. Sf: Simple Features for R; CRAN.R Project, 2021. Available online: https://CRAN.R-project.org/package=sf (accessed on 29 August 2021).
132. Natural Earth—Free Vector and Raster Map Data. Available online: https://www.naturalearthdata.com/ (accessed on 29 August 2021).
133. South, A. Rnaturalearth: World Map Data from Natural Earth; CRAN.R Project 2017. Available online: https://CRAN.R-project.org/package=rnaturalearth (accessed on 29 August 2021).
134. Smirnakou, S.; Ouzounis, T.; Radoglou, K. Effects of Continuous Spectrum LEDs Used in Indoor Cultivation of Two Coniferous Species *Pinus sylvestris* L. and *Abies Borisii-Regis* Mattf. *Scand. J. Res.* **2017**, *32*, 115–122. [CrossRef]
135. Soriano, T.; Suárez-Rey, E.M.; Morales, M.I.; Romero, M.; Hernández, J.; Montero, J.I.; Antón, A.; Castilla, N. Solar Radiation Transmission in Mediterranean Plastic Greenhouses. *Acta Hortic.* **2009**, *807*, 73–78. [CrossRef]
136. Critten, D.L. Greenhouse Light Transmission: The Shading Effect of Infinite Parallel Cylinders under Diffuse Light. *J. Agric. Eng. Res.* **1983**, *28*, 569–572. [CrossRef]
137. Kittas, C.; Baille, A.; Giaglaras, P. Influence of Covering Material and Shading on the Spectral Distribution of Light in Greenhouses. *J. Agric. Eng. Res.* **1999**, *73*, 341–351. [CrossRef]
138. Hemming, S.; Dueck, T.A.; Marissen, N.; Jongschaap, R.E.E.; Kempkes, F.L.K.; van de Braak, N.J. *Diffuus Licht: Het Effect van Lichtverstrooiende Kasdekmaterialen op Kasklimaat, Lichtdoordringing en Gewasgroei*; Rapport/Wageningen UR, Agrotechnology & Food Innovations; Agrotechnology & Food Innovations: Wageningen, The Netherlands, 2005; ISBN 978-90-6754-978-3.
139. Kotilainen, T.; Robson, T.M.; Hernández, R. Light Quality Characterization under Climate Screens and Shade Nets for Controlled-Environment Agriculture. *PLoS ONE* **2018**, *13*, e0199628. [CrossRef] [PubMed]
140. Cossu, M.; Ledda, L.; Urracci, G.; Sirigu, A.; Cossu, A.; Murgia, L.; Pazzona, A.; Yano, A. An Algorithm for the Calculation of the Light Distribution in Photovoltaic Greenhouses. *Sol. Energy* **2017**, *141*, 38–48. [CrossRef]
141. Bakker, J.C.; Bot, G.P.A.; Challa, H.; van de Braak, N.J. *Greenhouse Climate Control: An Integrated Approach*; Wageningen Academic Publishers: Wageningen, The Netherlands, 1995; ISBN 978-90-74134-17-0.

142. Graamans, L.; van den Dobbelsteen, A.; Meinen, E.; Stanghellini, C. Plant Factories; Crop Transpiration and Energy Balance. *Agric. Syst.* **2017**, *153*, 138–147. [CrossRef]
143. Wallace, C.; Both, A.J. Evaluating Operating Characteristics of Light Sources for Horticultural Applications. *Acta Hortic.* **2016**, 435–444. [CrossRef]
144. Radetsky, L.C. *LED and HID Horticultural Luminaire Testing Report*; Lighting Research Center, Rensselaer Polytechnic Institute: Troy, NY, USA, 2018.
145. Stober, K.; Lee, K.; Yamada, M.; Pattison, M. *Energy Savings Potential of SSL in Horticultural Applications*; EERE Publication and Product Library: Washington, DC, USA, 2017.
146. Harbick, K.; Albright, L.D. Comparison of Energy Consumption: Greenhouses and Plant Factories. *Acta Hortic.* **2016**, *1134*, 285–292. [CrossRef]
147. Molin, E.; Martin, M. *Assessing the Energy and Environmental Performance of Vertical Hydroponic Farming*; IVL Swedish Environmental Research Institute: Stockholm, Sweden, 2018; p. 36.
148. Liaros, S.; Botsis, K.; Xydis, G. Technoeconomic Evaluation of Urban Plant Factories: The Case of Basil (*Ocimum basilicum*). *Sci. Total Environ.* **2016**, *554–555*, 218–227. [CrossRef] [PubMed]

*agriculture*

MDPI

*Article*

# The Impact of LED Light Spectrum on the Growth, Morphological Traits, and Nutritional Status of 'Elizium' Romaine Lettuce Grown in an Indoor Controlled Environment

**Bożena Matysiak [1,*], Stanisław Kaniszewski [2], Jacek Dyśko [2], Waldemar Kowalczyk [3], Artur Kowalski [2] and Maria Grzegorzewska [4]**

1 Department of Applied Biology, The National Institute of Horticultural Research, Konstytucji 3 Maja 1/3, 96-100 Skierniewice, Poland
2 Laboratory of Vegetable Crops and Edible Mushroom, The National Institute of Horticultural Research, Konstytucji 3 Maja 1/3, 96-100 Skierniewice, Poland; stanislaw.kaniszewski@inhort.pl (S.K.); jacek.dysko@inhort.pl (J.D.); artur.kowalski@inhort.pl (A.K.)
3 Laboratory of Chemical Analysis, The National Institute of Horticultural Research, Konstytucji 3 Maja 1/3, 96-100 Skierniewice, Poland; waldemar.kowalczyk@inhort.pl
4 Fruit and Vegetables Storage and Processing Department, The National Institute of Horticultural Research, Konstytucji 3 Maja 1/3, 96-100 Skierniewice, Poland; maria.grzegorzewska@inhort.pl
* Correspondence: bozena.matysiak@inhort.pl; Tel.: +48-46-834-53-83

**Abstract:** The study examined the influence of light quality on the growth and nutritional status of romaine lettuce grown in deep water culture with a floating raft system using two different nutrient solutions. Four spectra of LED light were used with different ratios of R, G, and B lights (80:10:10, 70:10:20, 60:10:30, and 70:18:12). Two nutrient solutions with a low (A) and moderately high (B) nutrient content were used. Regardless of the nutrient solution, the RGB 70:18:12 light promoted the production of leaf biomass as well as inhibited the accumulation of K and Mg in the leaves. Moreover, those plants were characterized by a low Nitrogen Balance Index (NBI) and a high flavonol index. In the last week of cultivation, there was a strong decrease in K, P, and nitrates in the nutrient solution, and an increase in Ca. In the final stage of growth, symptoms of withering of the tips of young leaves (tipburn) were observed on the plants. The most damage was observed on the plants growing under 70:10:20, 70:18:12, and with the higher concentration of minerals in the solution (B).

**Keywords:** hydroponics; *Lactuca sativa* var. *longifolia*; nitrogen balance index; tipburn; flavonol index

**Citation:** Matysiak, B.; Kaniszewski, S.; Dyśko, J.; Kowalczyk, W.; Kowalski, A.; Grzegorzewska, M. The Impact of LED Light Spectrum on the Growth, Morphological Traits, and Nutritional Status of 'Elizium' Romaine Lettuce Grown in an Indoor Controlled Environment. *Agriculture* **2021**, *11*, 1133. https://doi.org/10.3390/agriculture11111133

Academic Editors: Athanasios Koukounaras and Filippos Bantis

Received: 27 September 2021
Accepted: 10 November 2021
Published: 12 November 2021



## 1. Introduction

The multi-level production of plants in plant factories is one of the strategies for adapting agriculture to the advancing climate change. The decreasing production potential of soils, and the shortage of raw materials for plant production as well as the growing demand for food in cities are caused by intensifying urbanization processes [1,2]. This technology is particularly suitable for the cultivation of small-sized plants with a short production cycle, such as leafy vegetables and herbs, and valuable medicinal plants [1,3,4]. The consumption of energy, water, carbon dioxide, and land area for producing a unit mass of lettuce in a plant factory is much lower than in greenhouse cultivation, because of the possibility of cultivating plants on many levels, using closed fertigation systems, and recovery of water lost due to transpiration [5]. The most important factors limiting the development of technologies of plant production in plant factories are the high costs of investment and energy for artificial lighting. The multi-level production of plants requires the use of light sources with high energy conversion to the light used by plants in the photosynthesis process and generating little heat [6]. A major advantage of LEDs lamps is their electrical efficiency and photosynthetic efficacy. The small size and low heat energy emission mean that LED lighting can be installed near plants. The spectrum of LEDs can be adjusted based on plant growth requirements.

Studies conducted on lettuce (*Lactuca sativa* L.) as a model plant in facilities without sunlight have shown that changes in the light spectrum significantly affect the growth and development of the plants, as both the morphology and physiological processes depend on the quality of light [7,8]. The greatest influence is exerted by red light (600–700 nm) in combination with blue light (400–500 nm) because it affects the process of photosynthesis [8–14] and the morphological features of leaves that facilitate the absorption of light quanta by plants [15]. A high proportion of red light stimulates the production of biomass of green-leaf lettuce, with its optimal percentage in the total spectrum being in a fairly wide range from 50% to 80% [8,13,16–19]. With the increase in the proportion of blue light, lettuce plants grow more slowly [8], but the amounts of bioactive compounds in them, e.g., flavonoids, increase [10–12,19,20]. In the case of red-leaf lettuce, blue light is more effective in stimulating growth than red light [21,22]. Positive effects on the production of lettuce leaf biomass have also been obtained with green light (500–600 nm) [23] The addition of green light has been shown to efficiently drive photosynthesis, however, this effect depends on the light intensity [24]. At low PPFD, green light compared to red and blue, has the lowest photosynthetic efficiency, because of its low absorptance; on the other hand, at high PPFD quantum yield of $CO_2$ assimilation under green light is the highest [25]. Supplementation of the spectrum with green light at moderate PPFD decreased the intensity of photosynthesis but did not limit the growth of lettuce [8]. Butterhead lettuce (*Lactuca sativa* var. *capitata*) is the basic leafy vegetable produced in plant factories. The few studies relating to the production of plants in plant factories have concerned the romaine lettuce *Lactuca sativa* var. *longifolia* [26,27]. Despite its high taste quality and nutritional value [28], this variety, especially the 'mini' type is cultivated on a small scale in controlled atmosphere environments due to problems with obtaining high-quality plants and the lack of information on the requirements in relation to environmental conditions.

It has been demonstrated that with an increase in the photosynthetic photon flux density (PPFD) in the range from 150 to 300 μmol $m^{-2}$ $s^{-1}$, the growth rate, fresh and dry leaf weight, and the number of leaves of butterhead lettuce increased, but the negative effect was a greater number of leaves with tipburn symptoms, which was associated with a reduced calcium content [29]. The tipburn problem affects mainly head-forming lettuces, such as romaine lettuce and crisphead lettuce [30]. Increased light intensity (PPFD from 100 to 400 mol $\mu m^{-2}$ $s^{-1}$) promoted the production of biomass while reducing the nitrate content in lettuce leaves [31], even when high levels of PPFD were used only at the end of the production period [4]. The few studies concerned with the production of plants in plant factories have shown a significant correlation between the light spectrum and the nutritional status of lettuce plants with respect to macro- and micronutrients, including nitrates [12]. For microgreens, an increasing percentage of blue light in the LED illumination spectrum had a positive effect on the accumulation of mostly macro- and micronutrients [32]. In turn, Kyriacou et al. [33] showed that nitrate accumulation in microgreens was higher under monochromatic red and blue compared to red-blue lights, moreover monochromatic lights tended to increase K and Na and decrease Ca and Mg concentrations.

Cultivation in plant factories is aimed at maximizing the efficiency of the production process, and in the case of leafy vegetables, at achieving rapid weight gain of the above-ground part. In the hydroponic cultivation of lettuce, the composition and concentration of the nutrient solution supplied to the plants play a very important role. In order to obtain good quality and high yield lettuce, the appropriate composition and concentration of the fertigation medium are required. In greenhouse studies with a hydroponic flooding system, the optimal EC of the nutrient solution in the cultivation of butterhead and loose-leaf lettuce has been found to be 2 mS $cm^{-1}$. Increasing the nutrient concentration to an EC of 3 mS $cm^{-1}$ did not increase the yield of lettuce, while a further increase to an EC of 4 mS $cm^{-1}$ resulted in a significant reduction in the yield [34,35]. Moreover, it was found that the EC of 4 mS $cm^{-1}$ increased the nitrate content above the permissible limit set by the European Commission [36]. Too rapid lettuce growth can lead to an unbalanced uptake of minerals from the nutrient solution and to the occurrence of deficiencies (e.g., calcium) or an excess of minerals (nitrates, potassium), which can lead to a reduction in plant quality and significant economic losses.

The aim of the study was to assess the influence of the LED light spectrum and the composition of the mineral nutrient solution on the production of biomass, morphological features, and nutritional status of 'Elizium' romaine lettuce in an indoor controlled environment. Changes in the content of basic nutrients in the hydroponic medium during plant growth were analyzed.

## 2. Materials and Methods

### 2.1. Plant Material and Growth Conditions

The study was conducted on romaine lettuce *Lactuca sativa* var. *longifolium* 'Elizium' type 'mini'. Seedlings of the lettuce were produced on trays, in cubes of mineral wool (0.02 × 0.02 m) in a phytotron (model FD 730 DD INOX, BIOSELL, Warsaw, Poland) in a laboratory building where constant temperature (22 °C) and humidity (65%) were maintained throughout the day, with PPFD at the plant level of 75 μmol $m^{-2}$ $s^{-1}$ and a 16 h photoperiod. Twenty-one-day-old seedlings were used in the study. The experiment was performed in an outdoor free-standing container (Weldon, Brzezówka, Poland) with dimensions 6.0 × 2.6 × 3.2 m adapted to a phytotron by BIOSELL (Warsaw, Poland) fitted with two two-shelf racks. On each of the four shelves (3.3 × 0.6 m), there were placed 6 styrofoam containers (0.4 × 0.6 × 0.2 m) for growing lettuce in a hydroponic system. The lettuce seedlings were mounted on floating polystyrene rafts with openings for the plants and placed on the mineral nutrient solution contained in the containers (5 plants per container, 20 plants per $m^2$). Each container contained 20 L of the solution; the solution was constantly aerated (2.3 L $min^{-1}$). The temperature in the phytotron was set at 20/18 °C day/night, and the relative air humidity at 65%. The experiment was set up on 25 January 2021 and lasted 30 days.

### 2.2. Experimental Combinations—Light

Each shelf was fitted with panels with LEDs emitting different lights: red (R)—Hyper Red 660 nm (Osram Osconique P 30–30), blue (B)—Deep Blue 440 nm (Osram Osconique P 30–30), and white (W)—6500 K (Samsung CRI 80). The study used 4 spectra of LED light with the following spectral composition: RGB 80:10:10, RGB 70:10:20, RGB 60:10:30, and RGB 70:18:12. The light spectra used in the study are shown in Table 1 and Figure 1. Photometric measurements were made with a GL Spectrolux VIS spectrometer (GL Optic, Puszczykowo, Poland, https://gloptic.com accessed on 23 January 2021). The intensity of photosynthetically active light (PPFD) at the plant height was 160 μmol $m^{-2}$ $s^{-1}$ for a 16-h photoperiod. The total daily amount of light (DLI) was 9.2 mol $m^{-2}$.

**Table 1.** Spectral photon flux PPF (μmol $s^{-1}$) for the 4 LED light spectra in vertical cultivation of romaine lettuce (fraction of integral photon flux ranging from 340 to 780 nm in ultraviolet, blue, green, red, and far-red).

| Light Spectrum R:G:B (Red:Green:Blue) | UV-A 340–399 nm | Blue 400–499 nm | Green 500–599 nm | Red 600–699 nm | Far-Red 700–780 nm | R:B Ratio | Total nm |
|---|---|---|---|---|---|---|---|
| 80:10:10 | 0 | 45.2 | 45.0 | 349.5 | 1.3 | 7.7 | 441.0 |
| 70:10:20 | 0 | 89.6 | 42.0 | 306.6 | 1.0 | 3.4 | 439.2 |
| 60:10:30 | 0 | 127.4 | 46.5 | 267.2 | 1.1 | 2.1 | 442.2 |
| 70:18:12 | 0 | 52.0 | 80.3 | 304.1 | 1.8 | 5.8 | 438.2 |

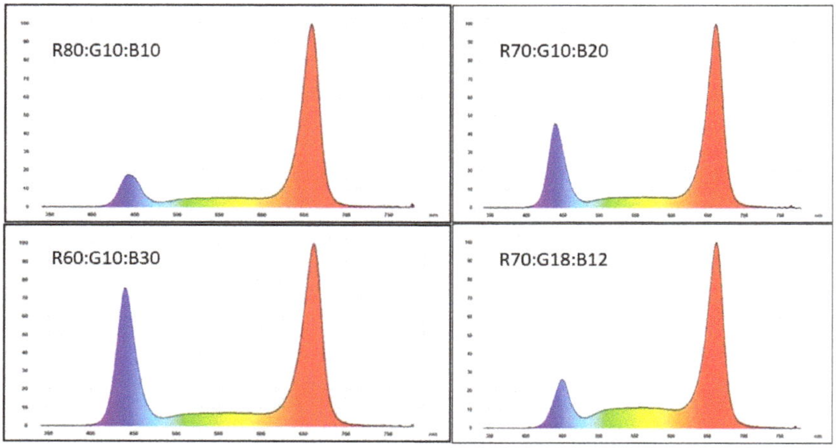


**Figure 1.** Spectral photon flux distribution for the 4 LED light treatments in vertical cultivation of romaine lettuce; fraction of integral photon flux ranging from 340 to 780 nm in red R 600–699 nm, green G 500–599 nm, and blue B 400–499 nm (*Y*-axis—relative values from 0 to 100%).

### 2.3. Experimental Combinations—Composition of Nutrient Solution

Two nutrient solutions with a different nutrient content (A and B) were used in the study. The concentrations (mg dm$^{-3}$) of macroelements in solution A (EC 1.6 mS cm$^{-1}$, pH 6.0) were: N-$NO_3$—130, N-$NH_4$—11, P—40, K—180, Ca—200, Mg—35; in solution B (EC 2.0 mS cm$^{-1}$, pH 6.0): N-$NO_3$—170, N-$NH_4$—10, P—50, K—210, Ca—210, Mg—45. The concentrations (mg dm$^{-3}$) of microelements in solutions A and B were the same: Fe—2.0, Mn—0.76, Zn—0.16, B—0.32, Cu—0.16, Mo—0.04.

During the cultivation of lettuce, the electrical conductivity EC, pH, and nutrient content in the nutrient solution were measured. The pH was determined with the potentiometric method and EC using the conductivity method. Mineral components in the nutrient solution, as N-$NO_3$, were analyzed by the potentiometric method; P, K, Ca, Mg, and $SO_4$ by the spectrophotometric method using a sequential emission spectrometer with inductively coupled plasma (ICP Perkin-Elmer model Optima 2000 DV, Boston, MA, USA). Plant samples (leaves) from each treatment were placed for 48 h in a forced-air dryer at 70 °C. They were analyzed after grinding and wet mineralization in a strong $HNO_3$ and $HClO_4$ acid mixture. The concentrations of macronutrients (P, K, Ca, Mg) and micronutrients (Fe, Mn, Cu, Zn, B) were determined in three replications using an ICP spectrometer. Selected elements were determined at their characteristic wavelengths [37]. The N content in plant samples was analyzed using the Kjeldahl method using Vapodest Kjeldahl apparatus, Gerhardt GmbH & Co., KG, Königswinter, Bonn, Germany [38]. All the nutrients were determined in three replications.

### 2.4. Growth and Morphological Characteristics of Plants

Lettuce plants were assessed after 30 days of cultivation. Leaf fresh weight, plant height and diameter, number of leaves (including leaves with tipburn), and head circumference (curled inner leaves) were determined.

### 2.5. Chlorophyll, Flavonol, and Nitrogen Balance Indices

An optical sensor was used for the assessment of chlorophyll and flavonol compounds, measuring the UV absorbance of the leaf epidermis by the double excitation of chlorophyll fluorescence (Dualex Scientific+ Instrument, Force-A, Orsay, Paris, France, https://www.force-a.com, accessed on 23 January 2021). The nitrogen balance index (NBI) was automatically calculated as a ratio of the chlorophyll index (Chl) to the flavonol index

(FLAV), i.e., NBI = Chl/FLAV. The device used in this study allows for non-destructive measurements of chlorophyll, flavonol content, and nitrogen balance in leaves, which makes it particularly suitable for photophysiological research. For each lighting combination, 30 young, fully expanded leaves were used for the determination of the flavonol and chlorophyll indices.

### *2.6. Experimental Design and Statistical Analysis*

The experiment used a two-factorial design of light spectrum × nutrient solution. Plants were subjected to illumination with four light spectra (R:G:B—80:10:10, 70:10:20, 60:10:30, 70:18:12) and two nutrient solutions (A and B). There were three container replicates for each of the eight treatments and thus 24 containers in total. In the study, eight experimental treatment groups were analyzed, with five samples (plants) in each treatment group. Two-way ANOVAs were used to test the effects of the light spectrum and nutrient solution on the growth traits of romaine lettuce. The treatment means were compared using Tukey's HSD. Statistical analysis was performed using the STATISTICA software, version 13.1 (StatSoft Inc., Tulsa, OK, USA).

## **3. Results**

### *3.1. Growth and Morphological Characteristics of Plants*

The quality of light significantly influenced the growth and development of 'Elizium' romaine lettuce grown in the hydroponic system without sunlight (Figure 2). However, the effects of light quality on the growth and morphological features of lettuce plants were not dependent on the type of nutrient solution used, and the interactions between the main factors tested (light spectrum × solution composition) were not significant (Table 2). A high proportion of red light R (600–699 nm), an increased proportion of green light G (500–599 nm), and a low proportion of blue light B (400–499 nm) in the spectrum of the light emitted by the LEDs (RGB 70:18:12) were favorable to biomass production, whereas the use of light with a reduced proportion of red light and a high proportion of blue light (RGB 60:10:30) was the least favorable for the growth of lettuce leaf biomass. Irrespective of the type of nutrient medium used, the lettuce plants grown under RGB 70:18:12 had the highest fresh weight of leaves (148 g), the highest number of leaves (24.4 cm), and the largest head circumference (29 cm), with these values being respectively 15%, 8%, and 7% higher than under RGB 60:10:30. By comparison, the plants growing under RGB 80:10:10 and RGB 70:10:20 had the largest diameter (20.2 and 19.7 cm, respectively), and those growing under RGB 70:10:20 were the tallest (20.2 cm). The plants growing under RGB 60:10:30 and RGB 70:18:12 had the lowest height and diameter.

The results of our study showed that a high ratio of red to blue light with a fairly high proportion of green light (RGB 70:18:12) stimulated the production of biomass in 'Elizium' romaine lettuce, and at the same time provided the most eye-friendly conditions, which is important when staying in sealed rooms with such lighting for longer periods of time [13]. Similarly, Mickens et al. [27] showed that the light emitted by LEDs with a spectrum similar to natural light, including the red, green, blue, and far-red bands (RGB 60:24:16 + FR), more strongly stimulated biomass production and the diameter of 'Outredgeous' romaine lettuce than the combination of only red light and blue light (RB 60:40). That study also showed that the requirements of lettuce plants at different stages of growth were different. In the initial period, white light combined with green light stimulated biomass production the most, while in the final stage—white light with red light. Monochromatic red light created unfavorable conditions for the growth of *Lactuca sativa* 'Grizzly' [39], and too much blue light resulted in the slower growth of lettuce plants [8,40]. It was been shown that the temporal shift of red light in relation to blue by 4 to 7 h gave a better effect than the simultaneous use of both types of LED light, which may, however, result from an extended photoperiod [26].

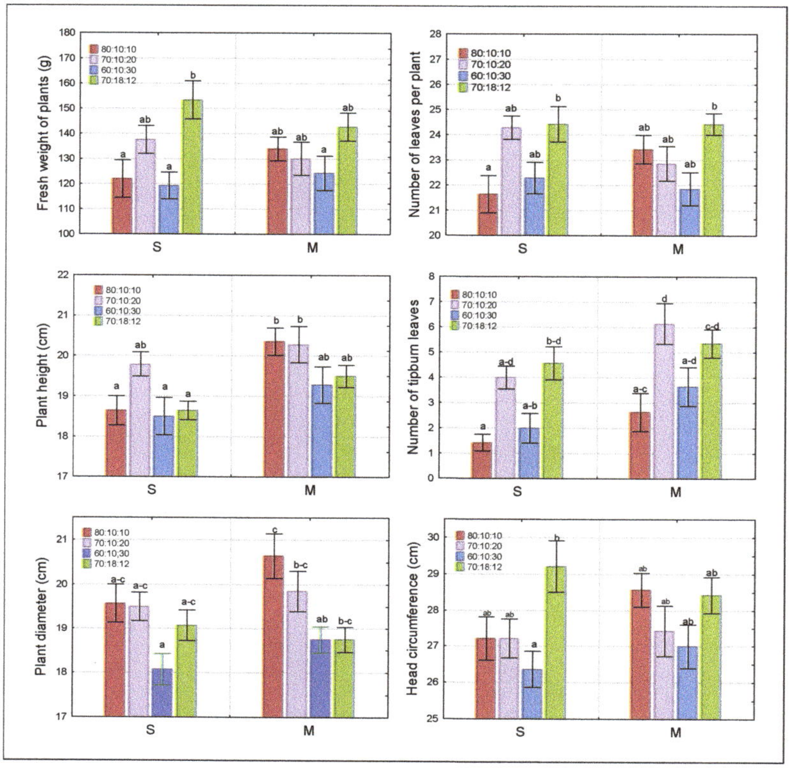

**Figure 2.** Fresh weight of leaves, plant height, plant diameter, number of leaves per plant, number of tipburn leaves per plant, and head circumference of 'Elizium' romaine lettuce grown in a hydroponic system in different nutrient solutions (S and M) under four different light spectra (R:G:B—80:10:10, 70:10:20, 60:10:30, and 70:18:12) in an indoor controlled environment. Bars represent means ± SE. Means followed by the same letter are not significantly different ($p < 0.05$) according to Tukey's HSD test.

**Table 2.** Significance of two-way ANOVA results ($p$-values) for the effects of light spectrum and nutrient solution on the measurements of biomass, morphological traits, chlorophyll, flavonol, and nitrogen balance indices (NBI) of 'Elizium' romaine lettuce.

| Growth and Morphological Trait | Light Spectrum | Nutrient Solution | Light Spectrum × Nutrient Solution |
|---|---|---|---|
| Fresh weight of leaves | **0.0005** | 0.9529 | 0.2375 |
| Plant height | **0.0119** | **0.0001** | 0.3899 |
| Plant diameter | **0.0001** | 0.1018 | 0.3182 |
| Head circumference | **0.0035** | 0.3910 | 0.3258 |
| Number of leaves per plant | **0.0009** | 0.9673 | 0.0734 |
| Percentage tipburn | **0.0001** | **0.0016** | 0.7371 |
| Chlorophyll index | **0.0001** | **0.0019** | 0.0651 |
| Flavonol index | **0.0001** | **0.0001** | 0.0720 |
| NBI | **0.0001** | **0.0001** | 0.5550 |

Significant $p$-values are shown in bold.

Our study showed that the quality of light influenced the development of physiological disorders manifested by the withering of the tips of young leaves (tipburn) of 'Elizium' romaine lettuce. The highest numbers of damaged leaves were observed in the plants growing under RGB 70:10:20 and RGB 70:18:12—respectively 5.3 and 5.0, which constituted 21% and 20% of all the leaves. The lowest numbers of damaged leaves were recorded in the plants growing under RGB 80:10:10 and RGB 60:10:30—respectively 9% and 12%. These results suggest that the faster the biomass production is, the more often tipburn symptoms occur on young romaine lettuce leaves. It is known that climatic factors such as high temperature and high light intensity, leading to the rapid growth of lettuce shoots, are conducive to the occurrence of tipburn [41–43] and that this is a genetically determined trait [30,42,44–46].

The type of nutrient solution influenced to only a small extent the growth and morphological features of 'Elizium' romaine lettuce, although the plants growing in solution B were 5% taller than those growing in solution A (on average for the tested light quality variants). Much stronger was the influence of nutrient solution on the occurrence of damage to the tips of young leaves (tipburn). In the case of plants growing in solution B, the percentage of leaves with tipburn symptoms was as high as 19%, whereas in solution A this percentage was 12%.

### 3.2. *Changes in the Composition of Nutrient Solution*

Weekly analyses of the composition of the hydroponic medium showed significant changes in pH, EC, and the concentrations of macronutrients during the 30-day growth period of 'Elizium' romaine lettuce (Figure 3). During the first 3 weeks of cultivation, the pH of the nutrient solution gradually decreased from 6.6 to 6.0, but in the last week, there was an increase in the pH value to 6.8 (on average for the two solutions). The EC value changed only slightly during the first three weeks and decreased in the last week of cultivation, reaching 1.5 and 1.7 mS $cm^{-1}$ for solutions A and B, respectively. Changes in individual macronutrients were similar for the two solutions used. As the plants grew, decreases in the concentrations of nitrates, phosphorus, and potassium, as well as increasing concentrations of Ca and sulphates in the nutrient solution were recorded, and these changes were especially significant in the last week. On average, for the two solutions (A and B), the concentrations of nitrates in them were lower by 6% after 3 weeks and by 20% after 4 weeks of growth in relation to their concentrations at the beginning of cultivation. The content of phosphorus in the medium after 3 weeks of cultivation was lower by 21%, and after 4 weeks by 41%. The greatest decreases were related to potassium; after 3 weeks of plant growth, the content of this component in the medium was lower by 20%, and after 4 weeks by as much as 64% in relation to the potassium content in the medium immediately after the start of cultivation. Ca and sulphate contents after 3 and 4 weeks of cultivation were higher than in the initial phase of plant growth.

The amount of water used in growing lettuce in an indoor controlled environment in a hydroponic system is very small. Pennisi et al. [47] showed that in the deep-water culture system for the production of *Lactuca sativa* cultivars 'Rebelina', 'Gautier', and 'Eyragues' with a biomass weight not exceeding 50 g, only 0.46–0.56 L of water was used per plant. They also showed that lighting conditions affected the efficiency of water use by lettuce plants. The higher the ratio of red to blue light, the higher was the water consumption. In our study, the average consumption of the nutrient solution during the 30 days of cultivation was 6.4 L per container, which gives the value of 1.29 L for the production of one 'Elizium' romaine lettuce with a leaf biomass of about 148 g and was not dependent on lighting conditions. The percentage of red light in the entire spectrum in all the lighting combinations used was quite high (60–80%).

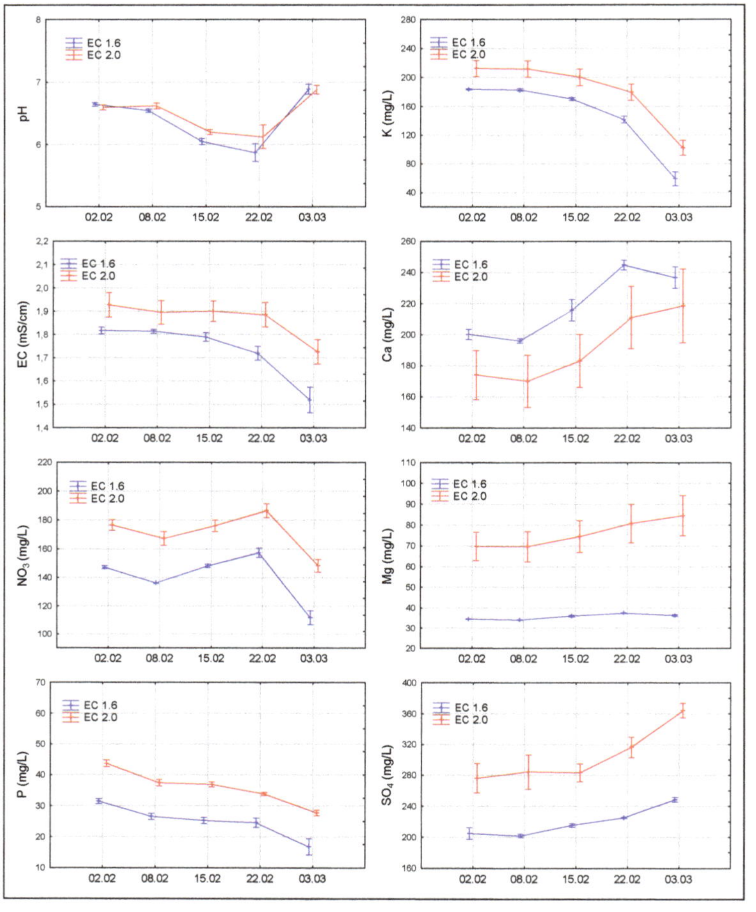


**Figure 3.** pH, electrical conductivity (EC), nitrate nitrogen (N-$NO_3$), phosphorus, potassium, calcium, magnesium, and sulfates contents in two different nutrient solutions (A and B) at weekly intervals from 25 January to 22 February 2021, for 'Elizium' romaine lettuce grown in a hydroponic system. Each data point is the average (±SE) for the four different light spectra (R:G:B).

### 3.3. Mineral Composition of Plants

Although the nutrient solutions contained small (A) or moderate (B) amounts of minerals, the concentrations of macro- and microelements in the lettuce plants were within the optimal range for most of the components and quite high for nitrates and potassium [48], which corresponded to a strong decrease in the concentrations of nitrates and K in the nutrient medium (Table 3, Figure 3). The mineral composition of the 'Elizium' romaine lettuce plants depended both on the lighting conditions in which the plants were grown, as well as on the nutrient solution used, with the influence of the nutrient solution being much stronger than that of the light quality (Table 3). The lettuce plants grown under RGB 70:18:12 had the lowest K (7.7%) and Mg (0.34%) contents, while under the other light spectra these amounts ranged from 8.1–8.4% for K, and reached the value of 0.41% (on average) for Mg. The amounts of other macronutrients (nitrates, N, P, and Ca) and micronutrients, except for B, did not depend on the quality of the light. The B content was the lowest (57 mg $kg^{-1}$ d.w.) at RGB 70:18:12. The lettuce plants grown in the hydroponic solution B contained more nitrates by 18%, total N by 4%, P by 5%, and Mg by 8%, but by

9% less K than in solution A. The Ca content in the plants was the same regardless of the solution used (1.2% on average). In the case of microelements, the type of solution did not affect the amounts of Fe and Mn, but the plants grown in solution B contained slightly less Cu and B, and more Zn.

**Table 3.** Concentrations of nitrate nitrogen (mg $kg^{-1}$ f.w.), macronutrients (N, P, K, Ca, and Mg, in %) and micronutrients (Fe, Mn, Cu, Zn and B, in mg $kg^{-1}$ d.w.) in the leaves of 'Elizium' romaine lettuce grown in a hydroponic system with different nutrient solutions (A and B) under illumination with four different light spectra (R:G:B—80:10:10, 70:10:20, 60:10:30, and 70:18:12) in an indoor controlled environment.

| Treatment | $N\text{-}NO_3$ | N | P | K | Ca | Mg | Fe | Mn | Cu | Zn | B |
|---|---|---|---|---|---|---|---|---|---|---|---|
| | mg $kg^{-1}$ f.w. | % | | | | | | | mg $kg^{-1}$ d.w. | | |
| Light spectrum R:G:B | | | | | | | | | | | |
| 80:10:10 | 3538 a | 4.40 a | 0.66 a | 8.38 b | 1.20 a | 0.41 b | 124 a | 149 a | 6.3 a | 51 a | 72 c |
| 70:10:20 | 3909 a | 4.34 a | 0.66 a | 8.37 b | 1.24 a | 0.41 b | 135 a | 126 a | 6.0 a | 45 a | 63 b |
| 60:10:30 | 3538 a | 4.36 a | 0.65 a | 8.10 ab | 1.18 a | 0.42 b | 135 a | 131 a | 5.8 a | 48 a | 58 ab |
| 70:18:12 | 3128 a | 4.24 a | 0.64 a | 7.77 a | 1.18 a | 0.34 a | 146 a | 126 a | 5.6 a | 43 a | 57 a |
| Nutrient solution | | | | | | | | | | | |
| A | 3218 a | 4.22 a | 0.64 a | 8.52 b | 1.20 a | 0.38 a | 141 a | 134 a | 6.5 b | 43 a | 69 b |
| B | 3899 b | 4.40 b | 0.67 b | 7.79 a | 1.20 a | 0.41 b | 130 a | 133 a | 5.4 a | 50 b | 55 a |
| Light spectrum × nutrient solution | n.s. | n.s. | n.s. | n.s | n.s. | n.s. | n.s. | n.s. | * | n.s. | * |
| Sufficient range (%) * | - | 2.1–5.6 | 0.5–0.9 | 4.0–8.0 | 0.9–2.0 | 0.4–0.8 | 50–200 | 25–200 | 5–18 | 30–200 | 25–65 |

Means followed by the same letter are not significantly different ($p < 0.05$) using Tukey's HSD test, ns = not significant, * Sufficient elemental ranges for the most recently matured leaf of greenhouse-grown lettuce, adapted from "Knott's Handbook for Vegetable Growers" [48].

Leafy vegetables, such as lettuce and spinach, contain the highest concentrations of nitrates [49]. The nitrate content in lettuce depends on the N content in the nutrient solution. In a study with flood fertigation of leaf lettuce [35], the nitrate content in the lettuce heads increased with the concentration of the nutrient solution, and at EC 3.0 mS $cm^{-1}$ exceeded the permissible limit imposed by the European Union. To protect human health, most European countries regulate the nitrate content in vegetables. For lettuce, different limits have been set for protected and open-grown crops [36]. No separate limits have been established for different types of lettuce, such as leaf lettuce and head lettuce. The maximum limits for nitrates in lettuce are 5000 in winter-grown plants and 4000 mg per kg of fresh product in other seasons of the year. The results of our study showed that the concentration of nitrates in the leaves of 'Elizium' romaine lettuce ('head' type) grown in the indoor controlled environment was quite high (3128–3909 mg $kg^{-1}$ f.w.) but did not exceed the limits for greenhouse winter crops. The study also showed that the concentration of nitrates in lettuce leaves in an indoor controlled environment could be managed by modifying the mineral composition of the nutrient solution. The nitrate content in the leaves of the lettuce plants grown in solution A with a low nitrate content (130 mg $L^{-1}$) was significantly lower (3218 mg $kg^{-1}$ f.w.) than of those grown in solution B with a higher nitrate content (170 mg $L^{-1}$). The modifications of the spectrum of the light emitted by LEDs at PPFD 160 µmol $m^{-2}$ $s^{-1}$ did not significantly affect the concentration of nitrates in the leaves of 'Elizium' romaine lettuce despite the wide range of red light to blue light ratio (2.1–7.7) in the spectra tested.

So far, little research has been conducted on the effect of light quality on the mineral composition of lettuce in indoor controlled environments, and the obtained results have been inconclusive [17,27,47,50,51]. The concentration of nitrates in the leaves of *Lactuca sativa* 'Grand Rapids' grown in greenhouse conditions was found to be significantly lower after short-term exposure of the plants to red light of high intensity (PPFD 500 µmol $m^{-2}$ $s^{-1}$) [52]. A similar effect was achieved by the alternating use of red light and blue light during the day [17] and a high ratio of red to blue light with a simultaneous periodic change in light intensity [51]. The addition of red light to white light generated by LEDs did not reduce the nitrate content in the leaves of *Lactuca sativa* cultivars 'Lvdie' and 'Ziya', with green leaves and purple leaves, although it stimulated the production of biomass [53]. By comparison, Liu et al. [54] showed that lamps generating a wide spectrum of light, such as fluorescent lamps and high-pressure sodium lamps (HPS), were more effective in reducing nitrates in lettuce than the combination of red light with blue light.

Amoozgar et al. [39] showed that *Lactuca sativa* 'Grizzly' grown in an indoor controlled environment accumulated much greater amounts of minerals in the leaves than when grown in a greenhouse. The concentration of macronutrients in the plants in the indoor controlled environment was on average 2 to 4 times higher than in greenhouse cultivation. 'Outredgeous' romaine lettuce had a much greater ability to accumulate K than other minerals [27]. Monochromatic red light increased the accumulation of K, P, and Fe, while red light combined with blue light increased the accumulation of N and Mg in the leaves [39]. Clavijo-Herrera et al. [55] and Pennisi et al. [47] showed that the accumulation of N in lettuce leaves did not depend on the ratio of red to blue light generated by LEDs. Our study showed that the tested ratios of red light to blue light, i.e., 80:10, 70:20, and 60:30 with the same proportion of green light (10%), did not significantly affect the accumulation of macro- or micronutrients in the leaves of 'Elizium' romaine lettuce, but with a higher proportion of green light (RGB 70:18:12) the plants contained less K and Mg, which may be due to the dilution effect, as these plants had the highest fresh weight. Increasing the blue-to-red light ratio from 0.1 to 4.5 had negatively affected biomass production and leaf growth of oakleaf lettuce *Lactuca sativa* 'Rouxai', but increased the concentrations of nitrogen, magnesium, zinc, and copper in the plants [56]. Similarly, in the case of 'Outredgeous' romaine lettuce, increasing the proportion of blue light relative to white light had increased the concentrations of K, Ca, Mg, and P in the plants, but the resulting plants were the smallest and had the lowest weight [27].

Our study showed that the quality of the LED-generated light with a red-to-blue ratio of 2.1–7.7, and also the concentrations of Ca in the nutrient solution, 170 and 200 mg $L^{-1}$, had no major effect on the accumulation of Ca in the leaves of 'Elizium' romaine lettuce. The average Ca content in the plants was 1.2%. At the same time, withering of the edges of young leaves (tipburn) was observed, and these symptoms were more common on the lettuce plants grown in the solution with the higher mineral content (B) and under RGB 70:10:20 and RGB 70:18:12. One of the main causes of the physiological disturbances causing tipburn is insufficient supply of Ca to young romaine lettuce leaves [29,57–59]. In our study, we observed increasing Ca concentrations in the hydroponic medium, and therefore the Ca concentration in the medium was not a direct cause of tipburn on lettuce leaves. Ca is not transported from older leaves to the younger ones, as a result of which the Ca content in mature lettuce leaves is higher than in young leaves [59–61]. Ca is transported from the roots to the leaves via the xylem and this process depends on the intensity of transpiration. Air humidity in an indoor controlled environment in hydroponic cultivation is usually high, which can create problems with adequately supplying Ca to young leaves [62], and this problem may especially concern the head-forming types of lettuce, such as crisphead lettuce and romaine lettuce. Lettuce plants grown in the DFT (Deep Flow Technique) hydroponic system have shown more severe tipburn symptoms than when grown in solid media [42]. The cause of the disturbances may also be an imbalance between the individual mineral components in the leaves, especially potassium and calcium [59]. There is little data on the accumulation of Ca in lettuce leaves depending on light quality. Increased Ca accumulation in plants has been obtained under LED lamps emitting white light compared to monochromatic red light and red light in combination with blue light [39], as well as under white light supplemented with blue light [27]. By contrast, Pennisi et al. [47] showed that the accumulation of Ca did not depend on the quality of light even if there was a relatively large variation in the ratio of red light to blue light (from 0.5 to 4).

### 3.4. Chlorophyll, Flavonol, and Nitrogen Balance Indices

Dualex Scientific+ is an innovative testing device designed for non-destructive measurements of the chlorophyll, flavonol, and nitrogen balance (NBI) indices in plants, and is used to monitor the nitrogen nutritional status of plants. Ouzounis et al. [63] confirmed the high correlation of the flavonol index determined with Dualex Scientific+ with the concentrations of flavonoids such as rutin and quercetin determined with the HPLC technique. Tremblay et al. [64], Padilla et al. [65], Agati et al. [66], and Kaniszewski et al. [67] confirmed the high correlation of the NBI index with the nutritional status of plants with respect to nitrogen.

The measurements made with the Dualex Scientific+ device when the plants of 'Elizium' romaine lettuce had obtained its marketable size showed that both the quality of light and the type of nutrient solution used significantly affected the chlorophyll, flavonol, and NBI indices in the leaves (Figure 4). However, no significant interaction was found between the quality of light and the nutrient solution. Irrespective of the nutrient solution used, the highest values of the flavonol index, as well as that of chlorophyll, were recorded for the lettuce plants grown with an increased proportion of blue light (RGB 60:10:30) and green light (RGB 70:18:12). The flavonol index in the leaves of the plants grown under these lighting conditions was 46% higher than under RGB 80:10:10 and 19% higher than under RGB 70:10:20. The highest value of the NBI index was shown by the plants growing under RGB 80:10:10, and this index was 31% higher than the values recorded for the other three spectra of light emitted by the LEDs. The measurements also revealed that the flavonol index in the leaves of the plants growing in solution A was 21% higher than in solution B, while the chlorophyll and NBI indices were lower by 4% and 25%, respectively, than in solution B.

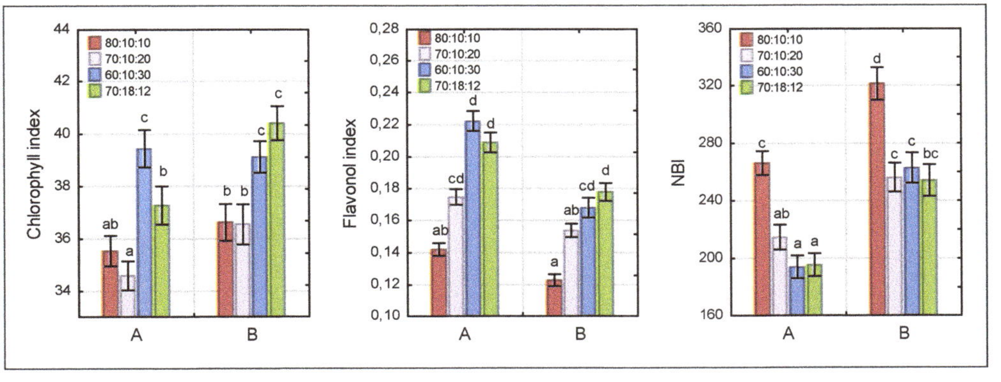


**Figure 4.** Chlorophyll index, flavonol index, and nitrogen balance index (NBI) of 'Elizium' romaine lettuce grown in a hydroponic system with different mineral concentrations of nutrient solution (A and B) under illumination with four different light spectra (R:G:B—80:10:10, 70:10:20, 60:10:30, and 70:18:12) in an indoor controlled environment. Bars represent the means ± SE. Means followed by the same letter are not significantly different ($p < 0.05$) according to Tukey's HSD test.

Our observations are generally consistent with the results of other authors relating to the various genotypes of lettuce, which indicate that blue light has a significant impact on the synthesis of bioactive compounds, including flavonoids [10–12,20,66,68]. In the case of red-leaf lettuce, supplementation of white light or red light with blue light has stimulated leaf pigmentation and the synthesis of secondary metabolites [27]. Increased levels of phytonutrients, including flavonoids, have been obtained after exposing lettuce plants to blue light with red light [69]. The biosynthesis of flavonoids is also affected by the nutritional status of plants with respect to nitrogen. Flavonoids, as nitrogen-free secondary metabolites, are considered indicators of nitrogen availability in the plant [70]. The concentrations of flavonoids increase with a low N availability. The highest level of flavonoids in lettuce leaves has been obtained with a medium containing the lowest tested mineral concentration [71]. It has also been shown that a high C/N ratio in plants stimulated the production of flavonoids, whereas a low C/N ratio inhibited their production [72]. In a study with cabbage [67], it was demonstrated that the chlorophyll index and the nitrogen balance index (NBI) were positively correlated with the N content in the leaves, whereas the flavonol index was negatively correlated. Similar relationships were evident in our study. The flavonol index was the highest for the nutrient solution with the low mineral content (A), while the high Chl and NBI indices corresponded to the low flavonol index. Our study also showed that the lettuce plants grown with an increased proportion of green light (RGB 70:18:12) were characterized by a high flavonol index and, at the same time, a low NBI index, while the nitrate concentration in the leaves was below the permissible limit. The resultant plants had the highest leaf fresh weight.

## 4. Conclusions

The quality of the light generated by LEDs significantly affects the rate of biomass production and the nutritional status of 'Elizium' romaine lettuce type 'mini' in an indoor controlled environment. Among the tested lighting combinations with different ratios of R, G, and B lights (80:10:10, 70:10:20, 60:10:30, and 70:18:12), the RGB 70:18:12 light promoted the production of leaf biomass, inhibited the accumulation of potassium in the leaves. Moreover, those plants were characterized by a low NBI index and a high flavonol index. In indoor cultivation, romaine lettuce accumulates significant amounts of minerals, especially nitrates and potassium. To achieve rapid growth of 'Elizium' romaine lettuce at a light intensity (PPFD) of 160 μmol $m^{-2}$ $s^{-1}$ and a 16-h photoperiod, it is sufficient for the nutrient solution to have a low concentration of minerals with the following composition (in mg $L^{-1}$): N-$NO_3$—130, N-$NH_4$—11, P—40, K—180, Ca—200, Mg—35, and EC 1.6 mS $cm^{-1}$. A relatively small increase in the concentration of minerals in the medium, on average by 25% (EC 2.0 mS $m^{-2}$ $s^{-1}$), significantly reduced the parameters related to food quality; there was a decrease in the flavonol index, an increase in the NBI index, and in the concentration of nitrates in plants, and at the same time the problem of withering of the tips of young leaves (tipburn) intensified.

**Author Contributions:** Conceptualization, B.M.; funding acquisition, S.K.; investigation, B.M., W.K. and A.K.; methodology, B.M., W.K. and J.D.; project administration, S.K. and M.G.; resources, B.M.; supervision, S.K.; writing—original draft, B.M.; writing—review and editing, B.M. and S.K. All authors have read and agreed to the published version of the manuscript.

**Funding:** This work was supported by funding through The National Centre for Research and Development in Poland, project number POIR. 01.01.01-00-0579/19, project title "Plantlab—innovative system of year-round production of romaine lettuce and freshwater fish in aquaponic technology".

**Institutional Review Board Statement:** Not applicable.

**Informed Consent Statement:** Informed consent was obtained from all subjects involved in the study.

**Data Availability Statement:** The data presented in this study are available on request from the corresponding author.

**Conflicts of Interest:** The authors declare no conflict of interest.

## References

1. Kozai, T.; Niu, G. Plant factory as a resource-efficient closed plant production system. In *Plant Factory: An Indoor Vertical Farming System for Efficient Quality Food Production*; Academic Press: Cambridge, MA, USA, 2016; pp. 69–90.
2. Kalantari, F.; Mohd Tahir, O.; Mahmoudi Lahijani, A.; Kalantari, S. A review of vertical farming technology: A guide for implementation of building integrated agriculture in cities. *Adv. Eng. Forum* **2017**, *24*, 76–91. [CrossRef]
3. Ohashi-Kaneko, K. Functional components in leafy vegetables. In *Plant Factory: An Indoor Vertical Farming System for Efficient Quality Food Production*; Academic Press: Cambridge, MA, USA, 2016; pp. 177–185.
4. Gómez, C.; Jiménez, J. Effect of end-of-production high-energy radiation on nutritional quality of indoor-grown red-leaf lettuce. *HortScience* **2020**, *55*, 1055–1060. [CrossRef]

5. Graamans, L.; Baeza, E.; van den Dobbelsteen, A.; Tsafaras, I.; Stanghellini, C. Plant factories versus greenhouses: Comparison of resource use efficiency. *Agric. Syst.* **2018**, *160*, 31–43. [CrossRef]
6. van Iersel, M.W. Optimizing LED lighting in controlled environment agriculture. In *Light Emitting Diodes for Agriculture*; Springer: Singapore, 2017; pp. 59–80.
7. Lin, K.-H.; Huang, M.-Y.; Huang, W.-D.; Hsu, M.-H.; Yang, Z.-W.; Yang, C.-M. The effects of red, blue, and white light-emitting diodes on the growth, development, and edible quality of hydroponically grown lettuce (*Lactuca sativa* L. var. *capitata*). *Sci. Hort.* **2013**, *150*, 86–91. [CrossRef]
8. Kang, W.H.; Park, J.S.; Park, K.S.; Son, J.E. Leaf photosynthetic rate, growth, and morphology of lettuce under different fractions of red, blue, and green light from light-emitting diodes (LEDs). *Hortic. Environ. Biotechnol.* **2016**, *57*, 573–579. [CrossRef]
9. Ohashi-Kaneko, K.; Takase, M.; Kon, N.; Fujiwara, K.; Kurata, K. Effect of light quality on growth and vegetable quality in leaf lettuce, spinach and komatsuna. *Environ. Control Biol.* **2007**, *45*, 189–198. [CrossRef]
10. Son, K.H.; Oh, M.M. Leaf shape, growth, and antioxidant phenolic compounds of two lettuce cultivars grown under various combinations of blue and red light-emitting diodes. *HortScience* **2013**, *48*, 988–995. [CrossRef]
11. Lee, J.S.; Lim, T.G.; Kim, Y.H. Growth and phytochemicals in lettuce as affected by different ratios of blue to red LED radiation. *Acta Hortic.* **2014**, *1037*, 843–848. [CrossRef]
12. Shen, Y.Z.; Guo, S.S.; Ai, W.D.; Tang, Y.K. Effects of illuminants and illumination time on lettuce growth, yield and nutritional quality in a controlled environment. *Life Sci. Space Res.* **2014**, *2*, 38–42. [CrossRef]
13. Han, T.; Vaganov, V.; Cao, S.; Li, Q.; Ling, L.; Cheng, X.; Pen, L.; Zhang, C.; Yakovlev, A.N.; Zhong, Y.; et al. Improving "color rendering" of LED lighting for the growth of lettuce. *Sci. Rep.* **2017**, *7*, 45944. [CrossRef]
14. Liu, H.; Fu, Y.; Hu, D.; Yu, J.; Liu, H. Effect of green, yellow and purple radiation on biomass, photosynthesis, morphology and soluble sugar content of leafy lettuce via spectral wavebands "knock out". *Sci. Hortic.* **2018**, *236*, 10–17. [CrossRef]
15. He, J.; Qin, L.; Chow, W.S. Impacts of LED spectral quality on leafy vegetables: Productivity closely linked to photosynthetic performance or associated with leaf traits? *Int. J. Agric. Biol. Eng.* **2019**, *12*, 16–25. [CrossRef]
16. Chang, C.L.; Chang, K.P. The growth response of leaf lettuce at different stages to multiple wavelength-band light-emitting diode lighting. *Sci. Hortic.* **2014**, *179*, 78–84. [CrossRef]
17. Chen, X.L.; Guo, W.Z.; Xue, X.Z.; Wang, L.C.; Qiao, X.J. Growth and quality responses of 'Green Oak Leaf' lettuce as affected by monochromic or mixed radiation provided by fluorescent lamp (FL) and light-emitting diode (LED). *Sci. Hortic.* **2014**, *172*, 168–175. [CrossRef]
18. Wang, J.; Lu, W.; Tong, Y.; Yang, Q. Leaf morphology, photosynthetic performance, chlorophyll fluorescence, stomatal development of lettuce (*Lactuca sativa* L.) exposed to different ratios of red light to blue light. *Front. Plant Sci.* **2016**, *7*, 250. [CrossRef]
19. Azad, O.K.; Kjaer, K.H.; Adnan, M.; Naznin, M.T.; Lim, J.D.; Sung, I.J.; Park, C.H.; Lim, Y.S. The evaluation of growth performance, photosynthetic capacity, and primary and secondary metabolite content of leaf lettuce grown under limited irradiation of blue and red LED light in an urban plant factory. *Agriculture* **2020**, *10*, 28. [CrossRef]
20. Son, K.H.; Lee, J.H.; Oh, Y.; Kim, D.; Oh, M.M.; In, B.C. Growth and bioactive compound synthesis in cultivated lettuce subject to light-quality changes. *HortScience* **2017**, *52*, 584–591. [CrossRef]
21. Johkan, M.; Shoji, K.; Goto, F.; Hashida, S.N.; Yoshihara, T. Blue light-emitting diode light irradiation of seedlings improves seedling quality and growth after transplanting in red leaf lettuce. *HortScience* **2010**, *45*, 1809–1814. [CrossRef]
22. Muneer, S.; Kim, E.-J.; Park, J.-S.; Lee, J.-H. Influence of green, red and blue light emitting diodes on multiprotein complex proteins and photosynthetic activity under different light intensities in lettuce leaves (*Lactuca sativa* L.). *Int. J. Mol. Sci.* **2014**, *515*, 4657–4670. [CrossRef]
23. Kim, H.H.; Goins, G.D.; Wheeler, R.M.; Sager, J.C. Green-light supplementation for enhanced lettuce growth under red-and blue light-emitting diodes. *HortScience* **2004**, *39*, 1617–1622. [CrossRef]
24. Terashima, I.; Fujita, T.; Inoue, T.; Chow, W.E.; Oguchi, R. Green light drives leaf photosynthesis more efficiently than red light in strong white light: Revisiting the enigmatic question of why leaves are green. *Plant Cell Physiol.* **2009**, *50*, 684–697. [CrossRef]
25. Liu, J.; van Iersel, M.W. Photosynthetic physiology of blue, green, and red light: Light intensity effects and underlying mechanisms. *Front. Plant Sci.* **2021**, *12*, 619987. [CrossRef]
26. Jishi, T.; Kimura, K.; Matsuda, R.; Fujiwara, K. Effects of temporally shifted irradiation of blue and red LED light on cos lettuce growth and morphology. *Sci. Hortic.* **2016**, *198*, 227–232. [CrossRef]
27. Mickens, M.A.; Skoog, E.J.; Reese, L.E.; Barnwell, P.L.; Spencer, L.E.; Massa, G.D.; Wheeler, R.M. A strategic approach for investigating light recipes for 'Outredgeous' red romaine lettuce using white and monochromatic LED. *Life Sci. Space Res.* **2018**, *19*, 53–62. [CrossRef]
28. Kim, M.J.; Moon, Y.; Kopsell, D.A.; Park, S.; Tou, J.C.; Waterland, N.L. Nutritional value of crisphead 'Iceberg' and romaine lettuces (*Lactuca sativa* L.). *J. Agric. Sci.* **2016**, *8*, 734. [CrossRef]
29. Sago, Y. Effects of light intensity and growth rate on tipburn development and leaf calcium concentration in butterhead lettuce. *HortScience* **2016**, *51*, 1087–1091. [CrossRef]
30. Jenni, S.; Hayes, R.J. Genetic variation, genotype × environment interaction, and selection for tipburn resistance in lettuce in multi-environments. *Euphytica* **2010**, *171*, 427–439. [CrossRef]
31. Fu, W.; Li, P.; Wu, Y.; Tang, J. Effects of different light intensities on anti-oxidative enzyme activity, quality and biomass in lettuce. *Sci. Hort.* **2012**, *39*, 129–134.

32. Brazaitytė, A.; Miliauskienė, J.; Vaštakaitė-Kairienė, V.; Sutulienė, R.; Laužikė, K.; Duchovskis, P.; Małek, S. Effect of different ratios of blue and red LED light on Brassicaceae microgreens under a controlled environment. *Plants* **2021**, *10*, 801. [CrossRef] [PubMed]
33. Kyriacou, M.C.; El-Nakhel, C.; Pannico, A.; Graziani, G.; Soteriou, G.A.; Giordano, M.; Zarrelli, A.; Ritieni, A.; De Pascale, S.; Rouphael, Y. Genotype-specific modulatory effects of select spectral bandwidths on the nutritive and phytochemical composition of microgreens. *Front. Plant Sci.* **2019**, *10*, 1501. [CrossRef]
34. Sabat, T.; Kaniszewski, S.; Dyśko, J. Flood fertigation of leaf lettuce grown in various substrates. *J. Elem.* **2019**, *24*, 19–29. [CrossRef]
35. Sabat, T.; Kaniszewski, S.; Dyśko, J. Effect of flood fertigation on yield of greenhouse lettuce grown in different substrates. *J. Elem.* **2015**, *20*, 407–416. [CrossRef]
36. European Commission. European Commission Regulation EC No. 1881/2006 of 19 December 2006 setting maximum levels for certain contaminants in foodstuffs. *Off. J. Eur. Union* **2006**, *364*, 5–24.
37. Boss, C.H.; Fredeen, K.J. *Concepts, Instrumentation, and Techniques in Inductively Coupled Plasma Optical Emission Spectrometry*, 3rd ed.; Perkin Elmer: Shelton, CT, USA, 2004. Available online: https://www.perkinelmer.com/lab-solutions/resources/docs/GDE_Concepts-of-ICP-OES-Booklet.pdf (accessed on 5 January 2021).
38. Latimer, G., Jr. *Official Methods of Analysis*, 19th ed.; AOAC International: Gaithersburg, MD, USA, 2012; ISBN 978-0-935584-83-7.
39. Amoozgar, A.; Mohammadi, A.; Sabzalian, M.R. Impact of light-emitting diode irradiation on photosynthesis, phytochemical composition and mineral element content of lettuce cv. Grizzly. *Photosynthetica* **2017**, *55*, 85–95. [CrossRef]
40. Meng, Q.; Runkle, E.S. Growth responses of red-leaf lettuce to temporal spectral changes. *Front. Plant Sci.* **2020**, *11*, 571788. [CrossRef]
41. Wissemeier, A.H.; Zühlke, G. Relation between climatic variables, growth and the incidence of tipburn in field-grown lettuce as evaluated by simple, partial and multiple regression analysis. *Sci. Hortic.* **2002**, *93*, 193–204. [CrossRef]
42. Assimakopoulou, A.; Kotsiras, A.; Nifakos, K. Incidence of lettuce tipburn as related to hydroponic system and cultivar. *J. Plant Nutr.* **2013**, *36*, 1383–1400. [CrossRef]
43. Carotti, L.; Graamans, L.; Puksic, F.; Butturini, M.; Meinen, E.; Heuvelink, E.; Stanghellini, C. Plant factories are heating up: Hunting for the best combination of light intensity, air temperature and root-zone temperature in lettuce production. *Front. Plant Sci.* **2021**, *11*, 592171. [CrossRef] [PubMed]
44. Jenni, S.; Truco, M.J.; Michelmore, R.W. Quantitative trait loci associated with tipburn, heat stress-induced physiological disorders, and maturity traits in crisphead lettuce. *Theor. Appl. Genet.* **2013**, *126*, 3065–3079. [CrossRef] [PubMed]
45. Holmes, S.C.; Wells, D.E.; Pickens, J.M.; Kemble, J.M. Selection of heat tolerant lettuce (*Lactuca sativa* L.) cultivars grown in deep water culture and their marketability. *Horticulturae* **2019**, *5*, 50. [CrossRef]
46. Macias-González, M.; Truco, M.J.; Bertier, L.D.; Jenni, S.; Simko, I.; Hayes, R.J.; Michelmore, R.W. Genetic architecture of tipburn resistance in lettuce. *Theor. Appl. Genet.* **2019**, *132*, 2209–2222. [CrossRef]
47. Pennisi, G.; Orsini, F.; Blasioli, S.; Cellini, A.; Crepaldi, A.; Braschi, I.; Spinelli, F.; Nicola, S.; Fernandez, J.; Stanghellini, C.; et al. Resource use efficiency of indoor lettuce (*Lactuca sativa* L.) cultivation as affected by red:blue ratio provided by LED lighting. *Sci. Rep.* **2019**, *9*, 14127. [CrossRef] [PubMed]
48. Maynard, D.N.; Hochmuth, G.J. *Knott's Handbook for Vegetable Growers*, 5th ed.; John Wiley and Sons: Hoboken, NJ, USA, 2013; p. 642, ISBN 978-1-118-68610-2.
49. Iammarino, M.; Taranto, A.; Cristino, M. Monitoring of nitrites and nitrates levels in leafy vegetables (spinach and lettuce): A contribution to risk assessment. *J. Sci. Food Agric.* **2014**, *94*, 773–778. [CrossRef]
50. Pinho, P.; Jokinen, K.; Halonen, L. The influence of the LED light spectrum on the growth and nutrient uptake of hydroponically grown lettuce. *Lighting Res. Technol.* **2017**, *49*, 866–881. [CrossRef]
51. Loconsole, D.; Cocetta, C.; Santoro, P.; Ferrante, A. Optimization of LED lighting and quality evaluation of Romaine Lettuce grown in an innovative indoor cultivation system. *Sustainability* **2019**, *11*, 841. [CrossRef]
52. Samuolienė, G.; Samuolienė, A.; Duchovskis, P.; Bliznikas, Z.; Vitta, P.; Žukauskas, A. Decrease in nitrate concentration in leafy vegetables under a solid-state illuminator. *HortScience* **2009**, *44*, 1857–1860. [CrossRef]
53. Yan, Z.; He, D.; Niu, G.; Zhou, Q.; Qu, Y. Growth, nutritional quality, and energy use efficiency of hydroponic lettuce as influenced by daily light integrals exposed to white versus white plus red light-emitting diodes. *Hortscience* **2019**, *54*, 1737–1744. [CrossRef]
54. Liu, H.; Fu, Y.; Yu, J.; Liu, H. Accumulation and primary metabolism of nitrate in lettuce (*Lactuca sativa* L. var. *Youmaica*) grown under three different light sources. *Commun. Soil Sci. Plant Anal.* **2016**, *47*, 1994–2002. [CrossRef]
55. Clavijo-Herrera, J.; van Santen, E.; Gómez, C. Growth, water-use efficiency, stomatal conductance, and nitrogen uptake of two lettuce cultivars grown under different percentages of blue and red light. *Horticulturae* **2018**, *4*, 16. [CrossRef]
56. Meng, Q.; Boldt, J.; Runkle, E.S. Blue radiation interacts with green radiation to influence growth and predominantly controls quality attributes of lettuce. *J. Am. Soc. Hortic. Sci.* **2020**, *145*, 75–87. [CrossRef]
57. Ashkar, S.A.; Ries, S.K. Lettuce tipburn as related to nutrient imbalance and nitrogen composition. *J. Am. Soc. Hortic. Sci.* **1971**, *96*, 448–452.
58. Saure, M.C. Causes of the tipburn disorder in leaves of vegetables. *Sci. Hortic.* **1998**, *76*, 131–147. [CrossRef]
59. Barta, D.J.; Tibbitts, T.W. Calcium localization and tipburn development in lettuce leaves during early enlargement. *J. Am. Soc. Hortic. Sci.* **2000**, *125*, 294–298. [CrossRef]

60. Gilliham, M.; Dayod, M.; Hocking, B.J.; Xu, B.; Conn, S.J.; Kaiser, B.N.; Leigh, R.A.; Tyerman, S.D. Calcium delivery and storage in plant leaves: Exploring the link with water flow. *J. Exp. Bot.* **2011**, *62*, 2233–2250. [CrossRef] [PubMed]
61. Birlanga, V.; Acosta-Motos, J.R.; Pérez-Pérez, J.M. Genotype-dependent tipburn severity during lettuce hydroponic culture is associated with altered nutrient leaf content. *Agronomy* **2021**, *11*, 616. [CrossRef]
62. Lee, J.G.; Choi, C.S.; Jang, Y.A.; Jang, S.W.; Lee, S.G.; Um, Y.C. Effects of air temperature and air flow rate control on the tipburn occurrence of leaf lettuce in a closed-type plant factory system. *Hortic. Environ. Biotechnol.* **2013**, *54*, 303–310. [CrossRef]
63. Ouzounis, T.; Fretté, X.; Rosenqvist, E.; Ottosen, C.-O. Spectral effects of supplementary lighting on the secondary metabolites in roses, chrysanthemums, and campanulas. *J. Plant Physiol.* **2014**, *171*, 1491–1499. [CrossRef]
64. Tremblay, N.; Wang, Z.; Cerovic, Z. Sensing crop nitrogen status with fluorescence indicators. A review. *Agron. Sustain. Dev.* **2012**, *32*, 451–464. [CrossRef]
65. Padilla, F.M.; Peña-Fleitas, M.T.; Gallardo, M.; Thompson, R.B. Proximal optical sensing of cucumber N status using chlorophyll fluorescence indices. *Eur. J. Agron.* **2015**, *73*, 83–97. [CrossRef]
66. Agati, G.; Tuccio, L.; Kusznierewicz, B.; Chmiel, T.; Bartoszek, A.; Kowalski, A.; Grzegorzewska, M.; Kosson, R.; Kaniszewski, S. Nondestructive optical sensing of flavonols and chlorophyll in white head cabbage (*Brassica oleracea* L. var. *capitata* subvar. *alba*) grown under different nitrogen regimens. *J. Agric. Food Chem.* **2016**, *64*, 85–94. [CrossRef]
67. Kaniszewski, S.; Kowalski, A.; Dyśko, J.; Agati, G. Application of a combined transmittance /fluorescence leaf clip sensor for the non-destructive determination of nitrogen status in white cabbage plants. *Sensors* **2021**, *21*, 482. [CrossRef] [PubMed]
68. Matysiak, B.; Kowalski, A. The growth, photosynthetic parameters and nitrogen status of basil, coriander and oregano grown under different led light spectra. *Acta Sci. Pol. Hortorum Cultus* **2021**, *20*, 13–22. [CrossRef]
69. Hooks, T.; Masabni, J.; Sun, L.; Niu, G. Effect of pre-harvest supplemental UV-A/Blue and Red/Blue LED lighting on lettuce growth and nutritional quality. *Horticulturae* **2021**, *7*, 80. [CrossRef]
70. Deng, B.; Li, Y.; Xu, D.; Qingqing, Y.; Guihua, L. Nitrogen availability alters flavonoid accumulation in Cyclocarya paliurus via the effects on the internal carbon/nitrogen balance. *Sci. Rep.* **2019**, *9*, 2370. [CrossRef]
71. Song, J.; Huang, H.; Hao, Y.; Song, S.; Zhang, Y.; Liu, H. Nutritional quality, mineral and antioxidant content in lettuce affected by interaction of light intensity and nutrient solution concentration. *Sci. Rep.* **2020**, *10*, 2796. [CrossRef]
72. Wan, H.; Zhang, J.; Song, T.; Tian, J.; Yao, Y. Promotion of flavonoid biosynthesis in leaves and calli of ornamental crab apple (*Malus* sp.) by high carbon to nitrogen ratios. *Front. Plant Sci.* **2015**, *6*, 673. [CrossRef]

agriculture

MDPI

*Article*

# Growth and Energy Use Efficiency of Grafted Tomato Transplants as Affected by LED Light Quality and Photon Flux Density

**Jianfeng Zheng [1], Peidian Gan [1], Fang Ji [1], Dongxian He [1,*] and Po Yang [2]**

[1] Key Laboratory of Agricultural Engineering in Structure and Environment, Ministry of Agriculture and Rural Affairs, China Agricultural University, Beijing 100083, China; zjf@cau.edu.cn (J.Z.); peppergan@cau.edu.cn (P.G.); jifang@cau.edu.cn (F.J.)
[2] Beijing Lighting Valley Technology Company, Beijing 100083, China; yangpo@139.com
* Correspondence: hedx@cau.edu.cn; Tel.: +86-010-62737550

**Abstract:** This study was conducted to compare the effects of broad spectrum during the whole seedling period and photon flux density (PFD) in the healing stage on the growth and energy use efficiency of grafted tomato (*Lycopersicon esculentum* Mill.) transplants in a plant factory. Fluorescent lights, white LED lights, and white plus red LED lights were applied at the growth processes of grafted tomato transplants from germination of rootstock and scion to post-grafting. Three levels of PFD (50, 100, 150 μmol $m^{-2}$ $s^{-1}$) were set in the healing stage under each kind of light quality. The results indicated that the growth and quality of grafted tomato transplants under different broad spectrums were influenced by the ratio of red to blue light (R/B ratio) and the ratio of red to far-red light (R/FR ratio). A higher R/B ratio was beneficial to total dry matter accumulation, but excessive red light had a negative effect on the root to shoot ratio and the seedling quality index. The higher blue light and R/FR ratio suppressed stem extension synergistically. The LED lights had good abilities to promote plant compactness and leaf thickness in comparison with fluorescent lights. The plant compactness and leaf thickness increased with the increase in daily light integral in the healing stage within a range from 2.5 to 7.5 mol $m^{-2}$ $d^{-1}$ (PFD, 50 to 150 μmol $m^{-2}$ $s^{-1}$). Compared to fluorescent lights, the LED lights showed more than 110% electrical energy saving for lighting during the whole seedling period. Higher PFD in the healing stage did not significantly increase the consumption of electric power for lighting. White plus red LED lights with an R/B ratio of 1.2 and R/FR ratio of 16 were suggested to replace fluorescent lights for grafted tomato transplants production considering the high quality of transplants and electrical energy saving, and PFD in the healing stage was recommended to be set to 150 μmol $m^{-2}$ $s^{-1}$.

**Keywords:** broad spectrum; white plus red LED; red to blue ratio; daily light integral; photosynthetic capacity; seedling quality

**Citation:** Zheng, J.; Gan, P.; Ji, F.; He, D.; Yang, P. Growth and Energy Use Efficiency of Grafted Tomato Transplants as Affected by LED Light Quality and Photon Flux Density. *Agriculture* **2021**, *11*, 816. https://doi.org/10.3390/agriculture11090816

Academic Editors: Athanasios Koukounaras and Filippos Bantis

Received: 19 July 2021
Accepted: 24 August 2021
Published: 27 August 2021



## 1. Introduction

Grafted tomato (*Lycopersicon esculentum* Mill.) transplants are the optimal combination of rootstock and scion with desirable production traits. They usually have the advantages of tolerance to soil-borne diseases and abiotic stresses, promotion of plant vigor, yield increases, and so on, compared to non-grafted ones [1]. The percentage of tomato cultivation area with grafted transplants has increased to 75% in the Netherlands, 50% in France, 40% in Japan, and 25% in Korea [2]. Although the proportion of tomato grafting is only around 1%, China is the country with the largest grafted tomato cultivation area of more than 20 thousand hectares [3].

The grafted tomato transplant production usually starts with the raising of rootstock and scion plantlets, followed with grafting and healing, and ends with the acclimation [1].

The management of light, temperature, and relative humidity during healing is very important for the survival and vigorous growth of tomato grafted transplants. Therefore, the professional nurseries usually utilize healing chambers to provide a controlled environment for the healing of grafted transplants [4]. Many studies have focused on environment optimization in the healing stage for grafted transplants production in the greenhouse [5–7]. It is noteworthy that grafted tomato transplants have been commercially produced in the plant factory with artificial lighting (PFAL) for the advantages of high-quality, pesticide-free, and annual production [8]. In comparison with the traditional production methods, production in PFAL is different in the preparation of rootstock and scion plantlets. To ensure the high-quality and pesticide-free production of transplants and stability of production, all the growth processes of grafted tomato transplants from sowing to post-grafting are in a fully controlled environment with artificial lighting. Generally, lighting accounts for 70%–80% of the total electricity consumption in a PFAL and greatly affects the growth and development of transplants [9]. It is meaningful to optimize the lighting environment during the whole seeding period to promote the growth of grafted tomatoes and reduce the electric consumption of lighting.

Recently, power-efficient and narrow-spectrum LED lights have been gradually replacing the fluorescent lights as the light sources in PFALs [10]. The chlorophyll absorption spectra were often used as the theoretical support for the design of horticultural commercial LED fixtures made by a combination of red and blue LEDs. However, the efficiency of red plus blue LED lights for plants is doubtful [11]. For a living plant, the pigments in vivo can absorb light with wavelengths from 400 nm to 700 nm and transfer the excitation energy to chlorophyll a for photoreaction [12]. Green light can be used for photosynthesis quite efficiently compared to blue light according to the relative quantum yield curve determined by McCree [13]. Kim et al. [14] reported that the growth of lettuce under red and blue LEDs was highly enhanced by an addition of 24% green lights, while the total photon flux density (PFD) and the ratio of red to blue light (R/B ratio) remained unchanged. The leaves can acclimate their photosystem composition to their growth light spectrum and quantum yields can be enhanced substantially under combined different wavelengths [15]. It has been proved that the phosphor-conversion white LEDs with continuous broad spectrum have great potential to promote high yields and energy-saving in PFALs compared to fluorescent lights and red and blue LED lights [16].

However, one thing to consider is that the common phosphor-based white LEDs that are widely used in architectural lighting applications [17] usually have a relative deficiency of red light in the spectrum considering the high use efficiency of red light for plants. One solution for the lack of red light in the spectrum of white LEDs is adding red LEDs to the white LED lights. The white LED light with additional red light emitted a more promising spectrum for increasing photosynthesis and energy efficiency of Chinese cabbage, lettuce, pepper, and tomato [18,19]. The quality of grafted tomato transplants was significantly promoted by supplementing lighting with white plus red plus blue LED lights compared to white LED lights in greenhouses [20].

The main aim of the present study is to compare the effects of the broad spectrum emitted by fluorescent lights, white LED lights, and white plus red LED lights during the whole seedling period on the growth and energy use efficiency of grafted tomato transplants in a PFAL. Considering that PFD in the healing stage affects the growth and quality of grafted transplants [21,22], three levels of PFD were set in the healing stage of grafted tomato transplants grown under each light quality to investigate the possible interactive effects of light quality and PFD. The results are expected to provide suggestions on replacing fluorescent lights with white LED lights for grafted tomato transplants production in a PFAL.

## 2. Materials and Methods

### *2.1. Plant Materials and Growth Conditions*

The experiment was conducted in a walk-in growth chamber equipped with an automatic environmental control system, the internal size of which is 2.8 m long, 2.9 m wide, and 2.4 m high (Figure 1). The tomatoes (*Lycopersicon esculentum* Mill.) "Zhezhan No.1" and "Dongfeng No.1" (Jinan Weili Seeds Co., Ltd., Jinan, China) were used as rootstock and scion, respectively. Rootstock seeds were sown 2 days after sowing scion seeds, in 72-cell and 128-cell plug trays filled with a mixture of vermiculite and perlite (3:1, *v*/*v*), respectively. The scion and rootstock seeds germinated 4 days after sowing the scion in the dark with a temperature of 28 °C ± 1 °C and relative humidity of 75% ± 8%.

(**A**) (**B**) (**C**)

**Figure 1.** Tomato transplants at pre-grafting stage (**A**), healing stage (**B**), and post-healing stage (**C**).

A total of 20 days after sowing the scion, the stem diameter of rootstock and scion was in the range between 2.0 mm and 3.0 mm. The grafting process was carried out manually by experienced workers using the cleft grafting method [2]. Air temperature in the growth chamber was maintained at 25 °C ± 1 °C and 20 °C ± 1 °C in photoperiod and dark period, respectively, and relative humidity was always controlled at 90% ± 5% within 7 days after grafting (in the healing stage). At pre-grafting and post-healing stage, air temperature and relative humidity in the growth chamber were maintained at 24 °C ± 1 °C and 60% ± 5%, respectively, in the light period, and 20 °C ± 1 °C and 65% ± 5%, respectively, in the dark period. During the whole seedling stage, the $CO_2$ concentration was maintained at 800 ± 50 μmol mol$^{-1}$ in photoperiod and no control in the dark period.

During the experiment, the tomato transplants were fully sub-irrigated every two or three days with the nutrient solution, which was prepared using purified water based on the Japanese horticultural formula (mg L$^{-1}$): $KNO_3$, 808; $Ca(NO_3)_2{\cdot}4H_2O$, 944; $MgSO_4{\cdot}7H_2O$, 492; $NH_4H_2PO_4$, 153; DTPA-Fe-7, 42.9; $H_3BO_3$, 2.82; $MnSO_4{\cdot}H_2O$, 1.54; $CuSO_4{\cdot}5H_2O$, 0.08; $ZnSO_4{\cdot}7H_2O$, 0.22; $(NH_4)_6Mo_7O_{24}{\cdot}4H_2O$, 0.03, respectively. One-third strength of standard nutrient solution was used after seed germination. One-half strength of standard nutrient solution was used after cotyledon unfolding. Full strength of standard nutrient solution was used after first true leaf unfolding, until to the end of experiment. The pH of nutrient solution was controlled at 6.0–6.5 during the experiment.

### *2.2. Lighting Treatments*

Three kinds of tubular LED lighting fixtures (W-LED-16W, WR-LED5/1-16W, WR-LED5/3-16W, Beijing Lighting Valley Technology Co., Beijing, China) were used for lighting treatments (Table 1). They were white LED light with R:B ratio of 0.9 (L0.9), white plus red LED light with white and red chips in number ratio 5:1 and R:B ratio of 1.2 (L1.2), white plus red LED light with white and red chips in number ratio 5:3 and R:B ratio of 2.2 (L2.2), respectively. Each light is 1.2 m long and has 120 chips arranged linearly with a spacing of 1 cm. All the white LED chips had a color temperature of 6500 K, and red chips had a peak at 660 nm. One kind of tri-phosphor fluorescent light (Shanghai Nonghui Biotechnology Corp., Shanghai, China) in a color temperature of 4200 K with R:B ratio of 1.8 (F1.8) was used as a control. The spectral distributions of PFD emitted by four lighting sources across the 300–800 nm wavelength were measured at 15 cm below the lights using a fiber spectrometer (AvaSpec-ULS2048, Avantes Inc., Apeldoorn, The Netherlands) (Figure 2). The photon flux fraction of ultraviolet light (300–399 nm), blue light (400–499 nm), green light (500–599 nm), red light (600–699 nm), and far-red light (700–800 nm) were calculated for light quality analysis. Lighting treatments using four kinds of lights lasted 15 days after germination of rootstock and scion with the PFD of 250 μmol $m^{-2}$ $s^{-1}$ and photoperiod of 14 h $d^{-1}$. On grafting day, all the lights were turned off to provide a dark environment for grafted seedlings. On the second day after grafting, the PFD was set to three levels, namely 50, 100, 150 μmol $m^{-2}$ $s^{-1}$, under each lighting source for 6 days until the healing stage was over. At post-healing stage, the PFD was set back to 250 μmol $m^{-2}$ $s^{-1}$ for 3 days. Different levels of PFD were achieved by changing the number and horizontal position of lights. Each lighting treatment included 24 plants and three replicates.

**Table 1.** Lighting treatments created by four kinds of light sources during the whole seedling period and three levels of photon flux density in the healing stage.

| Treatment | Light Source for the Whole Seedling Period | Photon Flux Density (μmol $m^{-2}$ $s^{-1}$) | | Photoperiod (h $d^{-1}$) | Daily Light Integral (mol $m^{-2}$ $d^{-1}$) | |
|---|---|---|---|---|---|---|
| | | Healing Stage (7 d) [Z] | Pre-Grafting/Post-Healing Stage (15 d/3 d) | | Healing Stage | Pre-Grafting/Post-Healing Stage |
| F1.8-P050 [Y] | | 50 | | | 2.5 | |
| F1.8-P100 | F1.8 | 100 | 250 | 14 | 5.0 | 12.6 |
| F1.8-P150 | | 150 | | | 7.6 | |
| L0.9-P050 | | 50 | | | 2.5 | |
| L0.9-P100 | L0.9 | 100 | 250 | 14 | 5.0 | 12.6 |
| L0.9-P150 | | 150 | | | 7.6 | |
| L1.2-P050 | | 50 | | | 2.5 | |
| L1.2-P100 | L1.2 | 100 | 250 | 14 | 5.0 | 12.6 |
| L1.2-P150 | | 150 | | | 7.6 | |
| L2.2-P050 | | 50 | | | 2.5 | |
| L2.2-P100 | L2.2 | 100 | 250 | 14 | 5.0 | 12.6 |
| L2.2-P150 | | 150 | | | 7.6 | |

[Z] The healing stage was 7 days. All the lights were turned off on the first day, and three levels of PFD were provided on the next 6 days. [Y] Symbol represents lighting source of F1.8 and photon flux density of 50 μmol $m^{-2}$ $s^{-1}$.

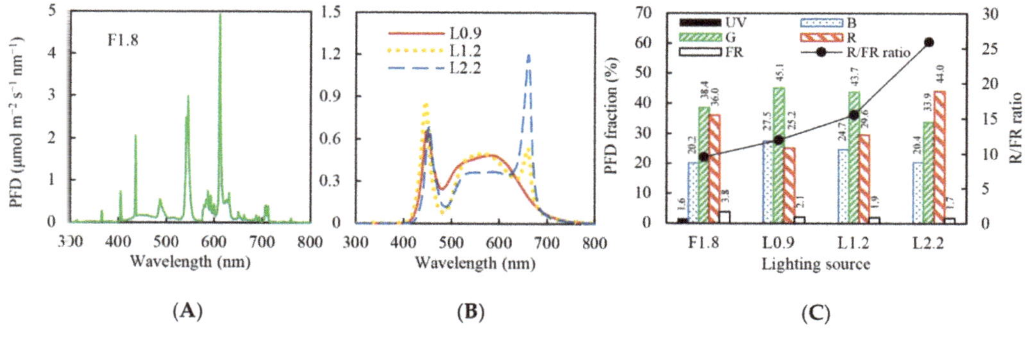


**Figure 2.** Spectral distribution of photon flux density (PFD) emitted by fluorescent lights (F1.8) with R:B ratio of 1.8 (**A**) and three LED lights (L0.9, L1.2 and L2.2) with R:B ratio of 0.9, 1.2 and 2.2, respectively (**B**). The total PFD was normalized to 100 μmol $m^{-2}$ $s^{-1}$. The PFD of ultraviolet light (UV, 300–399 nm), blue light (B, 400–499 nm), green light (G, 500–599 nm), red light (R, 600–699 nm) and far-red light (FR, 700–800 nm) were obtained by integral calculation (**C**).

### 2.3. Measurements and Calculations

At ten days after grafting, the general destructive measurements were carried out to determine the transplant characteristics. Six plants were randomly selected to measure in each treatment. Plant height (PH, cm) was measured from the substrate surface to the growing point of stem apex using a ruler. Stem diameters (SD, mm) were measured 1 cm above and below the joint using a vernier caliper, and the average value of them was used. All unfolded true leaves were counted as the leaf number, and the total leaf area was calculated according to the pixel value of the leaf image taken by a scanner (LiDE 110, Canon (China) Co., Ltd., Shenzhen, China). The fresh leaves, stems and root were weighed separately, then they were dried for 72 h in a drying oven at 80 °C, and the dry matter of leaves, stems and root were measured, respectively.

The root to shoot (R/S) ratio was the ratio of root dry matter (RDM) and shoot dry matter (SDM). The seedling quality index (SQI) was calculated as TDM/(PH/SD + SDM/RDM) [23], where TDM was a total dry matter in grams, and PH and SD are in centimeters and millimeters, respectively. The plant compactness was defined as SDM/PH. The specific leaf area (SLA) was the ratio of total leaf area per plant to total leaf dry matter per plant [24].

The net photosynthetic rate (Pn) of tomato leaf was measured using a portable photosynthesis system (LI-6400XT, LI-COR Biosciences Inc., Lincoln, NE, USA) equipped with a 6400-02B leaf chamber, in which PFD, air temperature, and $CO_2$ concentration were set at 250 μmol $m^{-2}$ $s^{-1}$, 25 °C, and 800 μmol $mol^{-1}$, respectively. The potential maximum quantum yield of primary PSII photochemistry (Fv/Fm) of the tomato leaf was measured using a multi-function plant efficiency analyzer (M-PEA, Hansatech Instruments Ltd., Norfolk, UK) after dark-adaptation for 30 min.

The fresh leaf tissue in approximately 70.0 mg ($W$) of each transplant was extracted in 80% ($v/v$) acetone for 48 h in the dark. The total volume of extract solution is 10 mL. The absorbance of the solution at 663 nm ($A_{663}$), 646 nm ($A_{646}$) and 470 nm ($A_{470}$) were measured by a spectrophotometer (UV-3150, Shimadzu Corporation, Kyoto, Japan). Chlorophyll a, chlorophyll b and total carotenoids content (mg $g^{-1}$) were calculated as $(122.5A_{663} - 27.9A_{646})/W$, $(215.0A_{646} - 51.0A_{663})/W$ and $(50.5A_{470} - 20.8A_{663} - 92.1A_{646})/W$, respectively [25].

The light energy use efficiency of the plant community ($LUE_p$) and electrical energy use efficiency of lighting ($EUE_L$) were calculated by $f{\cdot}D/PAR_p$ and $f{\cdot}D/A_L$, respectively [26]. $f$ is the conversion factor from dry mass to chemical energy fixed in dry matter (about 20 MJ $kg^{-1}$); $D$ is the average dry matter increase rate of grafted tomato plants during the whole seedling period (kg $m^{-2}$ $h^{-1}$); $PAR_P$ is average photosynthetically active radiation received at the plant community surface (MJ $m^{-2}$ $h^{-1}$); $A_L$ is electrical energy consumed by lights, which is measured by a power monitor (T8006, Shenzhen BeiDian Instrument Co., Shenzhen, China).

### 2.4. Data Statistics and Analysis

A two-way analysis of variance (ANOVA) was conducted to test the effects of light quality and PFD on plant growth and energy use efficiency using IBM SPSS Statistics 23 (IBM, Inc., Chicago, IL, USA). Duncan's test was used to make post-hoc multiple comparisons at $\alpha = 0.05$ level ($n = 6$). For the data in figures, means were separated across the four kinds of light quality and three levels of PFD, respectively, if there was no interaction between light quality and PFD.

## 3. Results

### 3.1. Growth Characteristics of Grafted Tomato Transplants

The morphology of grafted tomato transplants affected by light quality during the whole seedling period and PFD in the healing stage are shown in Figure 3. The PFD in the healing stage had interactive effects with light quality on plant height and Pn (Table 2). The height of grafted tomato transplants under F1.8-P100 was at the highest level, and that under L1.2-P100 was at the lowest level. Compared to F1.8-P100, the plant height under L1.2-P100 was 30% lower. The highest Pn of 14.1 μmol $m^{-2}$ $s^{-1}$ was observed on the transplants grown under F1.8-P150 and L1.2-P100, and the lowest Pn of 10.9 μmol $m^{-2}$ $s^{-1}$ was found on the transplants grown under L2.2-P50 and L2.2-P150. There were no interactive effects of light quality and PFD on the stem diameter, leaf number, leaf area, and TDM. The stem diameter was not affected by light quality during the whole seedling period. The PFD of 50 μmol $m^{-2}$ $s^{-1}$ in the healing stage resulted in the smallest stem diameters under F1.8 and L1.2. The leaf number and leaf area were not affected by PFD in the healing stage. The highest leaf number and leaf area were observed under F1.8. The TDM was affected by light quality and PFD. It was the highest under L2.2, and there were no significant differences under F1.8, L0.9, and L1.2. The PFD of 50 μmol $m^{-2}$ $s^{-1}$ in the healing stage led to the lowest TDM. There were no significant differences of TDM under 100 and 150 μmol $m^{-2}$ $s^{-1}$ in the healing stage.

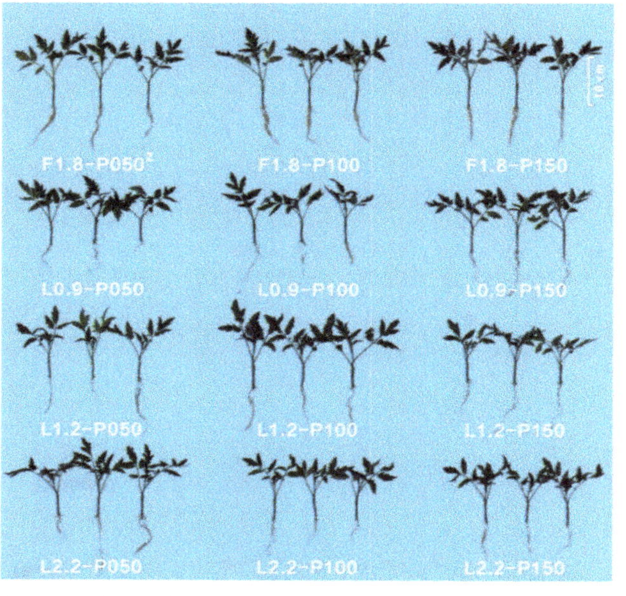


**Figure 3.** Effects of light quality during the whole seedling period and photon flux density in the healing stage on the morphology of grafted tomato transplants. $^{Z}$ Symbol represents lighting source of F1.8 and photon flux density of 50 μmol $m^{-2}$ $s^{-1}$.

No interactive effects of light quality and PFD were found on the Fv/Fm, ratio of chlorophyll a to chlorophyll b (chl a/b) and ratio of total chlorophylls to total carotenoids ((a + b)/(x + c)) (Figure 4). Fv/Fm of grafted tomato transplants grown under different light sources ranged from 0.811 to 0.822, not affected by PFD in the healing stage. The highest value of Fv/Fm 0.822 was observed on the transplants grown under L0.9, and the lowest value of Fv/Fm 0.811 was found on the transplants grown under L2.2. The PFD in the healing stage had no effects on the composition of photosynthetic pigments. Although there were statistical differences in chl a/b and (a + b)/(x + c) under different light quality, the chl a/b and (a + b)/(x + c) just varied within the range of 3.33 to 3.45 and 4.70 to 5.11, respectively. Moreover, the lowest values of chl a/b and (a + b)/(x + c) both emerged under L2.2.

**Table 2.** Morphological characteristics, biomass and net photosynthetic rate (Pn) of grafted tomato transplants as affected by light quality (LQ) during the whole seedling period and photon flux density (PFD) in the healing stage. Different letters in the same column indicate significant differences at $\alpha = 0.05$ level ($n = 6$) according to Duncan's test. NS and * represent nonsignificant and significant difference, respectively.

| Treatment | Plant Height (cm) | Stem Diameter (mm) | Leaf Number | Leaf Area (cm$^2$) | Total Dry Matter (g) | Net Photosynthetic Rate (μmol m$^{-2}$ s$^{-1}$) |
|---|---|---|---|---|---|---|
| F1.8-P050 [Z] | 7.6 ± 0.6 ab | 2.7 ± 0.1 b | 3.8 ± 0.4 a | 46.5 ± 4.3 a | 0.19 ± 0.02 b | 11.9 ± 1.4 ab |
| F1.8-P100 | 7.9 ± 0.4 a | 3.0 ± 0.2 a | 3.8 ± 0.4 a | 46.0 ± 5.4 a | 0.22 ± 0.03 ab | 12.2 ± 1.1 ab |
| F1.8-P150 | 7.2 ± 0.2 b | 2.9 ± 0.2 ab | 3.7 ± 0.5 ab | 46.9 ± 7.4 a | 0.22 ± 0.04 ab | 14.1 ± 1.2 a |
| L0.9-P050 | 6.2 ± 0.8 cd | 2.9 ± 0.2 ab | 3.5 ± 0.5 ab | 43.4 ± 4.7 ab | 0.20 ± 0.03 b | 12.0 ± 1.7 ab |
| L0.9-P100 | 6.0 ± 0.4 cd | 2.9 ± 0.2 ab | 3.2 ± 0.4 b | 41.9 ± 8.2 ab | 0.20 ± 0.03 b | 12.8 ± 1.7 ab |
| L0.9-P150 | 6.6 ± 0.7 c | 2.9 ± 0.2 ab | 3.8 ± 0.4 a | 46.6 ± 6.8 a | 0.24 ± 0.03 a | 12.3 ± 2.3 ab |
| L1.2-P050 | 5.4 ± 0.3 e | 2.7 ± 0.1 b | 3.0 ± 0.0 b | 36.6 ± 4.1 b | 0.19 ± 0.02 b | 13.8 ± 1.2 a |
| L1.2-P100 | 5.5 ± 0.3 e | 3.0 ± 0.2 a | 3.3 ± 0.5 ab | 44.2 ± 5.1 ab | 0.23 ± 0.03 a | 14.1 ± 1.3 a |
| L1.2-P150 | 5.6 ± 0.2 de | 3.0 ± 0.1 a | 3.3 ± 0.5 ab | 36.7 ± 1.7 b | 0.23 ± 0.02 a | 11.0 ± 1.0 b |
| L2.2-P050 | 6.3 ± 0.4 cd | 2.9 ± 0.1 ab | 3.2 ± 0.4 b | 46.4 ± 3.0 a | 0.24 ± 0.03 a | 10.9 ± 2.6 b |
| L2.2-P100 | 6.7 ± 0.5 c | 2.8 ± 0.2 ab | 3.2 ± 0.4 b | 42.4 ± 8.6 ab | 0.24 ± 0.05 a | 12.7 ± 1.6 ab |
| L2.2-P150 | 6.0 ± 0.6 d | 2.8 ± 0.2 ab | 3.6 ± 0.5 ab | 42.4 ± 6.4 ab | 0.24 ± 0.04 a | 10.9 ± 0.8 b |
| LQ | * | NS | * | * | * | * |
| PFD | NS | * | NS | NS | * | NS |
| LQ × PFD | * | NS | NS | NS | NS | * |

[Z] Symbol represents lighting source of F1.8 and photon flux density of 50 μmol m$^{-2}$ s$^{-1}$.

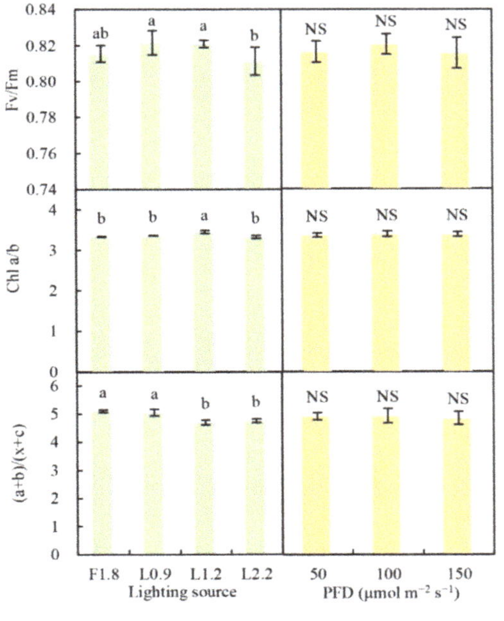


**Figure 4.** Effects of light quality during the whole seedling period and photon flux density (PFD) in the healing stage on Fv/Fm, ratio of chlorophyll a to chlorophyll b (chl a/b) and ratio of total chlorophylls to total carotenoids ((a + b)/(x + c)) of grafted tomato transplants. There were no interactive effects of light quality and PFD on Fv/Fm ($P = 0.056$), chl a/b ($P = 0.440$) and (a + b)/(x + c) ($P = 0.256$). Duncan's test was used to make post-hoc multiple comparisons at $\alpha = 0.05$ level ($n = 6$). Vertical bars represent standard deviations. Different letters, a and b, indicate significant differences and NS indicates nonsignificant differences.

There were no interactive effects of light quality and PFD on parameters of compactness and SLA (Figure 5). Compared to fluorescent lights, the LED lights led to higher compactness and lower SLA. The addition of red light to white LED lights significantly improved the compactness and reduced the SLA. However, no significant differences in compactness and SLA were found under two kinds of white plus red LED lights. The compactness increased linearly with an increase in daily light integral (DLI) in the healing stage at the range of 2.5 to 7.5 mol $m^{-2}$ $d^{-1}$. SLA decreased with the increase in DLI, responding in an opposite manner to compactness. Interactive effects of light quality and PFD were observed on the R/S ratio and SQI (Figure 6). There were no significant differences in R/S ratio between different PFD under F1.8 and L0.9 light quality. However, the R/S ratio had an increasing and decreasing trend, respectively, with the increase in PFD under L1.2 and L2.2 light quality. The SOI had a similar response as the R/S ratio to light quality and PFD. The R/S ratio and SQI of grafted tomato transplants under L1.2-P150 were both at comparable levels compared to that under F1.8.

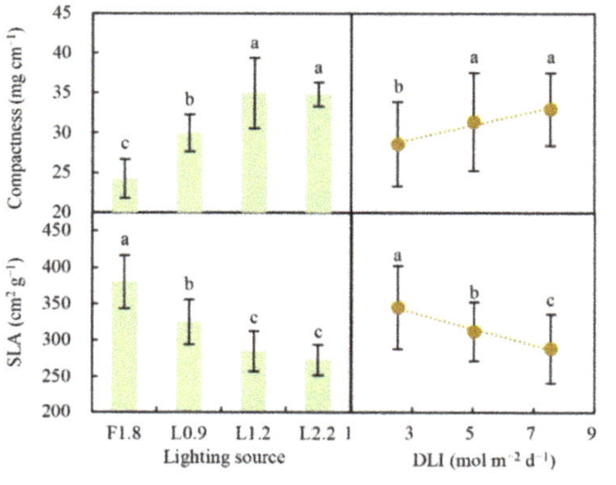


**Figure 5.** Effects of light quality during the whole seedling period and daily light integral (DLI) in the healing stage on compactness, specific leaf area (SLA) of grafted tomato transplants. There were no interactive effects of light quality and DLI on compactness ($P$ = 0.216) and SLA ($P$ = 0.716). Duncan's test was used to make post-hoc multiple comparisons at $\alpha$ = 0.05 level ($n$ = 6). Vertical bars represent standard deviations. Different letters, a–c, indicate significant differences.

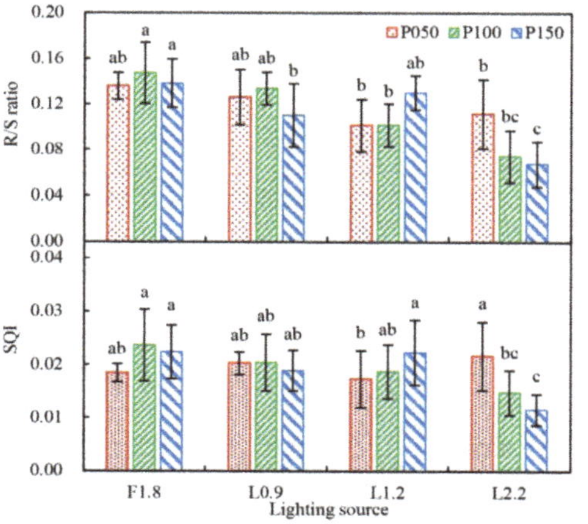


**Figure 6.** Root to shoot (R/S) ratio and seedling quality index (SQI) of grafted tomato transplants as affected by light quality during the whole seedling period and photon flux density in the healing stage. Means were separated by Duncan's test at $\alpha$ = 0.05 level ($n$ = 6). Vertical bars represent standard deviations. Different letters, a–c, indicate significant differences.

### 3.2. Energy Use Efficiency

There were no interactive effects on $LUE_P$ and $EUE_L$ between light quality during the whole seedling period and PFD in the healing stage (Figure 7). The $LUE_P$ and $EUE_L$ were both affected by light quality during the whole seedling period but not affected by PFD in the healing stage. Compared to F1.8, the $LUE_p$ of grafted tomato transplants under L0.9 and L1.2 were both at the same level, and that under L2.2 increased by 19% with the highest value of 0.025. There were no significant differences in EUE of grafted tomato transplants grown under L0.9, L1.2 and L2.2, which improved by 123%, 126%, and 110%, respectively, compared to that under F1.8 with the lowest value of 0.0031.

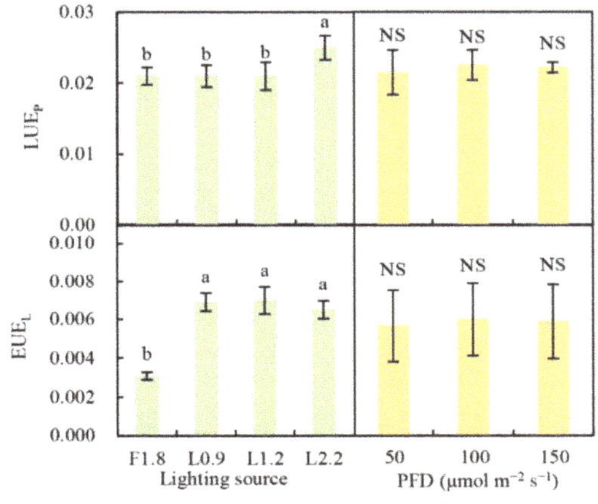


**Figure 7.** Effects of light quality during the whole seedling period and photon flux density (PFD) in the healing stage on energy use efficiency of grafted tomato transplants. There were no interactive effects of light quality and PFD on $LUE_P$ ($P$ = 0.212) and $EUE_L$ ($P$ = 0.063). Duncan's test was used to make post-hoc multiple comparisons at $\alpha$ = 0.05 level ($n$ = 6). Vertical bars represent standard deviations. Different letters, a and b, indicate significant differences and NS indicates nonsignificant differences.

## 4. Discussion

### 4.1. Growth of Grafted Tomato Transplants as Affected by Light Quality and PFD

Light quality is an essential factor for plant growth and development. Tubular fluorescent lights have been used as lighting sources for transplant production in commercialized PFALs since 2002 for their balanced spectrum and high efficiency [27]. Many studies aiming to optimize LED lighting for plant production in PFALs usually set fluorescent lights as the control in experiments [6,28,29]. They are willing to accept LED lights as lighting sources when the growth and quality of plants are comparable or better under LED lights in comparison with fluorescent lights considering the huge energy-saving advantages. In this study, high-quality transplants were evaluated according to the suggestions of Lee et al. [2] that they should have a proper size with healthy thick leaves and well-developed root systems, free from environmental stress during the growth stage. To understand the relationship between spectrum and growth and quality of grafted tomato transplants, integrated values of blue (400–499 nm), green (500–599 nm), red (600–699 nm), and far-red (700–800 nm) PFD in different lighting environments were used for discussion.

In the current study, light quality during the whole seeding period and PFD in the healing stage affected the TDM independently. Hernández et al. [29] compared the physiological responses of tomato seedlings to different R/B ratios using a combination of monochromatic blue LED and red LED, and found that R/B ratios in the range 1.0–2.3 were best for tomato seedling production in a PFAL, considering the higher dry matter accumulation. The R/B ratio of lights used in this study ranged from 0.9 to 2.2, similar to the ratio range mentioned above. Our results showed that there were no significant differences in TDM under an R/B ratio of 0.9–1.8, but a 14.2% increase in TDM was observed under an R/B ratio of 2.2. It seemed to indicate that grafted tomato transplants grown under L2.2 had the highest quantum yield for $CO_2$ fixation. However, it cannot be explained by the Pn of grafted tomato transplants for the reason that it decreased significantly under L2.2, especially under PFD of 50 and 100 μmol $m^{-2}$ $s^{-1}$ in the healing stage. Correspondingly, a slight decrease in Fv/Fm was also found under L2.2. It has been reported that red light alone led to a lower Fv/Fm value but higher TDM on lettuces [30]. Hernández et al. [29] also observed a similar response on non-grafted tomato seedlings. A similar phenomenon was also reported by Bantis et al. [6], who found that grafted watermelon seedlings treated by a higher R/B ratio in the healing stage had a higher TDM but showed a lower Pn and Fv/Fm value. Hernández et al. [29] explained that this phenomenon was caused by the different light intersections due to different leaf areas and leaf numbers. However, the leaf area and leaf number under L2.2 were not significantly higher than those under other treatments in this study.

The photosynthetic quantum yield of leaves is wavelength-dependent, and the leaves can acclimate their photosystem composition to their growth light spectrum [15]. Adaptation to different light environment is usually characterized by the adjusted composition of photosynthetic pigments and various Chl fluorescences in the leaf [31]. In this study, Fv/Fm of grafted tomato transplants grown under different light sources ranged from 0.811 to 0.822. It indicated that the grafted tomato transplants under different lighting environments did not suffer from obvious environmental stress [32]. The chl a/b and (a + b)/(x + c) showed a mildly decreased value under L2.2, similar to Fv/Fm. These results showed that the leaves of grafted tomato transplants acclimated their photosystem to adapt to the higher red lights of L2.2 by increasing chlorophyll b and total carotenoids fraction to resist possible photoinhibition. In this study and that conducted by Bantis et al. [6], the Pn of plants grown under different light qualities was measured under the same lighting environment provided by red and blue LED built in the leaf chamber; the measured Pn cannot represent the real Pn during the growth period, since the leaves have adapted to their growth light spectrum. The Pn measured under the growth light spectrum may be a good indicator to explain the response of TDM. McCree [33] reported that effects of light of different wavelengths could be treated as independent and additive and that the relative photosynthetic rate of a leaf under white light could be roughly calculated as the sum of the products of action spectrum by the spectral energy flux distribution of the white light, which is also known as yield photon flux of lighting. Based on this method and action spectrum of tomato leaf measured by McCree [13], the yield photon flux of F1.8, L0.9, L1.2, and L2.2 can be estimated as 9.5, 9.4, 9.6, and 10.0 (relative value), respectively, which could explain the TDM response in this study.

It is necessary to prevent excessive elongation of stem for tomato transplant production in high planting densities. Wollaeger and Runkle [34] reported that tomato transplants grown under blue light alone or in combination with red reduced the height compared to that under pure red light and red plus green light. The effect of blue light on plant height was highly species-dependent [35]. Tomato was classified as the blue-inhibition type according to the stem elongation in response to monochromatic light [28]. In the current study, the grafted tomato transplants under L0.9 and L1.2 had a lower height than those under F1.8 and L2.2. It was a result of a higher fraction of blue light in L0.9 and L1.2. However, F1.8 showed a significant promotion effect on plant height compared to L2.2, although they had the same dose of blue light. This phenomenon may be caused by a lower ratio of red to far-red light (R/FR ratio) of F1.8 compared to L2.2, since the low R/FR ratio can induce the shade-avoidance response [36]. There were shreds of evidence that the high R/FR ratio and blue light could suppress stem extension synergistically; in other words, the full expression of shade avoidance required both low R/FR and reduced blue light [6,29,37]. What's more, the higher fraction of green light of F1.8 might be another reason for the taller height. Green light can also induce the shade-avoidance response and promote the elongation of the stem [38,39]. This study indicated that the white plus red LED light of L1.2 with an R/B ratio of 1.2 and an R/FR ratio of 16 had a good ability of synergistically reducing shade-avoidance response for grafted tomato transplants.

In this study, LED lights showed good abilities to promote the compactness of grafted tomato transplants compared to fluorescent lights. The compactness of grafted tomato transplants was improved by a higher R/B ratio under LED lighting, but no significant differences were found between R/B ratios of 1.2 and 2.2. Excessive addition of red light to white LED light did not have a positive effect on improving transplant compactness. Similar results were also observed on grafted watermelon transplants: the plant compactness was not further promoted by a higher R/B ratio as the red-light fraction was over 64% under red plus blue LED lighting [6]. However, these results were not supported by Hernández et al. [29], who reported that the compactness of non-grafted tomato transplants was promoted by a higher fraction of blue light under red plus blue LED lighting, and no significant differences were observed when blue-light fraction was over 50%. This difference was caused by the different responses of SDM and plant height to R/B ratio. The SLA is the leaf area per unit of dry matter, and the lower SLA indicates a thicker leaf [24]. It was reported that a higher R/FR ratio led to the thicker leaves of tropical tree seedlings [40]. In this study, the far-red fraction of fluorescent light was two times that of LED light; therefore, it led to the lowest SLA. The grafted tomato transplants grown under L1.2 and L2.2 had the thickest leaves.

The healing process after grafting was influenced by PFD in the healing stage. Vu et al. [41] reported that grafted tomato seedlings under red light in approximately 15 $\mu$mol $m^{-2}$ $s^{-1}$ in the healing stage were significantly more compact than those in dark. The compactness and SLA had obvious responses to DLI in the healing stage in this study. Within the range of 2.5 to 7.5 mol $m^{-2}$ $d^{-1}$ in the healing stage, the compactness increased, and SLA decreased proportionally with the increase in DLI. This result agreed with the results of Jang et al. [21] who reported that the SLA of grafted cucumber seedlings decreased as DLI in the healing stage increased from 0 to 10.2 mol $m^{-2}$ $d^{-1}$. Hu et al. [22] also reported that compactness and SLA of grafted tomato transplants increased and decreased, respectively, as DLI in the healing stage increased from 2.2 to 6.5 mol $m^{-2}$ $d^{-1}$. Moreover, the compactness and SLA of grafted watermelon transplants also had similar responses to increasing supplemental DLI of 0–5.8 mol $m^{-2}$ $d^{-1}$ in greenhouses [42]. Our results indicated that high DLI in the healing stage was beneficial to the formation of compact plant shape. The provision of the PFD of 150 $\mu$mol $m^{-2}$ $s^{-1}$ is recommended at the healing stage.

A higher R/S ratio indicates that more carbohydrates are distributed to the roots, thus favoring the development of strong transplants [43]. In the current study, the R/S ratio of grafted tomato transplant was interactively affected by the light quality and PFD in the healing stage. A higher PFD in the healing stage promoted the R/S ratio under L1.2 but reduced the R/S ratio under L2.2. The L2.2 was not conducive to the allocation of dry matter to roots, especially under high PFD in the healing stage. Similarly, the R/S ratio of watermelon transplants significantly decreased when the R/B ratio of the broad LED spectrum was more than 1.7 [23]. It may be caused by excessive red light in photon fluxes, since it has been reported that red light was more beneficial to the growth rate of shoots than roots of lettuces, and the R/S ratio of lettuces decreased with increasing red LED ratio [30]. Higher PFD in the healing stage under L1.2 led to a higher R/S ratio. A similar result was also observed on the grafted cucumber seedlings [21]. The SQI includes parameters of height, stem diameter, and R/S ratio, and was suggested as a valuable indicator alongside the R/S ratio and compactness to evaluate the quality of grafted watermelon seedling [23]. In this study, the response of SQI to light quality and PFD in the healing stage was similar to the R/S ratio. Increased PFD in the healing stage under L1.2 promotes the increase in SQI. Sun et al. [44] also reported that the SQI of grafted cucumber seedlings with pumpkin rootstock increased with the increase in PFD within a range from 0 to 200 $\mu$mol $m^{-2}$ $s^{-1}$.

Moreover, L1.2 showed the same ability as F1.8 to promote the quality of grafted tomato transplants when PFD in the healing stage was 150 μmol $m^{-2}$ $s^{-1}$.

*4.2. Energy Use Efficiency as Affected by LED Quality*

There were no effects of PFD in the healing stage on $LUE_P$ and $EUE_L$ of grafted tomato transplants, which may be caused by the little differences in dry matter accumulation under different PFD in the short healing time of only six days. The differences caused by different PFD in the healing stage were narrowed to a negligible level when compared to the differences in $LUE_P$ and $EUE_L$ during the whole seedling stage.

Compared to TDM and $LUE_P$ of grafted tomato transplants grown under F1.8, those under L2.2 increased by 14% and 19%, respectively. The difference between the two increase rates was due to the difference in light energy input. According to Planck's equation, photon energy is inversely proportional to wavelength. It means that a greater fraction of red light and a lesser fraction of blue light contribute to less light energy when total photon flux is fixed. In this study, the fractions of red, green, and blue light photon flux of L2.2 were 44.0%, 33.9%, and 20.4%, respectively. The highest fraction of red light and the lowest fraction of green and blue light can explain why the increase rate of $LUE_P$ is higher than that of TDM of grafted tomato transplants under L2.2. In the current experiment, the average $LUE_P$ of grafted tomato transplants was approximately in the range between 0.021 and 0.025, which was about half times lower than that of non-grafted tomato seedling production in the PFAL reported by previous studies [45,46]. It is largely because half of the biomass of rootstock and scion was discarded during grafting, which was not regarded as the usable biomass of grafted transplants, and the decreased growth rate in the healing stage was another possible reason.

The $EUE_L$ of grafted tomato transplants under LED lights improved by 110%–126% compared to that under fluorescent lights. The LED lights showed good energy-saving ability. It is expected that replacing part of white LED chips with red LED chips can promote the photon efficacy of white LED fixtures for the reason that the photon efficiency of red LEDs is usually higher than that of white LEDs [47]. However, the $EUE_L$ of grafted tomato transplants under L2.2 did not increase compared to that under L0.9 and L1.2, although the $LUE_P$ of L2.2 was the highest among three kinds of LED lights. It indicated that L2.2 had a poor photon efficiency, which may be related to the manufacturing process of LED fixtures. Regardless, white LEDs are widely used in architectural lighting applications, the cost of which is now only 20% that of red LEDs [47]. Therefore, it is not necessary to add red light in excess considering the $EUE_L$ and cost of LED fixtures.

## 5. Conclusions

Compared to fluorescent lights, the LED lights show more than 110% electrical energy saving for lighting during the whole seedling period. The growth and quality of grafted tomato transplants under different broad LED light spectrums are influenced by the R/B ratio. The addition of an appropriate amount of red light to white LED lights can promote the total dry matter and enhance plant compactness and leaf thickness of grafted tomato transplants. However, excessive red light brings about a negative effect on the seedling quality and potentially increases costs of white plus red LED light fixtures. Higher PFD in the healing stage had a positive impact on the growth and quality of grafted tomato transplants and did not significantly increase the consumption of electric power for lighting. White plus red LED lights with R/B ratio of 1.2 and R/FR ratio of 16 are suggested to replace fluorescent lights for grafted tomato transplants production considering the high quality of transplants and electrical energy saving, and PFD in the healing stage is recommended to be set to 150 μmol $m^{-2}$ $s^{-1}$.

**Author Contributions:** Conceptualization, D.H.; Data curation, P.G.; Formal analysis, J.Z.; Funding acquisition, D.H.; Investigation, P.G.; Methodology, D.H.; Project administration, D.H.; Resources, P.Y.; Supervision, D.H.; Validation, F.J. and D.H.; Visualization, J.Z.; Writing—Original draft, J.Z.; Writing—Review and editing, F.J. and D.H. All authors have read and agreed to the published version of the manuscript.

**Funding:** This work was funded by the National Key Research and Development Program of China (Grant No. 2017YFB0403901).

**Conflicts of Interest:** The authors declare no conflict of interest.

## References

1. Colla, G.; Pérez-Alfocea, F.; Schwarz, D. *Vegetable Grafting: Principles and Practices*; CAB International: Oxfordshire, UK, 2017.
2. Lee, J.M.; Kubota, C.; Tsao, S.J.; Bie, Z.; Hoyos Echevarria, P.; Morra, L.; Oda, M. Current status of vegetable grafting: Diffusion, grafting techniques, automation. *Sci. Hortic.* **2010**, *127*, 93–105. [CrossRef]
3. Huang, Y.; Kong, Q.S.; Chen, F.; Bie, Z.L. The history, current status and future prospects of vegetable grafting in China. *Acta Hortic.* **2015**, *1086*, 31–40. [CrossRef]
4. Dong, W.; Zhou, Z.C.; Bu, Y.L.; Zhuo, J.Q.; Chen, L.Z.; Li, Y.Z. Research and application of grafted seedlings healing room. *Acta Hortic.* **2015**, *1086*, 51–58. [CrossRef]
5. Lee, K.M.; Lim, C.S.; Muneer, S.; Jeong, B.R. Functional vascular connections and light quality effects on tomato grafted unions. *Sci. Hortic.* **2016**, *201*, 306–317. [CrossRef]
6. Bantis, F.; Koukounaras, A.; Siomos, A.S.; Fotelli, M.N.; Kintzonidis, D. Bichromatic red and blue LEDs during healing enhance the vegetative growth and quality of grafted watermelon seedlings. *Sci. Hortic.* **2020**, *261*, 109000. [CrossRef]
7. Lang, K.M.; Nair, A.; Litvin, A.G. An alternative healing method for grafted tomato transplants: The effect of light exclusion and substrate temperature on plant survival and growth. *HortTechnology* **2020**, *30*, 677–684. [CrossRef]
8. Yamaguchi, K.; Yang, C.S. *Plant Factory: An Indoor Vertical Farming System for Efficient Quality Food Production*; Kozai, T., Niu, G.H., Takagaki, M., Eds.; BERG EARTH Co. Ltd.: Uwajima, Japan; Academic Press: London, UK, 2016; pp. 382–386.
9. Kozai, T.; Niu, G.H. Introduction. In *Plant Factory: An Indoor Vertical Farming System for Efficient Quality Food Production*; Kozai, T., Niu, G.H., Takagaki, M., Eds.; Academic Press: London, UK, 2020; pp. 3–6.
10. Kozai, T. PFAL business and R&D in Asia and North America: Status and perspectives. In *Plant Factory: An Indoor Vertical Farming System for Efficient Quality Food Production*; Kozai, T., Niu, G.H., Takagaki, M., Eds.; Academic Press: London, UK, 2020; pp. 35–39.
11. Sharakshane, A. White LED lighting for plants. *BioRxiv* **2016**, e215095, (Preprint).
12. Croce, R.; Müller, M.G.; Bassi, R.; Holzwarth, A.R. Carotenoid-to-chlorophyll energy transfer in recombinant major light-harvesting complex (LHCII) of higher plants. I. Femtosecond transient absorption measurements. *Biophys. J.* **2001**, *80*, 901–915. [CrossRef]
13. McCree, K.J. The action spectrum, absorptance and quantum yield of photosynthesis in crop plants. *Aagric. Meteorol.* **1972**, *9*, 191–216. [CrossRef]
14. Kim, H.H.; Goins, G.D.; Wheeler, R.M.; Sager, J.C. Green-light supplementation for enhanced lettuce growth under red-and blue-light-emitting diodes. *HortScience* **2004**, *39*, 1617–1622. [CrossRef]
15. Hogewoning, S.W.; Wientjes, E.; Douwstra, P.; Trouwborst, G.; Van Ieperen, W.; Croce, R.; Harbinson, J. Photosynthetic quantum yield dynamics: From photosystems to leaves. *Plant Cell* **2012**, *24*, 1921–1935. [CrossRef]
16. Nozue, H.; Shirai, K.; Kajikawa, K.; Gomi, M.; Nozue, M. White LED light with wide wavelength spectrum promotes high-yielding and energy-saving indoor vegetable production. *Acta Hortic.* **2018**, *1227*, 585–592. [CrossRef]
17. Cho, J.; Park, J.H.; Kim, J.K.; Schubert, E.F. White light-emitting diodes: History, progress, and future. *Laser Photonics Rev.* **2017**, *11*, 1600147. [CrossRef]
18. Avercheva, O.V.; Berkovich, Y.A.; Konovalova, I.O.; Radchenko, S.G.; Lapach, S.N.; Bassarskaya, E.M.; Kochetova, G.V.; Zhigalova, T.V.; Yakovleva, O.S.; Tarakanov, I.G. Optimizing LED lighting for space plant growth unit: Joint effects of photon flux density, red to white ratios and intermittent light pulses. *Life Sci. Space Res.* **2016**, *11*, 29–42. [CrossRef] [PubMed]
19. Urrestarazu, M.; Nájera, C.; Gea, M.D.M. Effect of the spectral quality and intensity of light-emitting diodes on several horticultural crops. *HortScience* **2016**, *53*, 268–271. [CrossRef]
20. Wei, H.; Hu, J.; Liu, C.; Wang, M.; Zhao, J.; Kang, D.I.; Jeong, B.R. Effect of supplementary light source on quality of grafted tomato seedlings and expression of two photosynthetic genes. *Agronomy* **2018**, *8*, 207. [CrossRef]
21. Jang, Y.; Goto, E.; Ishigami, Y.; Mun, B.; Chun, C. Effects of light intensity and relative humidity on photosynthesis, growth and graft-take of grafted cucumber seedlings during healing and acclimatization. *Hort. Environ. Biotechnol.* **2011**, *52*, 331–338. [CrossRef]
22. Hu, B.; Ling, P.; Kleinhenz, M. Grafted tomato propagation: Effects of light intensity and temperature on graft healing and plant regrowth. *Acta Hortic.* **2016**, *1140*, 327–330. [CrossRef]
23. Bantis, F.; Koukounaras, A.; Siomos, A.; Radoglou, K.; Dangitsis, C. Optimal LED wavelength composition for the production of high-quality watermelon and interspecific squash seedlings used for grafting. *Agronomy* **2019**, *9*, 870. [CrossRef]
24. Hunt, R. *Basic Growth Analysis: Plant Growth Analysis for Beginners*; Unwin Hyman Ltd.: London, UK, 1990.

25. Lichtenthaler, H.K. Chlorophylls and carotenoids: Pigments of photosynthetic biomembranes. *Methods Enzymol.* **1987**, *148*, 350–382.
26. Kozai, T. Resource use efficiency of closed plant production system with artificial light: Concept, estimation and application to plant factory. *Proc. Jpn. Acad. Ser. B* **2013**, *89*, 447–461. [CrossRef]
27. Goto, E. Plant production in a closed plant factory with artificial lighting. *Acta Hortic.* **2012**, *956*, 37–49. [CrossRef]
28. Jang, Y.; Mun, B.; Seo, T.; Lee, J.; Chun, C. Effects of light quality and intensity on the carbon dioxide exchange rate, growth, and morphogenesis of grafted pepper transplants during healing and acclimatization. *Kor. J. Hort. Sci. Technol.* **2013**, *31*, 260–266. [CrossRef]
29. Hernández, R.; Eguchi, T.; Deveci, M.; Kubota, C. Tomato seedling physiological responses under different percentages of blue and red photon flux ratios using LEDs and cool white fluorescent lights. *Sci. Hortic.* **2016**, *213*, 270–280. [CrossRef]
30. Son, K.H.; Oh, M.M. Leaf shape, growth, and antioxidant phenolic compounds of two lettuce cultivars grown under various combinations of blue and red light-emitting diodes. *HortScience* **2013**, *48*, 988–995. [CrossRef]
31. Lichtenthaler, H.K.; Babani, F. Light adaptation and senescence of the photosynthetic apparatus. changes in pigment composition, chlorophyll fluorescence parameters and photosynthetic activity. In *Chlorophyll a Fluorescence: A Signature of Photosynthesis*; Papageorgiou, G.C., Govindjee, Eds.; Springer: Dordrecht, The Netherlands, 2004; pp. 713–736.
32. Björkman, O.; Demmig, B. Photon yield of $O_2$ evolution and chlorophyll fluorescence characteristics at 77 K among vascular plants of diverse origin. *Planta* **1987**, *170*, 489–504. [CrossRef] [PubMed]
33. McCree, K.J. Significance of enhancement for calculations based on the action spectrum for photosynthesis. *Plant Physiol.* **1972**, *49*, 704–706. [CrossRef]
34. Wollaeger, H.M.; Runkle, E.S. Growth of impatiens, petunia, salvia, and tomato seedlings under blue, green, and red light-emitting diodes. *HortScience* **2014**, *49*, 734–740. [CrossRef]
35. Cope, K.R.; Snowden, M.C.; Bugbee, B. Photobiological interactions of blue light and photosynthetic photon flux: Effects of monochromatic and broad-spectrum light sources. *Photochem. Photobiol.* **2014**, *90*, 574–584. [CrossRef] [PubMed]
36. Clifford, S.C.; Runkle, E.S.; Langton, F.A.; Mead, A.; Foster, S.A.; Pearson, S.; Heins, R.D. Height control of poinsettia using photoselective filters. *HortScience* **2004**, *39*, 383–387. [CrossRef]
37. Ciolfi, A.; Sessa, G.; Sessi, M.; Possenti, M.; Salvucci, S.; Carabelli, M.; Morelli, G.; Ruberti, I. Dynamics of the shade-avoidance response in arabidopsis. *Plant Physiol.* **2013**, *163*, 331–353. [CrossRef]
38. Wang, Y.; Folta, K.M. Contributions of green light to plant growth and development. *Am. J. Bot.* **2013**, *100*, 70–78. [CrossRef] [PubMed]
39. Schenkels, L.; Saeys, W.; Lauwers, A.; Proft, M.P.D. Green light induces shade avoidance to alter plant morphology and increases biomass production in *Ocimum basilicum* L. *Sci. Hortic.* **2020**, *261*, e109002. [CrossRef]
40. Kwesiga, F.; Grace, J. The role of the red/far-red ratio in the response of tropical tree seedlings to shade. *Ann. Bot.* **1986**, *57*, 283–290. [CrossRef]
41. Vu, N.T.; Kim, Y.S.; Kang, H.M.; Kim, I.S. Influence of short-term light irradiation under pre-and post-grafting period on the graft-take ratio and quality of tomato seedlings. *Hort. Environ. Biotechnol.* **2014**, *55*, 27–35. [CrossRef]
42. Wei, H.; Wang, M.; Jeong, B.R. Effect of supplementary lighting duration on growth and activity of antioxidant enzymes in grafted watermelon seedlings. *Agronomy* **2020**, *10*, 337. [CrossRef]
43. Currey, C.J.; Hutchinson, V.A.; Lopez, R.G. Growth, morphology, and quality of rooted cuttings of several herbaceous annual bedding plants are influenced by photosynthetic daily light integral during root development. *HortScience* **2012**, *47*, 25–30. [CrossRef]
44. Sun, F.; Ma, S.; Gao, L.; Qu, M.; Tian, Y. Enhancing root regeneration and nutrient absorption in double-rootcutting grafted seedlings by regulating light intensity and photoperiod. *Sci. Hortic.* **2020**, *264*, 109192. [CrossRef]
45. Yokoi, S.; Kozai, T.; Ohyama, K.; Hasegawa, T.; Chun, C.; Kubota, C. Effects of leaf area index of tomato seedling populations on energy utilization efficiencies in a closed transplant production system. *J. Soc. High Technol. Agr.* **2003**, *15*, 231–238. [CrossRef]
46. Ji, F.; Gan, P.D.; Liu, N.; He, D.X.; Yang, P. Effects of LED spectrum and daily light integral on growth and energy use efficiency of tomato seedlings. *Trans. Chin. Soc. Agric. Eng.* **2020**, *36*, 231–238.
47. Kusuma, P.; Pattison, P.M.; Bugbee, B. From physics to fixtures to food: Current and potential LED efficacy. *Hortic. Res.* **2020**, *7*, e56. [CrossRef] [PubMed]

agriculture

Article

# Enhancing Growth and Glucosinolate Accumulation in Watercress (*Nasturtium officinale* L.) by Regulating Light Intensity and Photoperiod in Plant Factories

**Vu Phong Lam [1,2], Jaeyun Choi [1] and Jongseok Park [1,3,*]**

[1] Department of Horticultural Science, Chungnam National University, Daejeon 34134, Korea; phonglamdhtaybac@gmail.com (V.P.L.); teuk0207@naver.com (J.C.)
[2] Department of Agronomy, Tay Bac University, Son La 360000, Vietnam
[3] Graduate School of Bio-AI Convergence, Chungnam National University, Daejeon 34134, Korea
* Correspondence: jongseok@cnu.ac.kr; Tel.: +82-42-821-5737

**Abstract:** Recent advancements in light-emitting diode technology provide an opportunity to evaluate the correlation between different light sources and plant growth as well as their secondary metabolites. The aim of this study was to determine the optimal light intensity and photoperiod for increasing plant growth and glucosinolate concentration and content in watercress. Two-week-old seedlings were transplanted in a semi-deep flow technique system of a plant factory for 28 days under four photoperiod–light intensity treatments (12 h—266 μmol· $m^{-2}{\cdot}s^{-1}$, 16 h—200 μmol· $m^{-2}{\cdot}s^{-1}$, 20 h—160 μmol· $m^{-2}{\cdot}s^{-1}$, and 24 h—133 μmol· $m^{-2}{\cdot}s^{-1}$) with the same daily light integral. The mean values of shoot fresh and dry weights were the highest under the 20 h—160 μmol· $m^{-2}{\cdot}s^{-1}$ treatment, although there was no significant difference. Net photosynthesis and stomatal conductance gradually decreased with decreasing light intensity and increasing photoperiod. However, total glucosinolate concentration was significantly higher under 20 h—160 μmol· $m^{-2}{\cdot}s^{-1}$ and 24 h—133 μmol· $m^{-2}{\cdot}s^{-1}$ compared with 12 h—266 μmol· $m^{-2}{\cdot}s^{-1}$ and 16 h—200 μmol· $m^{-2}{\cdot}s^{-1}$. The total glucosinolate content was the greatest under 20 h—160 μmol· $m^{-2}{\cdot}s^{-1}$ treatment. These data suggest that the 20 h—160 μmol· $m^{-2}{\cdot}s^{-1}$ treatment promoted the maximum shoot biomass and glucosinolate content in watercress. This study supplies the optimal light strategies for the future industrial large-watercress cultivation.

**Keywords:** deep flow technique; glucosinolate; light-emitting diode; net photosynthesis; shoot biomass; watercress

**Citation:** Lam, V.P.; Choi, J.; Park, J. Enhancing Growth and Glucosinolate Accumulation in Watercress (*Nasturtium officinale* L.) by Regulating Light Intensity and Photoperiod in Plant Factories. *Agriculture* **2021**, *11*, 723. https://doi.org/10.3390/agriculture11080723

Academic Editors: Athanasios Koukounaras, Filippos Bantis and Éva Darkó

Received: 22 June 2021
Accepted: 27 July 2021
Published: 30 July 2021



## 1. Introduction

Watercress (*Nasturtium officinale* L.; Brassicaceae) is a semi-aquatic or aquatic perennial herb mainly cultivated in Asia, North and South America, and Europe [1]. Watercress is evaluated as an aquatic weed in some regions. It is used in soups (as garnish), fresh salads, and in other dishes [2]. The European Food Safety Authority has indicated that watercress is a safe vegetable of the group "herbs, edible flowers, and leaf vegetables" [3]. The US Centers for Disease Control and Prevention selected watercress as one of the crops containing the highest nutrient content per calorie [4]. It contains compounds such as vitamins, polyphenols, carotenoids, and isothiocyanates, and glucosinolates are the most crucial components present in watercress [2]. Watercress is a known medicine for treating cough, bronchial problems, and asthma [5]. Watercress has pharmacological actions such as antioxidant, anti-inflammatory, cardioprotective actions, antipsoriatic, antibacterial, and anticancer properties [1,2,6,7]. Because of the abundance of chemical components, watercress can be used in the food, medicine, and cosmetics industries.

Growing plants in a plant factory using artificial light is an efficient method of agricultural cultivation for combating climate change and worldwide food shortage [8]. Water shortages, unusual weather, and depletion of the agricultural land area result in a decrease in field crop production globally [9,10]. Nevertheless, these environmental problems do not affect crop production in a closed plant factory as the conditions for growth are controlled using temperature regulation, air conditioning, air circulation fans, artificial light, nutrient solutions, and $CO_2$ enrichment [11,12]. Crops cultivated in a plant factory always depend on artificial light (light intensity, photoperiod, and light quality), which controls the photosynthetic process, plant physiology, biochemistry, and morphology [13,14]. Thus, upgrading the light source efficiency would significantly decrease the expense of the plant factory system, which in turn would promote sustainable cultivation because the impacts of ecological and costs could be decreased.

The lighting can be supplied for plants at uniform and fixed times with specific-wavelength illumination in a plant factory because lighting schedules can be controlled, and particular photoperiods can be adjusted to promote plant growth and quality. For example, longer photoperiods increased the fresh weight of lettuce [15]. The growth of lettuce was increased with longer photoperiods and lower photosynthetic photon flux density (PPFD) at the same daily light integral (DLI) because the longer photoperiods compensated for a lower PPFD [16]. The growth of *Achimenes* cultivars grown under a low light intensity and longer photoperiods was higher as compared to that of those grown under a high light intensity and shorter photoperiods at the same DLI [17]. In addition, plant growth and morphology changes were reported due to changes in light intensity and photoperiod [18]. In general, these reports indicated that plant growth can be promoted under longer photoperiods with the same DLI. There have been several studies on the influences of different light intensities [19–21] and a combination of photoperiods and light intensities on plant biomass and secondary metabolites [15,22,23]. The optimum plant growth, yield, and quality can be obtained by controlling the light-emitting diode photoperiod and light quality [24,25]. Therefore, establishing a light regime that provides a favorable light photoperiod and intensity for plant growth and development is an essential step in cultivating plants in hydroponic systems in plant factories.

To date, several studies have been conducted on the influence of different light qualities, photoperiods, and light intensities on the growth and quality of watercress [26,27]. However, the effects of different light intensities in combination with different photoperiods on the growth and glucosinolate concentration and content of watercress grown in a plant factory have not yet been reported. For year-round production of good quality watercress in plant factories, it is important to understand the growth and quality responses to combined conditions of two light factors including photoperiod and light intensity. Thus, the aim of this study was to find the optimal light intensity and photoperiod treatment to increase plant growth and glucosinolate content in watercress. We hypothesized that plant growth and glucosinolate content in watercress increase with the increase of light intensity and photoperiod treatment.

## 2. Materials and Methods

### 2.1. Seedling Conditions

Watercress seeds were sown in rockwool cubes (240 holes; UR Rockwool, Suwon, Korea) and grown in a plant factory for 2 weeks. The air temperature and relative humidity in the plant factory were controlled at $20 \pm 2$ °C and $60 \pm 10\%$, respectively. White fluorescent lamps (TL5 14W/865 Philips, Amsterdam, Netherlands) were used for illumination. The photoperiod and PPFD were adjusted to 16 h per day and 150 $\mu mol \cdot m^{-2} \cdot s^{-1}$, respectively. The Hoagland solution for watercress seedlings (electrical conductivity 0.8 $dS \cdot m^{-1}$; pH 6.0) was supplied from 1 week after sowing.

### 2.2. Treatments

Two weeks after sowing, the seedlings were transplanted into four lighting treatments in a plant factory with the same daily light integral (11.52 mol $m^{-2}$ $d^{-1}$). Each treatment had 10 plants. The 12 h—266 µmol treatment photoperiod was set to 12 h per day with a PPFD of 266 $\mu mol \cdot m^{-2} \cdot s^{-1}$. The 16 h—200 µmol treatment photoperiod was set to 16 h per day with a PPFD of 200 μmol· $m^{-2}$·$s^{-1}$. The 20 h—160 μmol treatment photoperiod was set to 20 h per day with a PPFD of 160 μmol· $m^{-2}$·$s^{-1}$. The 24 h—133 μmol treatment photoperiod was set to 24 h per day with a PPFD of 133 μmol· $m^{-2}$·$s^{-1}$. Plants were cultivated under a 7:3 ratio of red/blue light-emitting diodes (LEDs) for 28 days under 400 μmol· $mol^{-1}$ $CO_2$ concentration and 60% ± 10% relative humidity. The blue and red LEDs had a peak wavelength of 450 and 660 nm, respectively. The experiment was conducted with two replicates for each treatment. The Hoagland nutrient solution for watercress plants (EC 2.0 dS·$m^{-1}$; pH 6.0) was supplied for 28 days. The day and night air temperatures were controlled at 22 and 20 °C, respectively.

### 2.3. *Measurement of Photosynthetic Parameters and SPAD Value*

The net photosynthetic rate and stomatal conductance of fully expanded leaves were measured with a portable photosynthesis system (LI-6400; Li-Cor, Lincoln, NE, USA) at 28 days after transplantation. The measurement conditions in the leaf chamber, namely $CO_2$ concentration, leaf temperature, airflow rate, and PPFD, were maintained at 400 μmol· $mol^{-1}$, 25 °C, 500 $cm^3$·$s^{-1}$ and 500 μmol·$m^{-2}$·$s^{-1}$, respectively. The SPAD values were measured with a portable chlorophyll meter (502, Minolta Camera Co., Ltd., Tokyo, Japan). All parameters were recorded for 6 plants ($n$ = 6) in each replication.

### 2.4. *Measurement of Plant Growth Parameters*

After 28 days of transplanting, the shoot fresh and dry weights and stem length were measured. The stem length and shoot fresh weight were determined using a ruler and an electronic scale (EW220-3NM, Kern & Sohn GmbH., Balingen, Germany), respectively. For determination of the shoot dry weight, samples were dried for one week in an oven (HB-502M; Hanback Sci, Suwon, Korea) at 70 °C and then weighed.

### 2.5. *Determination of the Individual Glucosinolate Concentration in Watercress*

The individual glucosinolate concentrations in the watercress were analyzed according to previous literature [28] but with some modifications. The shoots of watercress plants were collected at 28 days after transplanting, kept in a deep freezer at −70 °C after soaking in liquid nitrogen, then moved to a dry freezer (TFD5503, IL Shinbiobase Co., Ltd., Seoul, Korea) at −50 °C for 3 days and ground to powder. Glucosinolate was extracted in 70% ($v/v$) methanol with watercress powder (0.1 g) in a water bath for 5 min. Afterward, the samples were centrifuged at 12,000 × $g$ for 10 min, and the supernatant was analyzed as described by Lam et al. (2019) and Cuong et al. (2019) [28,29] to determine individual glucosinolate concentrations (glucobrassicin, 4-methoxyglucobrassicin, glucohirsutin, glucosiberin, and gluconasturtiin; Table 1). Desulfoglucosinolates were measured using a high-performance liquid chromatography (HPLC) system (1200 Infinity, Agilent Technologies, Santa Clara, CA, USA). An Inertsil ODS-3 (C18) column 150 × 3.0 $mm^2$ i.d., particle size 3 μm (GL Science, Tokyo, Japan), was used with a column temperature of 40 °C, a wavelength of 227 nm, and a flow rate of 0.4 mL·$min^{-1}$. The individual glucosinolates were measured by response factors (ISO 9167-1, 1992) [30] and the HPLC peak area ratios and with reference to a desulfosinigrin external standard. Glucosinolate contents (μmol/plant DW) were presented as total glucosinolate concentration in the shoot (μmol·$g^{-1}$ DW) multiplied by shoot dry weight (g).

### 2.6. *Statistical Analysis*

The growth and SPAD values were measured for six plants per replication. Photosynthetic parameters were measured for four plants per replication. The individual glucosinolate concentrations were measured for three plants per replication. For statistical analysis, one-way ANOVA was performed using SPSS 20.0 (SPSS, Inc., Chicago, IL, USA). Significant differences among treatments were verified at $p \leq 0.05$, using Tukey's multiple range test.

**Table 1.** Relative response factor values of the desulfoglucosinolates from watercress shoot extracts and their retention times on C18 column [28].

| Common Name | Side Chain Structure | Retention Time (min) | Response Factor |
|---|---|---|---|
| Glucobrassicin | indol-3-ylmethyl | 17.16 | 0.29 |
| 4- Methoxyglucobrassicin | 4-methoxyindol-3-ylmethyl | 16.05 | 0.25 |
| Glucohirsutin | 8-methylsulfinyloctyl | 13.78 | 1.1 |
| Glucosiberin | 7-methylsulfinylheptyl | 16.77 | 1 |
| Gluconasturtiin | 2-phenylethyl | 15.68 | 0.95 |

## 3. Results

### 3.1. Plant Growth Parameters, Chlorophyll Content, and Photosynthetic Parameters

The 12 h—266 µmol, 16 h—200 µmol, and 20 h—160 µmol treatments resulted in higher growth parameters relative to the 24 h—133 µmol treatment. The shoot fresh and dry weights were significantly higher (1.15 and 1.42 times, respectively) under the 20 h—160 µmol treatment compared to the 24 h—133 µmol treatment (Figure 1C,D). However, the stem length and SPAD value of the watercress were not significantly affected by photoperiod and light intensity combinations (Figure 1A,B).

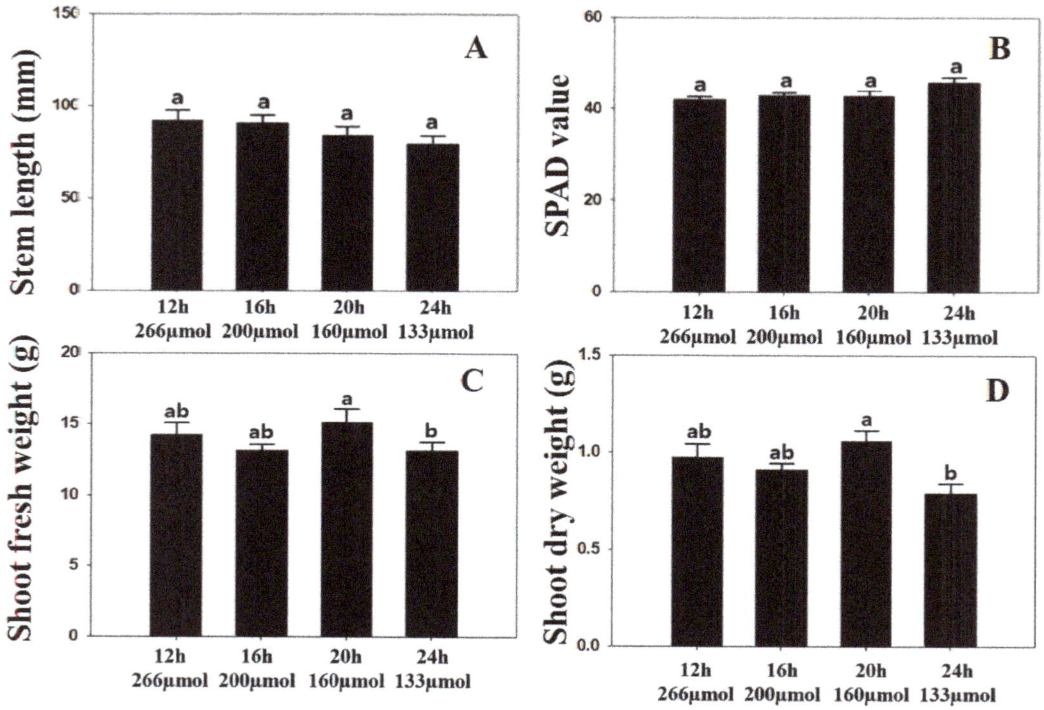


**Figure 1.** Stem length (**A**), SPAD value (**B**), shoot fresh weight (**C**), and shoot dry weight (**D**) under different lighting treatments of 12 h—266 µmol· $m^{-2}$·$s^{-1}$, 16 h—200 µmol· $m^{-2}$·$s^{-1}$, 20 h—160 µmol· $m^{-2}$·$s^{-1}$, and 24 h—133 µmol· $m^{-2}$·$s^{-1}$. Different letters above bars show significant differences at $p \leq 0.05$, using Tukey's multiple range test ($n = 6$).

The net photosynthesis and stomatal conductance were reduced with increasing photoperiod and decreasing light intensity (Figure 2). Specifically, the net photosynthesis under 16 h—200 µmol, 20 h—160 µmol, and 24 h—133 µmol treatments was 1.38, 1.52, and 3.32 times lower than that of 12 h—266 µmol treatment in the study, respectively. The stomatal conductance under 16 h—200 µmol, 20 h—160 µmol, and 24 h—133 µmol treatments was 1.41, 1.92, and 2.08 times lower than that of 12 h—266 µmol treatment in this study, respectively (Figure 2). Moreover, the shoot fresh and dry weights of the watercress were significantly low under lower light intensity and longer photoperiod treatments (24 h—133 µmol) compared with 12 h—266 µmol, 16 h—200 µmol, and 20 h—160 µmol treatments. There was no significant difference in shoot fresh and dry weights among 12 h—266 µmol, 16 h—200 µmol, and 20 h—160 µmol treatments (Figure 1).

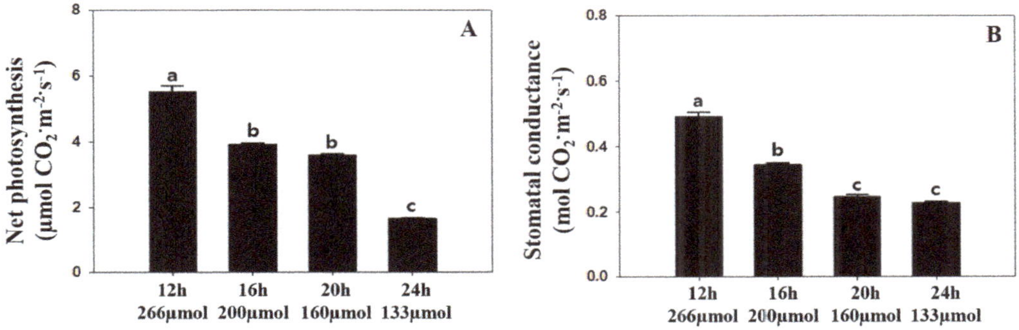


**Figure 2.** The net photosynthesis (**A**) and stomatal conductance (**B**) under different lighting treatments of 12 h—266 µmol· $m^{-2}$·$s^{-1}$, 16 h—200 µmol· $m^{-2}$·$s^{-1}$, 20 h—160 µmol· $m^{-2}$·$s^{-1}$, and 24 h—133 µmol· $m^{-2}$·$s^{-1}$. Different letters above bars show significant differences at $p \leq 0.05$, using Tukey's multiple range test ($n$ = 4).

### 3.2. Total Glucosinolate Concentration and Content

These analyses indicate that the watercress contained five different desulfoglucosinolates (glucohirsutin, 4-methoxyglucobrassicin, glucoiberin, glucohirsutin, and gluconasturtiin). Among the five desulfoglucosinolates identified, gluconasturtiin presented the highest concentration (Table 2). Gluconasturtiin accumulation in the shoot increased under 24 h—133 µmol treatment and had the highest concentration (82.51% of the total glucosinolate concentration). However, there was no significant difference in gluconasturtiin concentration among 24 h—133 µmol, 16 h—200 µmol, and 20 h—160 µmol treatments or between 20 h—160 µmol and 12 h—266 µmol treatments. The highest glucosiberin, 4-methoxyglucobrassicin, and glucohirsutin concentrations (13.61%, 3.79%, and 2.39% of the total glucosinolates, respectively) were observed under 20 h—160 µmol treatment. There was no significant difference in glucosiberin concentration among 24 h—133 µmol, 16 h—200 µmol, and 12 h—266 µmol treatments. There was no significant difference in glucohirsutin concentration among four treatments (24 h—133 µmol, 16 h—200 µmol, 12 h—266 µmol, and 20 h—160 µmol). There was no significant difference in 4-methoxyglucobrassicin concentration among 24 h—133 µmol, 16 h—200 µmol, and 20 h—160 µmol treatments or among 24 h—133 µmol, 16 h—200 µmol, and 12 h—266 µmol treatments. The highest glucobrassicin concentration (8.80% of the total glucosinolate) was recorded under 24 h—133 µmol treatment. There was no significant difference in glucobrassicin concentration among 24 h—133 µmol, 16 h—200 µmol, and 12 h—266 µmol treatments (Table 2). Overall, the total glucosinolate concentration was the greatest at 24 h—133 µmol treatment and was 1.28-fold higher than that of the 12 h—266 µmol treatment. There were no significant differences in total glucosinolate concentration in shoots between the 12 h and 16 h treatments or 20 h and 24 h treatments (Figure 3A). However, the total glucosinolate content in the shoot was the highest under 20 h—160 µmol treatment because glucosinolate contents (µmol/plant DW) were presented as total glucosinolate concentration in the shoot (µmol $g^{-1}$ DW) multiplied by shoot dry weight (g). There were no significant differences in total glucosinolate content in shoot among 24 h—133 µmol, 16 h—200 µmol, and 12 h—266 µmol treatments (Figure 3B).

**Table 2.** The individual glucosinolate concentration of watercress under different lighting treatments of 12 h—266 µmol· $m^{-2}·s^{-1}$, 16 h—200 µmol· $m^{-2}·s^{-1}$, 20 h—160 µmol· $m^{-2}·s^{-1}$, and 24 h—133 µmol· $m^{-2}·s^{-1}$.

| Lighting Treatment | Individual Glucosinolate Concentration in Watercress Shoots (mg·$g^{-1}$ DW) [z] | | | | |
|---|---|---|---|---|---|
| | Siber | Hirsu | Brassi | Metho | Nastur |
| 12 h—266 µmol | 1.04b | 0.53 | 1.83ab | 0.58b | 15.69b |
| 16 h—200 µmol | 0.63b | 0.45 | 1.75ab | 0.68ab | 18.92a |
| 20 h—160 µmol | 3.36a | 0.59 | 1.40b | 0.94a | 18.81ab |
| 24 h—133 µmol | 0.46b | 0.53 | 2.21a | 0.79ab | 20.71a |
| Significance [y] | *** | NS | ** | *** | * |

[z] Siber: glucosiberin, Hirsu: glucohirsutin, Brassi: glucobrassicin, Metho: 4-methoxyglucobrassicin, Nastur: gluconasturtiin. [y] Different letters show a significant difference within each treatment according to Tukey's multiple range test at Not Significant (NS) ($p > 0.05$), * $p \leq 0.05$; ** $p < 0.01$; and *** $p < 0.001$ ($n = 3$).

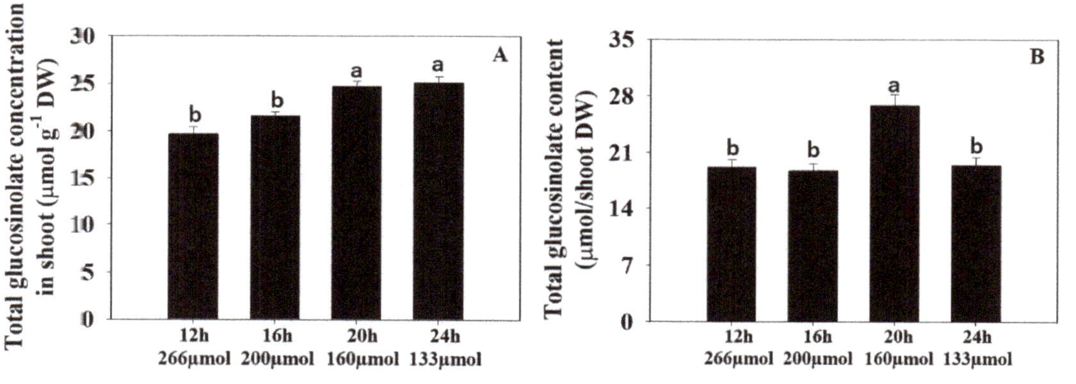


**Figure 3.** The glucosinolate concentration (**A**) and content (**B**) in watercress shoots under different lighting treatments of 12 h—266 µmol· $m^{-2}·s^{-1}$, 16 h—200 µmol· $m^{-2}·s^{-1}$, 20 h—160 µmol· $m^{-2}·s^{-1}$, and 24 h—133 µmol· $m^{-2}·s^{-1}$. Different letters above bars show significant differences at $p \leq 0.05$, using Tukey's multiple range test ($n = 3$).

## 4. Discussion

### *4.1. Plant Growth Parameters, Chlorophyll Content, and Photosynthetic Parameters*

Previous reports have indicated that higher light intensities enhanced growth and promoted crop production [31,32]. This was possible because of the wider expansion of the leaf under the higher light intensity treatment. A larger leaf leads to more light interception, which might have resulted in a significant increment in the shoot fresh and dry weights under higher light intensity [31]. The growth of ice plants under a fluorescent lamp, red LEDs, and blue LEDs was significantly higher under a higher light intensity treatment (150 µmol· $m^{-2}·s^{-1}$) than under a lower light intensity treatment (120 µmol· $m^{-2}·s^{-1}$) [20]. The biomass, stem diameter, and root/shoot ratio of soybean were higher under 400 and 500 µmol· $m^{-2}·s^{-1}$ than under 100 µmol· $m^{-2}·s^{-1}$ [33]. The leaf area was reduced by shade conditions [34]. Similarly, plant dry matter production of soybean decreased with decreasing light intensity [35]. Leaf fresh weight of *Arabidopsis thaliana* was significantly higher under a higher light intensity as compared to that of plants grown under a low light intensity at 6 weeks after transplanting [36]. Moreover, the fresh weight of watercress subjected to a long day (16 h) was significantly higher than that subjected to a short day (8 h) [26]. The fresh and dry weights of quinoa increased under a short photoperiod and high light intensity treatment [37]. Likewise, the biomass of the watercress increased under high light intensity and short photoperiod treatments.

Reductions in light intensity may influence the carbon balance in the plant. Rates of physiological process increase, while the photosynthetic yield decreases [33]. Normally, it is expected that the shading conditions or lower light intensities restrict leaf growth and result in smaller leaf areas with thinner leaves, reduced chlorophyll content, and thinner palisade tissues, leading to lower light-harvesting [38,39]. Furthermore, there is a reduction in stomatal conductance and density, which leads to poor $CO_2$ transportation under low light conditions. The electron transition from PSII to PSI is obstructed, whereas the activity and number of enzymes that participate in the Calvin cycle undergo a change. All of this results in a reduced carbon dioxide assimilation rate and a reduced net photosynthetic rate under low light conditions [33]. Previous reports have indicated that the main biochemical restraint related to shadow-associated down-adjustment of net photosynthetic rate is a decrease in the activity or amount of rubisco [33,40]. Photosynthetic capacity was reduced under low light conditions because carbon was restricted [41]. For example, low light intensity reduced the photosynthesis rate in pak choi [42] and soybean [33]. Thus, the net photosynthesis and stomatal conductance were reduced with increasing photoperiod and decreasing light intensity.

### 4.2. Total Glucosinolate Concentration and Content

Glucosinolates are bioactive compounds typically found in cruciferous group plants. It has been indicated that long days could enhance glucosinolate accumulation in *Arabidopsis* [43] and watercress [26]. Low light intensity increased the concentrations of 4-methoxyglucobrassicin, glucobrassicin, and neoglucobrassicin in pak choi [44]. The antioxidant activity and the total phenolic content of *Orthosiphon stamineus* under an open environment were higher than shade-grown conditions [45]. Total phenolic content in the leaves of *Ipomoea batatas* was higher under 16 than 8 h at a light intensity of 150 μmol· $m^{-2}$·$s^{-1}$ [46]. Antioxidant capacity and total phenolic content in lettuce continuously increased with increasing photoperiods in 150 μmol· $m^{-2}$·$s^{-1}$ conditions. Specifically, the phenolic content in lettuce was highest at 24 h under 150 μmol· $m^{-2}$·$s^{-1}$ and was 5.3-fold higher than under a 12 h period treatment [47]. Likewise, in the results of this experiment, total glucosinolate concentration was significantly higher under 20 h—160 μmol and 24 h—133 μmol compared with 12 h—266 μmol and 16 h—200 μmol. This indicates that the long photoperiods had a more photomorphogenic effect than a photosynthetic one. However, the total glucosinolate concentration was not significantly different between the 20 and 24 h photoperiods. The results indicate that the total glucosinolate concentration could increase with increasing photoperiods under low light intensity. However, it might reach a saturation point under low light intensity and long photoperiod. Expanding the photoperiod in weak light intensity conditions has a slight compensatory effect because it can decrease the negative influences of the weak light stress.

## 5. Conclusions

The results indicated that a photoperiod of 20 h at 160 μmol· $^{-2}$·$s^{-1}$ enhanced total glucosinolate content and plant biomass of watercress grown in a plant factory. Further studies can investigate the influence of light quality from LEDs on the productivity and bioactive compounds of watercress grown in a plant factory. Moreover, the results also suggested that longer photoperiod induction was a potential method for watercress glucosinolate production. There is great potential to apply these results to improve the quality of watercress plants and enhance the efficiency in watercress cultivation in plant factories.

**Author Contributions:** V.P.L. conducting the experiment, data collection and analysis, writing—original manuscript, writing—review and editing. J.C. conducting the experiment, data collection and analysis. J.P. Project administration, supervision, constructing the idea, experimental design, data analysis, writing—original manuscript, writing—review, and editing. All authors have read and agreed to the published version of the manuscript.

**Funding:** This research work was supported by Institute of Information & communications Technology Planning & Evaluation (IITP) grant funded by the Korean government (MSIT) (No. 2020-0-01441, Artificial Intelligence Conver-gence Research Center (Chungnam National University)).

**Institutional Review Board Statement:** Not applicable.

**Informed Consent Statement:** Not applicable.

**Data Availability Statement:** Not applicable.

**Conflicts of Interest:** The authors declare no conflict of interest.

## References


1. Zeb, A. Phenolic profile and antioxidant potential of wild watercress (*Nasturtium officinale* L.). *Springerplus* **2015**, *4*, 714. [CrossRef] [PubMed]
2. Klimek-Szczykutowicz, M.; Szopa, A.; Ekiert, H. Chemical composition, traditional and professional use in medicine, application in environmental protection, position in food and cosmetics industries, and biotechnological studies of *Nasturtium officinale* (watercress)—A review. *Fitoterapia* **2018**, *129*, 283–292. [CrossRef] [PubMed]
3. EFSA European Food Safety Authority (EFSA). Available online: http://www.efsa.europa.eu/ (accessed on 25 January 2020).
4. Noia, J.D. Defining powerhouse fruits and vegetables: A nutrient density approach. *Prev. Chronic. Dis.* **2014**, *11*, 130390. [CrossRef] [PubMed]
5. Suroowan, S.; Mahomoodally, M.F. A comparative ethnopharmacological analysis of traditional medicine used against respiratory tract diseases in Mauritius. *J. Ethnopharmacol.* **2016**, *177*, 61–80. [CrossRef]
6. Zafar, R.; Zahoor, M.; Shah, A.; Majid, F. Determination of antioxidants and antibacterial activities, total phenolic, polyphenol and pigment contents in *Nasturtium officinale*. *Pharmacologyonline* **2017**, *1*, 11–18.
7. Klimek-Szczykutowicz, M.; Dziurka, M.; Blazevic, I.; Dulovic, A.; Granica, S.; Korona-Glowniak, I.; Ekiert, H.; Szopa, A. Phytochemical and biological activity studies on *Nasturtium officinale* (watercress) microshoot cultures grown in rita(r) temporary immersion systems. *Molecules* **2020**, *25*, 5257. [CrossRef]
8. Al-Kodmany, K. The vertical farm: A review of developments and implications for the vertical city. *Buildings* **2018**, *8*, 24. [CrossRef]
9. Ali, S.; Liu, Y.; Ishaq, M.; Shah, T.; Abdullah; Ilyas, A.; Din, I.U. Climate change and its impact on the yield of major food crops: Evidence from pakistan. *Foods* **2017**, *6*, 39. [CrossRef]
10. Zhao, C.; Liu, B.; Piao, S.L.; Wang, X.H.; Lobell, D.B.; Huang, Y.; Huang, M.T.; Yao, Y.T.; Bassu, S.; Ciais, P.; et al. Temperature increase reduces global yields of major crops in four independent estimates. *Proc. Natl. Acad. Sci. USA* **2017**, *114*, 9326–9331. [CrossRef]
11. Kozai, T. Resource use efficiency of closed plant production system with artificial light: Concept, estimation and application to plant factory. *Proc. Jpn. Acad. Ser.* **2013**, *89*, 447–461. [CrossRef]
12. Kozai, T. Sustainable plant factory: Closed plant production systems with artificial light for high resource use efficiencies and quality produce. *Acta Hortic.* **2013**, *1004*, 27–40. [CrossRef]
13. Zakurin, A.O.; Shchennikova, A.V.; Kamionskaya, A.M. Artificial-light culture in protected ground plant growing: Photosynthesis, photomorphogenesis, and prospects of LED application. *Russ. J. Plant Physl.* **2020**, *67*, 413–424. [CrossRef]
14. Liu, Y.; Ren, X.X.; Jeong, B.R. Supplementary Light source affects growth, metabolism, and physiology of *Adenophora triphylla* (thunb.) A.DC. seedlings. *Biomed. Res. Int.* **2019**, *2019*. [CrossRef]
15. Mao, H.P.; Hang, T.; Zhang, X.D.; Lu, N. Both multi-segment light intensity and extended photoperiod lighting strategies, with the same daily light integral, promoted *lactuca sativa* l. growth and photosynthesis. *Agronomy* **2019**, *9*, 857. [CrossRef]
16. Kitaya, Y.; Niu, G.; Kozai, T.; Ohashi, M. Photosynthetic photon flux, photoperiod, and $CO_2$ concentration affect growth and morphology of lettuce plug transplants. *HortScience* **1998**, *33*, 988–991. [CrossRef]
17. Vlahos, J.; Heuvelink, E.; Martakis, G. A growth analysis study of three Achimenes cultivars grown under three light regimes. *Sci. Hortic.* **1991**, *46*, 275–282. [CrossRef]
18. Zha, L.Y.; Liu, W.K. Effects of light quality, light intensity, and photoperiod on growth and yield of cherry radish grown under red plus blue LEDs. *Hortic. Environ. Biote* **2018**, *59*, 511–518. [CrossRef]
19. Lu, N.; Bernardo, E.L.; Tippayadarapanich, C.; Takagaki, M.; Kagawa, N.; Yamori, W. Growth and accumulation of secondary metabolites in perilla as affected by photosynthetic photon flux density and electrical conductivity of the nutrient solution. *Front. Plant Sci.* **2017**, *8*, 708. [CrossRef]
20. Kim, Y.J.; Kim, H.M.; Kim, H.M.; Jeong, B.R.; Lee, H.J.; Kim, H.J.; Hwang, S.J. Ice plant growth and phytochemical concentrations are affected by light quality and intensity of monochromatic light-emitting diodes. *Hortic. Environ. Biote* **2018**, *59*, 529–536. [CrossRef]
21. Fan, X.X.; Xu, Z.G.; Liu, X.Y.; Tang, C.M.; Wang, L.W.; Han, X.L. Effects of light intensity on the growth and leaf development of young tomato plants grown under a combination of red and blue light. *Sci. Hortic.* **2013**, *153*, 50–55. [CrossRef]

22. Kang, J.H.; KrishnaKumar, S.; Atulba, S.L.S.; Jeong, B.R.; Hwang, S.J. Light intensity and photoperiod influence the growth and development of hydroponically grown leaf lettuce in a closed-type plant factory system. *Hortic. Environ. Biote* **2013**, *54*, 501–509. [CrossRef]
23. Yin, Y.H.; Yu, C.J.; Yu, L.; Zhao, J.S.; Sun, C.J.; Ma, Y.B.; Zhou, G.K. The influence of light intensity and photoperiod on duckweed biomass and starch accumulation for bioethanol production. *Bioresour. Technol.* **2015**, *187*, 84–90. [CrossRef] [PubMed]
24. Rehman, M.; Ullah, S.; Bao, Y.; Wang, B.; Peng, D.; Liu, L. Light-emitting diodes: Whether an efficient source of light for indoor plants? *Environ. Sci. Pollut. Res. Int.* **2017**, *24*, 24743–24752. [CrossRef] [PubMed]
25. Taulavuori, E.; Taulavuori, K.; Holopainen, J.K.; Julkunen-Tiitto, R.; Acar, C.; Dincer, I. Targeted use of LEDs in improvement of production efficiency through phytochemical enrichment. *J. Sci. Food Agric.* **2017**, *97*, 5059–5064. [CrossRef] [PubMed]
26. Engelen-Eigles, G.; Holden, G.; Cohen, J.D.; Gardner, G. The effect of temperature, photoperiod, and light quality on gluconasturtiin concentration in watercress (*Nasturtium officinale* R. Br.). *J. Agric. Food Chem.* **2006**, *54*, 328–334. [CrossRef] [PubMed]
27. Choi, J.Y.; Kim, S.J.; Bok, K.J.; Lee, K.Y.; Park, J.S. Effect of different nutrient solution and light quality on growth and glucosinolate contents of watercress in hydroponics. *J. Bio-Environ. Control* **2018**, *27*, 371–380. [CrossRef]
28. Lam, V.P.; Choi, J.; Kim, S.; Park, J.; Hernandez, R. Optimizing plant spacing and harvest time for yield and glucosinolate accumulation in watercress (*Nasturtium officinale* L.) grown in a hydroponic system. *Hortic. Sci. Technol.* **2019**, *37*, 733–743. [CrossRef]
29. Cuong, D.M.; Park, C.H.; Bong, S.J.; Kim, N.S.; Kim, J.K.; Park, S.U. Enhancement of glucosinolate production in watercress (*Nasturtium officinale*) hairy roots by overexpressing cabbage transcription factors. *J. Agric. Food Chem.* **2019**, *67*, 4860–4867. [CrossRef]
30. ISO Norm. Rapeseed-Determination of glucosinolate content. Part 1: Method using high-performance liquid chromatography. *ISO* **1992**, *9167*, 1–9.
31. Li, Q.; Kubota, C. Effects of supplemental light quality on growth and phytochemicals of baby leaf lettuce. *Environ. Exp. Bot.* **2009**, *67*, 59–64. [CrossRef]
32. Solovchenko, A.E.; Khozin-Goldberg, I.; Didi-Cohen, S.; Cohen, Z.; Merzlyak, M.N. Effects of light intensity and nitrogen starvation on growth, total fatty acids and arachidonic acid in the green microalga *Parietochloris Incisa*. *J. Appl. Phycol.* **2008**, *20*, 245–251. [CrossRef]
33. Feng, L.Y.; Raza, M.A.; Li, Z.C.; Chen, Y.K.; Bin Khalid, M.H.; Du, J.B.; Liu, W.G.; Wu, X.L.; Song, C.; Yu, L.; et al. The influence of light intensity and leaf movement on photosynthesis characteristics and carbon balance of soybean. *Front. Plant Sci.* **2019**, *9*, 1952. [CrossRef] [PubMed]
34. Gommers, C.M.M.; Visser, E.J.W.; Onge, K.R.S.; Voesenek, L.A.C.J.; Pierik, R. Shade tolerance: When growing tall is not an option. *Trends Plant Sci.* **2013**, *18*, 65–71. [CrossRef]
35. Yang, F.; Huang, S.; Gao, R.C.; Liu, W.G.; Yong, T.W.; Wang, X.C.; Wu, X.L.; Yang, W.Y. Growth of soybean seedlings in relay strip intercropping systems in relation to light quantity and red:far-red ratio. *Field. Crop Res.* **2014**, *155*, 245–253. [CrossRef]
36. Schumann, T.; Paul, S.; Melzer, M.; Dormann, P.; Jahns, P. Plant growth under natural light conditions provides highly flexible short-term acclimation properties toward high light stress. *Front. Plant Sci.* **2017**, *8*. [CrossRef] [PubMed]
37. Austin, J.; Jeon, Y.A.; Cha, M.-K.; Park, S.; Cho, Y.-Y. Effects of photoperiod, light intensity and electrical conductivity on the growth and yield of quinoa (*Chenopodium quinoa* Willd.) in a closedtype plant factory system. *Hortic. Sci. Technol.* **2016**, *34*, 405–413. [CrossRef]
38. Gong, W.Z.; Qi, P.F.; Du, J.B.; Sun, X.; Wu, X.L.; Song, C.; Liu, W.G.; Wu, Y.S.; Yu, X.B.; Yong, T.W.; et al. Transcriptome analysis of shade-induced inhibition on leaf size in relay intercropped soybean. *PLoS ONE* **2014**, *9*, e98465. [CrossRef] [PubMed]
39. Shafiq, I.; Hussain, S.; Raza, M.A.; Iqbal, N.; Ahsan Asghar, M.; Raza, A.; Fan, Y.F.; Mumtaz, M.; Shoaib, M.; Ansar, M.; et al. Crop photosynthetic response to light quality and light intensity. *J. Integr. Agric.* **2021**, *20*, 4–23. [CrossRef]
40. Cen, Y.P.; Sage, R.F. The regulation of rubisco activity in response to variation in temperature and atmospheric $CO_2$ partial pressure in sweet potato. *Plant Physiol.* **2005**, *139*, 979–990. [CrossRef] [PubMed]
41. Su, B.Y.; Song, Y.X.; Song, C.; Cui, L.; Yong, T.W.; Yang, W.Y. Growth and photosynthetic responses of soybean seedlings to maize shading in relay intercropping system in Southwest China. *Photosynthetica* **2014**, *52*, 332–340. [CrossRef]
42. Zhu, H.F.; Li, X.F.; Zhai, W.; Liu, Y.; Gao, Q.Q.; Liu, J.P.; Ren, L.; Chen, H.Y.; Zhu, Y.Y. Effects of low light on photosynthetic properties, antioxidant enzyme activity, and anthocyanin accumulation in purple pak-choi (*Brassica campestris* ssp *Chinensis* Makino). *PLoS ONE* **2017**, *12*, e0179305. [CrossRef] [PubMed]
43. Huseby, S.; Koprivova, A.; Lee, B.R.; Saha, S.; Mithen, R.; Wold, A.B.; Bengtsson, G.B.; Kopriva, S. Diurnal and light regulation of sulphur assimilation and glucosinolate biosynthesis in *Arabidopsis*. *J. Exp. Bot.* **2013**, *64*, 1039–1048. [CrossRef]
44. Yang, J.; Zhu, Z.J.; Gerendas, J. Interactive effects of phosphorus supply and light intensity on glucosinolates in pakchoi (*Brassica campestris* L. ssp *chinensis* var. *communis*). *Plant Soil.* **2009**, *323*, 323–333. [CrossRef]
45. Farhan, M.; Razak, S.A.; Pin, K.Y.; Chuah, A.L. Antioxidant activity and phenolic content of different parts of *Orthosiphon stamineus* grown under different light intensities. *J. Trop. Sci.* **2012**, *24*, 173–177.
46. Carvalho, I.S.; Cavaco, T.; Carvalho, L.M.; Duque, P. Effect of photoperiod on flavonoid pathway activity in sweet potato (*Ipomoea batatas* (L.) Lam.) leaves. *Food Chem.* **2010**, *118*, 384–390. [CrossRef]
47. Cho, J.Y.; Yoo, K.S.; Kim, J.; Choi, B.J.; Oh, W. Growth and bioactive compounds of lettuce as affected by light intensity and photoperiod in a plant factory using external electrode fluorescent lamps. *Hortic. Sci. Technol.* **2020**, *38*, 645–659. [CrossRef]

*agriculture*

MDPI

*Article*

# Effect of Colour of Light on Rooting Cuttings and Subsequent Growth of Chrysanthemum (*Chrysanthemum* × *grandiflorum* Ramat./Kitam.)

**Anita Schroeter-Zakrzewska * and Faisal Anggi Pradita**

Department of Ornamental Plants, Dendrology and Pomology, Poznan University of Life Sciences, Dąbrowskiego 159, 60-594 Poznań, Poland; praditaf@gmail.com
* Correspondence: anita.schroeter-zakrzewska@up.poznan.pl

**Abstract:** A closed system for plant production with artificial light is an innovative method of plant cultivation. The objective of this study was to investigate the effect of light colour on rooting cuttings and subsequent growth of chrysanthemum (*Chrysanthemum* × *grandiflorum* Ramat./Kitam.) During the experiments, the following conditions were maintained: photoperiod 16 h or 10 h, temperature 22 °C, relative humidity of 65–70%. LED lamps emitted the following light colours: white, blue, white + blue (50:50), and red + blue (75:25). For all light spectra, the photosynthetic photon flux density (PPFD) was 50 µmol $m^{-2}$ $s^{-1}$. The effectiveness of exposure to different light colours was measured with parameters: cutting weight (g), cutting length (cm), length of roots, and index of leaf greenness (SPAD). The measurements referred to plant features determining plant quality, i.e., the number of flower buds and flower head, the diameter of the flower head, height of plants, index of leaf greenness (SPAD), the number of leaves, and the fresh and dry weights of aboveground parts of plants. The rooting of cuttings and subsequent growth are integral processes in the cultivation of potted chrysanthemums. Both were differently affected by the colour of light from LED lamps. The exposure to red + blue light resulted in the highest leaf greenness index (SPAD) value and the shortest cuttings with the longest roots. White + blue light significantly influenced most of the growth parameters, except the height of the plants and the number of leaves.

**Keywords:** light colour; LED; rooting cuttings; chrysanthemum; growth room

**Citation:** Schroeter-Zakrzewska, A.; Pradita, F.A. Effect of Colour of Light on Rooting Cuttings and Subsequent Growth of Chrysanthemum (*Chrysanthemum × grandiflorum* Ramat./Kitam.). *Agriculture* **2021**, *11*, 671. https://doi.org/10.3390/agriculture11070671

Academic Editors: Athanasios Koukounaras and Filippos Bantis

Received: 1 June 2021
Accepted: 8 July 2021
Published: 15 July 2021



## 1. Introduction

Chrysanthemum is one of the most popular ornamental plants worldwide. Various methods have been used to cultivate and breed this plant in its long history. Chrysanthemum is a widely known quantitative short-day ornamental plant, which means that the length of the day and night significantly influences its growth. The plant can form flower buds in a daytime of 13.5 h or less. It can elongate its internodes and stem when the daytime is longer than 14 h under supplemental light replacing daylight. Both are necessary for different purposes. The former is for potted flowers, whereas the latter is for cut flowers [1]. Similar to other crops, chrysanthemum can normally be cultivated both indoors (in greenhouses and plastic tunnels) and outdoors (in fields). Potted chrysanthemums are usually grown indoors or in plastic tunnels and growth chambers. Ornamental plants, which strongly depend on the photoperiod, benefit from the cultivation in a plastic tunnel, where lighting, pests, diseases, watering, and harvesting are controlled. Lighting can be divided into two phases, i.e., the long-day phase and the short-day one. The aim of the former phase is to improve vegetative growth, mainly the stem length and width, and to increase the number of leaves. Plants are exposed to light for more than 12 h per day for 10–25 days. The aim of the latter phase is to promote generative growth, such as the formation of flower buds for the anthesis period. Plants are exposed to light less than 12 h per day for 6–11 weeks [2].

Fast propagation techniques are necessary to meet the market demand and produce plants for different purposes. Cutting is a simple and low-cost propagation method. Practically, it can be applied at the onset of the vegetative or reproductive phase [3]. Apart from that, common cutting methods have several advantages: simplicity, the general uniformity of results, faster root formation (even by applying auxin), and higher income. The economic reason tends to dominate mainly due to the increasing demand for cut and potted chrysanthemums worldwide [4]. Several factors may influence successful cutting propagation. These are: the application of phytohormone (auxin), the substrate for cuttings, stem parts cut, nutritional status, and light requirements. Light is crucial for plants, including short-day ones such as chrysanthemum [5]. Light, which is necessary for photosynthesis, has the characteristics of both a particle and a wave. The optimum light wavelength for photosynthesis is 400–700 nm. However, outside that range, the plants still sphotosynthesise at a low rate [6,7]. According to Zhen and Bugbee [8], far-red photons (701–750 nm) are abundant in sunlight but are considered inactive for photosynthesis and are thus excluded from the definition of photosynthetically active radiation (PAR; 400–700 nm). The consistent response among diverse species indicates that the mechanism is common in higher plants. These results suggest that far-red photons (701–750 nm) should be included in the definition of PAR.

A specific wavelength represents the colours of light affecting physiological processes, which are responsible for the plant's growth from the vegetative to the generative phase. For instance, the interaction of far-red (FR) and blue (B) light in a specific red (R): far-red (FR) spectrum increases the photosynthetic rate of chrysanthemum plants because chlorophyll can intensely absorb blue light (about 430 nm) and far-red light (about 660 nm). Both can enhance the electron excitation level during photosynthesis, activate some enzymes to catalyse reactions on the electron chain, and soptimise photosystem activities [9].

Studies on potted chrysanthemums cultivated in growth chambers showed that light affected the growth and flowering of the plants. Different colours of light may have different influences on the chrysanthemum flowering time. Zalewska et al. [10] report that the cultivar Baja from the Sombrero group of chrysanthemums was grown under short-day conditions and exposed to artificial blue light and daylight, which constituted the control. The blue light caused the plants to flower earlier than in the case of daylight. Jerzy et al. [11] observed that the source of light, such as a fluorescent lamp or LED lamp, had different effects on the duration of cultivation. Moreover, the intensity of light can manipulate some morphological features of plants, such as flower development and leaf elongation. The cultivation in growth chambers was found to have a better influence on the post-harvest quality of potted chrysanthemums. As light is a limiting factor, specific colours and wavelengths of light can be used for various purposes. Blue and white light can accelerate flowering, so both are suitable to meet market demands quickly. Red light delays flowering, so it can be applied to plants in storage rooms to keep them longer for later production [11].

The increasing demand for plant-based products such as fruit, flowers, and chemical substances caused the need to enhance plant growth and flowering by light supplementation. The use of light-emitting diode (LED) lamps positively affected protected cultivation in growth chambers, plastic tunnels, and greenhouses. At the same wavelength, the LED lamp produces lesser heat than a fluorescent lamp by different heat dissipation mechanisms, so it can save a higher amount of energy, and it is more eco-friendly. It is also beneficial for plants because it promotes germination, soptimises the photosynthetic rate, triggers stomatal conductance, creates plant compactness, extends roots and shoots, promotes flowering, and enhances fruit taste. These effects heavily depend on the plant species as well as the colour and wavelength of the light generated by LED lamps, which may either positively or negatively affect plant growth [12].

Kozai [13] observed that although there are numerous advantages of the use of LED lamps in agriculture, they are not applied in conventional farming in developing countries. On the contrary, developed countries such as Japan started using LED lamps in urban agriculture in 2010. This idea was implemented due to the limited number of fields for cultivation and insufficient sun exposure. LED lamps were applied to solve these problems. This trend also increases the use of LEDs in urban agriculture because they meet all requirements for high effectiveness (only specific spectra are used), safety, and healthy production. LED characteristics surpass the characteristics of other unused wavelengths and have low irradiance. They generate low amounts of heat, which is environmentally friendly.

Light is also important for plants to build biomass and activate the genes responsible for the activation of phytochromes. Several light colours can interact with cryptochromes and phototropins. Blue light affects stomatal conductance and thus increases photosynthesis. Potted chrysanthemums cultivated under 100% blue light had longer shoots than the plants grown under white light with shade avoidance response. Moreover, higher intensity of blue light can significantly affect rhizogenesis during adventitious root formation, which may be manifested by an increase in the dry root mass and the number of adventitious roots [9].

However, blue light may inversely affect root development in some species cultivated in vitro. For instance, Moon et al. [14] observed that birch pine and tsuru-rindo (*Triptospermum japonicum*) grown in blue light were scharacterised by low chlorophyll content, a small number of roots, and short internodal length and plant height.

Akbarian et al. [15] found that blue LEDs increased the emergence of seedlings and root length of *Zinnia* and *Impatiens* during the germination period. Furthermore, LEDs can make some ornamental plants (gerbera, marigold, petunia, and zinnia) sturdier than others, which were not illuminated with an LED lamp [16].

A closed plant production system with artificial light is an innovative plant cultivation method. The aim of the study was to assess the effect of different colours of light on the rooting of chrysanthemum (*Chrysanthemum* × *grandiflorum* Ramat./Kitam) cv. 'Nova Lime' cuttings and their subsequent growth in a room with no access to sunlight.

## 2. Materials and Methods

### 2.1. Plant Material

A total of 120 cuttings of medium-flowered pot chrysanthemum (*Chrysanthemum* × *grandiflorum* Ramat./Kitam.) cv. 'Nova Lime' were placed in a a controlled environment growth room in a three-layer shelf system) on 120 × 50 cm shelves lined with felt (Figure 1). The shelves were equipped with a LED Tube (LeuchTek, Ahrensburg, Germany). The lamps emitted light of different colours, i.e., white (cool white—5000 K), blue (460 nm), and as well as combinations of two colours: blue + white (50:50) (400–500 nm), and red + blue (75:25) (460–660 nm). The cuttings were rooted for 2 weeks. They were regularly sprayed with water in order to maintain high humidity. The experiment was completed when most of the cuttings formed roots. During the rooting, the day length was 16 h.

Next, the rooted cuttings were replanted into pots (1 $dm^3$) (5 cuttings per pot) with commercial peat substrate TS1 (pH 6.0). Then, the plants were placed in the same growth room and irradiated with the same light condition with the LED lamps. For 2 weeks, the rooted cuttings were grown in the growth room, where they were illuminated for 16 h a day (from 6 a.m. to 10 p.m.) and pinched off above the fifth leaf. After 7 days, the pinched off parts were sprayed with a retardant daminozide at concentration of 2550 $mg{\cdot}dm^{-3}$. Then, they were illuminated for 10 h (from 6 a.m. to 4 p.m.) for around 3 months to promote flower opening. They were watered every second day.

For different spectra of LED light, the photosynthetic photon flux density (PPFD) was 50 $\mu mol\ m^{-2}s^{-1}$ and was measured by means of Optel phytophotometer FR- 10 (Sonopan, Białystok, Poland). The lighting system was placed at 50 cm distances over the plant in order to adjust the photosynthetic photon flux density (PPFD) to the same value for all treatments.

Figure 1. Three-layer shelf system in growth room.

The air temperature in the growth room was maintained at 20 °C, and the relative humidity was 65–70%. The light spectrum characteristics measured with a spectroradiometer (USB 4000, Ocean Optics Inc., Dunedin, FL, USA) is shown in Figures 2–5.

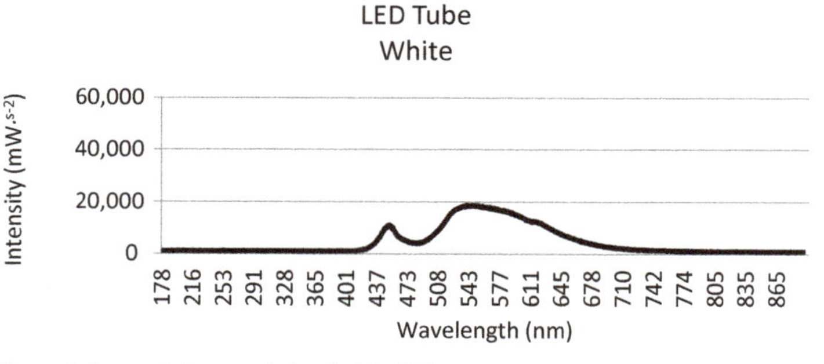


Figure 2. Spectral characteristic of white light.

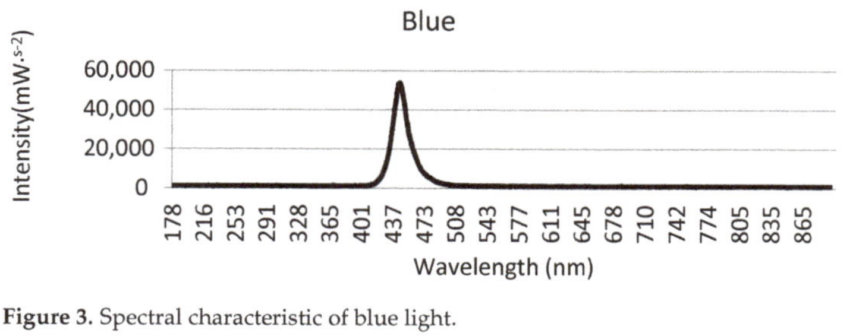


Figure 3. Spectral characteristic of blue light.

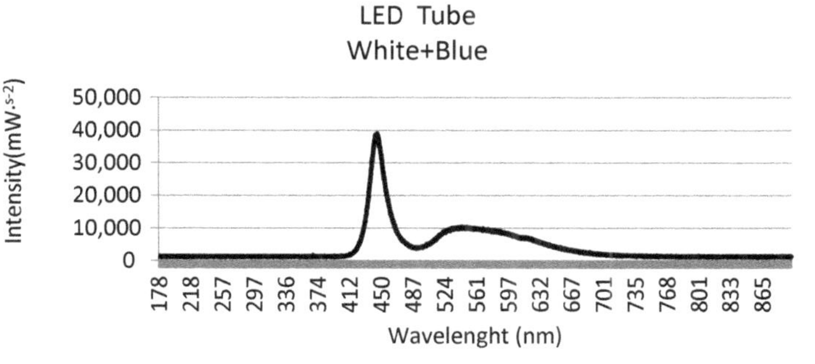


**Figure 4.** Spectral characteristic of white + blue light.

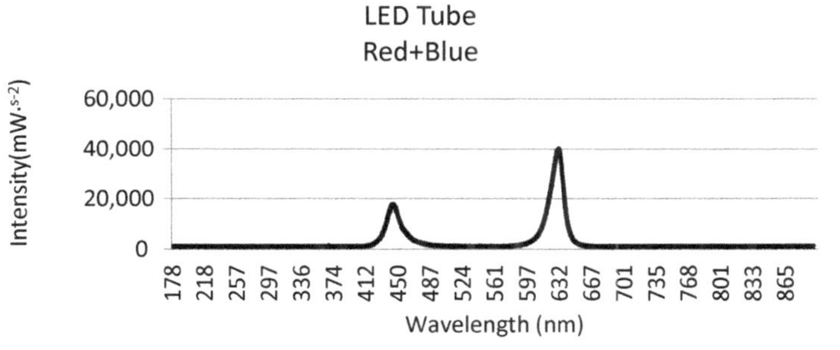


**Figure 5.** Spectral characteristic of red + blue light.

### 2.2. Experimental Design

The experiment was conducted in a completely srandomised design, where the exposure of the plants to different light colours was analysed, i.e., white light (W), blue light (B), red + blue light (RB), and white + blue light (WB). Rooted chrysanthemum cuttings were set in 4 treatments with 6 pots in each. There were 5 cuttings in each pot. The experiment was conducted at the Marcelin Experimental Station of the Faculty of Agricultural, Horticultural and Bioengineering Poznań University of Life Sciences, Poland, between July and November 2020.

The following morphological traits of the plants were measured: the number of flower buds and flower heads, the diameter of the flower heads, the height of the plants, the number of leaves, the fresh and dry weights of aboveground parts of plants. After the biometric measurements, the plants were dried in a drier 48 h at 60 °C.

The leaf greenness index (SPAD) was measured with an N-Tester apparatus (Yara International ASA, Norway). This measurement is used to determine the intensity of green colour in leaves and consists of the determination of the light absorption coefficient connected with the presence of chlorophyll at a wavelength of 650 nm and absorption by the leaf tissue at a wavelength of 940 nm.

### 2.3. Statistical Analysis

The results were analysed statistically with one-way analysis of variance and Duncan's multiple range test at a significance level of $\alpha = 0.05$.

## 3. Results

### 3.1. Cuttings Rooting

The experiment showed that the rooting of the cuttings was influenced by the light colour (Table 1). The parameters were measured before or at the beginning of the rooting process and 2 weeks later, after the cuttings rooted optimally. The exposure of the chrysanthemum cuttings to blue and white lights significantly increased their weight, i.e., from 0.7 and 0.8 g to 3.0 and 2.9 g, respectively. On the other hand, the cuttings exposed to white + blue and red + blue light combinations had the same average weight of 2.7 g.

**Table 1.** The effect of light colour to rooting cuttings of chrysanthemum.

| Colour of Light | Before Rooted Cuttings | | | After 2 Weeks Rooting Process | | | |
|---|---|---|---|---|---|---|---|
| | Weight of Cutting (g) | Length of Cutting (cm) | Index of Greening Leaves (SPAD) | Weight of Cutting (g) | Length of Cutting (cm) | Index of Greening Leaves (SPAD) | Length of Roots (cm) |
| White | 0.8 a * | 7.0 a | 35.2 a | 2.9 b | 9.1 b | 44.4 b | 8.9 a |
| Blue | 0.7 a | 7.5 a | 37.1 a | 3.0 b | 9.7 b | 40.4 a | 11.0 b |
| White + Blue | 0.7 a | 7.2 a | 36.2 a | 2.7 a | 9.8 b | 47.0 c | 11.2 b |
| Red + Blue | 0.8 a | 7.4 a | 36.7 a | 2.7 a | 8.7 a | 51.6 d | 13.1 c |

* Means followed by the same letters are not significantly different at $\alpha = 0.05$.

The length of the cuttings exposed to white, white + blue, and blue lights increased considerably and amounted to 9.1, 9.7, and 9.8 cm, respectively. Regardless of the length of individual cuttings, the exposure to light increased their length by about 2.1–2.6 cm. The smallest increase in the length of the chrysanthemum cuttings was observed under red + blue light—1.7 cm, approximately from 7.4 to 8.7 cm.

The statistical analysis showed that the colour of light also influenced the leaf greenness index (SPAD). Three of the light treatments increased the index after 2 weeks of rooting. The highest SPAD values (51.0) were noted for the cuttings grown under red + blue light. The other two treatments increased the index value as follows: white light—44.4 and white + blue light—47.0.

Regardless of the three main parameters, the chrysanthemum cuttings successfully formed numerous roots of different sizes and lengths to absorb nutrients and water. Therefore, the length of the roots was measured from the base of the main shoot to the main root apex. As shown in Table 1, the chrysanthemum plants developed the longest roots (average length—about 13.1 cm) under red + blue light. The exposure to white + blue and blue lights resulted in shorter roots, i.e., 11.2 and 11.0 cm, respectively. The chrysanthemum cuttings grown under white light had the shortest roots, i.e., 8.9 cm.

### 3.2. Subsequent Growth of Rooting Cutting

After the rooting of the cuttings, the chrysanthemum plants were transferred into pots (diameter 14 cm) in order to observe their subsequent growth under different colours of light from the LED lamps. Eight parameters were measured at the end of the generative period: the height of the plants, the number of flowers, the flower diameter, the number of flower buds, the leaf greenness index, the number of leaves, and fresh and dry weights. The results showed that the colour of light significantly affected the morphological traits of the chrysanthemums.

The plants exposed to white and red + blue lights grew shorter. By contrast, the tallest plants were produced under blue light. They were on average 17–41% taller than the other plants (Table 2).

**Table 2.** The effect of light colour on vegetative features of chrysanthemum.

| Colour of Light | Height of Plants (cm) | Index of Greening Leaves (SPAD) | Number of Leaves | Fresh Weight of Above-Ground Parts of Plants (g) | Dry Weight of Above-Ground Parts of Plants (g) |
|---|---|---|---|---|---|
| White | 9.2 a * | 57.4 b | 242.6 a | 52.2 b | 28.9 b |
| Blue | 15.4 c | 55.5 b | 265.1 a | 52.4 b | 32.3 c |
| White + Blue | 12.7 b | 65.9 c | 281.3 a | 65.2 c | 38.1 d |
| Red + Blue | 10.6 a | 52.8 a | 366.0 b | 46.9 a | 26.3 a |

* Means followed by the same letters are not significantly different at $\alpha = 0.05$.

Regardless of the average number of leaves in one pot per treatment, during the subsequent growth, the rooted cuttings produced the most leaves under red + blue light, i.e., about 366 leaves. The plants exposed to the other lights produced about 84–123 fewer leaves than those cultivated under red + blue light.

The colour of light significantly influenced the leaf greenness index. The leaves of the plants exposed to white + blue light were darker than the other leaves—their SPAD value was 65.9. Conversely, the plants grown under red + blue light had the lowest SPAD values (52.8). The SPAD values of the plants grown under white and blue lights were 57.4 and 55, respectively.

The chrysanthemum plants exposed to white + blue light had the highest fresh and dry weights, i.e., about 65.2 and 38.1 g on average. The plants exposed to white and blue lights did not differ significantly in the fresh weight. Their average fresh weights were 52.2 and 52.4 g, respectively. The plants exposed to red + blue light had the lowest fresh (46.9 g) and dry (26.3 g) weights.

Interestingly, only the chrysanthemum plants cultivated under white + blue light had almost two times more flowers—about five flowers/pot. The other treatments resulted in similar numbers of flowers—not more than three flowers/pot (Table 3).

**Table 3.** The effect of light colour on generative feature of chrysanthemum.

| Colour of Light | Number of Flower Heads | Flower Head Diameter (cm) | Number of Flower Buds |
|---|---|---|---|
| White | 2.6 a * | 4.4 a | 9.8 a |
| Blue | 2.8 a | 6.2 b | 15.8 c |
| White + Blue | 5.0 b | 6.9 b | 17.3 d |
| Red + Blue | 2.3 a | 4.2 a | 11.1 b |

* Means followed by the same letters are not significantly different at $\alpha = 0.05$.

The flower diameter also depended significantly on the light colour. White and red + blue lights noticeably inhibited the growth of flowers. The plants grown under blue and white + blue lights developed the largest flowers.

The statistical analysis showed that the colour of light also influenced the number of flower buds. The plants produced the most flowers and buds under white + blue light. On the other hand, the plants developed the fewest flowers and buds under white light—2.4 flowers and 7.5 buds on average.

## 4. Discussion

### *4.1. Rooted Cutting Activity*

Root formation, which is affected by several factors, is a crucial point for the survival of cuttings. The number of adventitious roots formed during rhizogenesis determines plants' abilities to absorb soil nutrients. Being short-day plants, chrysanthemum cuttings can form adventitious roots when stimulated by light and hormone activities. Light is a major factor responsible for the expression of genes resulting in phytohormone and phytochrome activity. Light promotes or inhibits the growth pattern of chrysanthemum cuttings depending on its wavelength, irradiance, and colour [17]. In our experiment, the colour of light affected the formation of roots and subsequent growth of chrysanthemums.

Two weeks of observation of root development in the cuttings showed that the colours of light differently affected the weight and length of the cuttings, their leaf greenness index (SPAD), and root length. Red + blue light (RB) accelerated the growth of rooted cuttings and induced the greenness of leaves but resulted in the low fresh weight of the cuttings. An earlier study on cherry rootstock microcuttings of the Colt cultivar showed that exposure to dichromatic blue and red light caused a remarkable increase in the root length. Root elongation was related to the phytochrome photoequilibrium value and photoreceptors. Blue and red light interacted strongly and regulated elongation activities more intensely than a single light colour. This effect was caused by the fact that red light stimulated root elongation more intensely than far-red and blue light and thus compensated for the blue light root elongation deficiency [18]. The exposure of chrysanthemum cuttings to dichromatic red + blue light with the same PPFD ratio (30:30 $\mu$mol $m^{-2}s^{-1}$) gave similar results. After 3 weeks, the cuttings exposed to red + blue light had the highest percentage of developed root [6].

On the contrary, the exposure of *Jatropha curcas* rooted cuttings to red + blue light generated by LED lamps (50:50 photon flux density) inhibited root formation, as opposed to blue, red, and white lights [19]. Kurilcik et al. [20] observed that blue light added to red and far-red light affected the rhizogenesis of chrysanthemum microcuttings. The blue light component was found to inhibit the rooting rate, but it increased the ratio of the fresh and dry weight of the explants due to the interaction between cryptochromes and phytochromes. Both of them affected the development of adventitious roots, but the influence of phytochromes was more pronounced. Phytochromes also regulate phytohormones such as auxin. However, this scheme requires further investigations, as plants respond differently to each photon flux density of the light colour. The growth of plants may be differently affected by the colour of light.

Matysiak [21] observed that the combination of light characteristics did not always result in optimal rooting percentage. The exposure of roses of the 'Konstancin' cultivar to dichromatic white + blue light resulted in a lower rooting percentage than the exposure to the monochromatic white light of high photosynthetic photon flux density (PPFD) 100 $\mu$mol $m^{-2}$ $s^{-1}$. The dichromatic light did not increase the fresh weight of the cuttings (roots and shoots) and the length of shoots significantly. The rooting of rose cuttings was induced by high irradiance and delayed by the blue light component during the exposure. These findings were consistent with the results of our experiment, in which the cuttings rooted under blue light were characterised by an intermediate increase in weight and length.

As Table 1 shows, the chrysanthemum cuttings in our experiment poorly responded to the exposure to monochromatic lights, i.e., white and blue lights, as expressed by the length of roots, cuttings, and the leaf greenness index (SPAD), except the weight of the cuttings. Baque and Hahn [22] observed a moderate effect of blue light on the length of roots of *Morinda citrifolia* leaf microcuttings grown in vitro. However, the root growth under red and blue lights were faster than under red + blue light. The fresh weight of the cuttings was also positively affected by monochromatic light. A lower amount of $H_2O_2$ applied under blue LEDs accelerated root formation triggered by rapid cell division.

This was caused by blue light slightly induced the activity of the genes expressing superoxide dismutase (SOD), and subsequently activated ascorbate peroxidase (APX) and catalase (CAT) mRNA to function at the same time in order to metabolically convert toxic $H_2O_2$ into $H_2O$. The lower amount of $H_2O_2$ and higher amount of $H_2O$ promoted cell elasticity. Further, the elasticity triggered a division of cells. As a result, the root cells divided rapidly because there was a low level of toxic free radicals.

The illumination with white LEDs had a less significant effect on the root length, but in comparison with the other light treatments, it slightly increased the weight, length, and greenness index of the cuttings. The research on the sensitivity of rooting behaviour of *Wiekstromia gemmate* microcuttings showed that specific light wavelengths and other qualities were reflected by various growth patterns. Consequently, the plants grown under white LED lamps had shorter roots than those grown under warm white fluorescent lamps and subsequently, they had epinastic and greener leaves. Further investigations showed that that other light spectral could not support auxin transport basipetally, so the endogenous auxin level may be high in the apical zone but low in the root elongation zone [23].

The rooted cuttings exposed to blue light had the lowest leaf greenness index (SPAD) value after 2 weeks. The second-lowest leaf greenness index (SPAD) value was noted after the white-light treatment. Interestingly, both of the light treatments positively affected the leaf greenness index. This result was consistent with the findings of the experiments conducted by Zheng and Van Labeke [24] and Schroeter-Zakrzewska et al. [25]. The researchers observed that the blue light component caused the shaded leaf effect, so physically, the leaves were darker than under the other treatments.

### 4.2. Subsequent Growth of Rooted Cuttings

The successful transferring of rooted chrysanthemum cuttings to their subsequent growth was determined by light characteristics such as the type of light and its colour. Nonetheless, some light colours had little or no effect on the vegetative and generative growth of potted chrysanthemums. Blue light combined with white and red lights considerably increased the height of chrysanthemums of the Covington cultivar. Monochromatic blue light had a moderate influence on the plant height [25]. Our experiment showed that the plants exposed to blue light had different characteristics. They were taller than the other plants. However, there were similar results of an earlier study on plants exposed to white + blue light. The plants grew optimally under exposure to blue light. The other treatments did not noticeably enhance the height of the plants.

The next vegetative parameter, i.e., the number of leaves, was related to the plant height. As light spectra enabled the manipulation of the chrysanthemum shoot architecture, they caused the shade avoidance syndrome. The cuttings grown only in the growth chamber with exposure to red and blue light were characterised by numerous bud outgrowths and the shortest height. Due to low apical dominance, these plants were short and compact. The application of red light either as a single light source or its combination with other light spectra [26]. Similar to earlier studies, our experiment also showed that the exposure to blue and red lights resulted in the shortest and most compact plants with numerous axillary vegetative buds. The buds then formed numerous thick and small leaves.

The quality of ornamental plants is also determined by leaf colour. It was shown in the experiments that the light colour significantly affected the SPAD index value. It increased after the exposure to white + blue light as well as white and blue lights, but it did not increase after the exposure to red + blue light. This finding was similar to the results of the study on lettuce conducted by Kleiber et al. [27]. The SPAD index value of the lettuce cultivated under white + blue light was lower than that of the lettuce grown under blue light.

The LED lamps emitting white + blue light resulted in more flower buds, flowers, and subsequently in greater flower diameters than the exposure to blue light, which resulted in the highest SPAD index value. The higher leaf greenness index (SPAD) value was positively correlated with the photosynthetic rate and thus affected the generative stage. Nissim-Levi et al. [28], Schroeter-Zakrzewska et al. [25], and Jerzy et al. [11] made similar observations for LED lamps emitting monochromatic blue light, which triggered rapid flower bud development. The induction of blue light itself depended strongly on the duration of exposure. Partch and Sancar [29] found that the blue light component activated cryptochromes so that they controlled flowering photoperiodically. The cryptochrome activity pathways were complex due to the interaction between auxin biosynthesis and light regulation [30].

On the other hand, Kaiser et al. [31] studied tomato plants and found that the combination of red + blue light induced the development of flower buds and increased the yield and biomass more than single blue light. This effect may have been caused by the plants' different responses to the light spectra, which depended not only on physiological activities but also on the cellular and gene levels. Further investigations on other crops are necessary for this field to find appropriate artificial or supplemental light treatments.

In order to analyse the correlation between light colour induction and each growth stage of chrysanthemums, the fresh and dry weight of the aboveground parts of plants were measured at the end of the experiment. The strongest correlation was observed between the high SPAD index value caused by the illumination with white + blue light and the fresh and dry weights of the aboveground parts of plants.The favourable effect of the white and blue light onto the fresh and dry weights in Scarlet sage was reported by Schroeter-Zakrzewska [32]. In the conducted experiment, the plants exposed to red + blue light were characterised by the lowest fresh and dry weights. They obtained different results reported by Heo et al. [33] in African marigold and Scarlet sage which were characterised by higher dry weight when the plants were cultivating under red- and blue- coloured lights.

## 5. Conclusions

The aim of the study was to assess the effect of different colours of light on the rooting of chrysanthemums (*Chrysanthemum* × *grandiflorum* Ramat./Kitam) cv. 'Nova Lime' cuttings and their subsequent growth in a growth room with no access to sunlight. During this experiment, morphological characteristics were measured. The cuttings rooted under white and blue lights were heavier than those exposed to white + blue and red + blue light combinations. The cuttings rooted under red + blue light were the shortest. The exposure to red + blue light resulted in the highest index of greening leaves (SPAD) value and the shortest cuttings with longest roots. White + blue light significantly influenced most of the growth parameters of chrysanthemum plants, except the height of the plants and the number of leaves. The blue light treatment optimally affected the plant height. Red + blue light emitted by the LED lamps had a significant influence on the number of leaves.

A closed system for plant production with artificial light is an innovative method of plant cultivation. The growing trend of closed cultivation system is still applied in vegetable crops widely and in ornamental crops; in turn, it had already been started a few years ago. Thus, more information about this area is needed for farmers of ornamental plants. The conducted experiment proved that good quality chrysanthemum could be obtained at a low light intensity, which can significantly reduce energy costs. Lower electrical energy costs can increase profits.

**Author Contributions:** Conceptualisation, A.S.-Z.; methodology, A.S.-Z.; software, A.S.-Z. and F.A.P.; validation, A.S.-Z. and F.A.P.; formal analysis, A.S.-Z. and F.A.P.; investigation, A.S.-Z. and F.A.P.; resources, F.A.P. and A.S.-Z.; data curation, A.S.-Z. and F.A.P.; writing—original draft preparation, F.A.P. and A.S.-Z.; writing—review and editing, A.S.-Z. and F.A.P.; visualisation, A.S.-Z. and F.A.P.; supervision, A.S.-Z.; project administration, A.S.-Z. and F.A.P.; funding acquisition, A.S.-Z. and F.A.P. All authors have read and agreed to the published version of the manuscript.

**Funding:** The APC was co-funded within the framework of the Ministry of Science and Higher Education program 'Regional Initiative Excellence' in years 2019–2022, (No.005/RID/2018/19), financing amount 1,200,000 PLN.

**Institutional Review Board Statement:** Not applicable.

**Informed Consent Statement:** Not applicable.

**Data Availability Statement:** Not applicable.

**Acknowledgments:** Authors thank the horticultural farm A.A. Kłak in Poznań, for donating cuttings.

**Conflicts of Interest:** The authors declare no conflict of interest. The funders had no role in the design of the study; in the collection, analyses, or interpretation of data; in the writing of the manuscript, or in the decision to publish the results.

## References


1. Sung, J.W.; Sander, W.H.; Wim, V.L. Responses of supplemental blue light on flowering and stem extension growth of cut chrysanthemum. *Sci. Hortic.* **2014**, *165*, 69–74.
2. Ploeg, A.V.D.; Heuvelink, E. The influence of temperature on growth and development of chrysanthemum cultivars: A review. *J. Hortic. Sci. Biotechnol.* **2006**, *81*, 174–182. [CrossRef]
3. De Vier, C.L.; Geneve, R.L. Flowering influences adventitious root formation in chrysanthemum cuttings. *Sci. Hortic.* **1997**, *70*, 309–318. [CrossRef]
4. Blythe, E.K.; Sibley, J.L.; Tilt, K.M.; Reuter, J.M. Methods of auxin application in cutting propagation: A review of 70 years of scientific discovery and commercial practice. *J. Environ. Hortic.* **2007**, *25*, 166–185. [CrossRef]
5. Thakur, T.; Grewal, H.S. Influence of photoperiodic night interruption on sustainable potted flower production of chrysanthemum cv. Snowball. *Int. J. Curr. Microbiol. Appl. Sci.* **2018**, *72*, 1282–1287. [CrossRef]
6. Taiz, L.; Zeiger, E. *Plant Physiology*, 3rd ed.; Sinauer Associates: Sunderland, UK, 2002; pp. 111–143.
7. McCree, K. Test of current definitions of photosynthetically active radiation against leaf photosynthesis data. *Agric. Meteorol.* **1972**, *10*, 443–453. [CrossRef]
8. Zhen, S.; Bugbee, B. Far-red photons have equivalent efficiency to traditional photosynthetic photons: Implications for re-defining photosynthetically active radiation. *Plant Cell Environ.* **2020**, *43*, 1259–1272. [CrossRef]
9. Christiaens, A.; Gobin, B.; Huylenbroeck, J.V.; Labeke, M.V. Adventitious rooting of Chrysanthemum is stimulated by a low red: Far-red ration. *J. Plant Physiol.* **2019**, *236*, 117–123. [CrossRef] [PubMed]
10. Zalewska, M.; Wożny, A.; Piszczek, P. Effect of blue light on the flowering of Dendranthema grandiflora Tzvelev Sambrero group. *Zesz. Probl. Postępu Nauk. Rolniczych.* **2002**, *483*, 305–310. (In Polish)
11. Jerzy, M.; Zakrzewski, P.; Schroeter-Zakrzewska, A. Effect of colour light on the opening of inflorescence buds and post-harvest longevity of pot chrysanthemums (*Chrysanthemum x grandiflorum* (Ramat.) Kitam). *Acta Agrobot.* **2011**, *64*, 13–18. [CrossRef]
12. Ouzounis, T.; Rosenqvist, T.E.; Ottosen, C.O. Spectral effects of artificial light on plant physiology and secondary metabolism: A review. *HortScience* **2015**, *50*, 1128–1135. [CrossRef]
13. Kozai, T. Why LED Lighting for Urban Agriculture? In *LED Lighting for Urban Agriculture*; Kozai, T., Fujiwara, K., Runkle, E.S., Eds.; Springer Science + Business Media Singapore: Singapore, 2016; pp. 3–18.
14. Moon, H.K.; Park, S.Y.; Kim, Y.W.; Kim, C.S. Growth of tsuru-rindo (*Triptospermum japonicum*) cultured in-vitro under various sources of light-emitting diode (LED) irradiation. *J. Plant Biol.* **2006**, *49*, 1–6. [CrossRef]
15. Akbarian, B.; Maltoobi, M.; Mahna, N. Effects of LED light on seed emergence and seedling quality of four bedding flowers. *J. Ornam. Plants* **2016**, *6*, 115–123.
16. Craver, J.K.; Boldt, J.K.; Lopez, R.G. Comparison of supplemental lighting provided by high-pressure sodium lamps or light-emitting diodes for the propagation and finishing of bedding plants in a commercial greenhouse. *HortScience* **2019**, *54*, 52–59. [CrossRef]
17. Gill, C.S.; Jung, H.Y.; Lee, C.; Eom, S.H. Blue light and NAA treatment significantly improve rooting on single leaf-bud cutting of chrysanthemum via upregulated rooting-related genes. *Sci. Hortic.* **2020**, *274*, 1–9. [CrossRef]
18. Lacona, M.; Muleo, R. Light quality affects in vitro adventitious rooting and ex vitro performance of cherry rootstock colt. *Sci. Hortic.* **2010**, *125*, 630–636. [CrossRef]
19. Daud, N.; Faizal, A.; Geelen, D. Adventitious rooting of *Jatropha curcas* L. is stimulated by phloroglucinol and by red LED light. *In Vitro Cell. Dev. Biol. Plant* **2013**, *49*, 183–190. [CrossRef]
20. Kurilcik, A.; Miklusyte-Canova, R.; Dapkuniene, S.; Zilinskaite, S.; Kurilcik, G.; Tamulaitis, G.; Duchovskis, P.; Zukauskas, A. In vitro culture of chrysanthemum plantlets using light-emitting diodes. *Cent. Eur. J. Biol.* **2008**, *3*, 161–167. [CrossRef]
21. Matysiak, B. Vertical production of Konstancin rose cuttings in the growth chamber under LED light. *J. Hortic. Res.* **2020**, *28*, 29–34. [CrossRef]
22. Baque, M.A.; Hahn, E.J. Induction mechanism of adventitious root from leaf explants of *Morinda citrifolia* as affected by auxin and light quality. *In Vitro Cell. Dev. Biol. Plant* **2010**, *46*, 71–80. [CrossRef]
23. Verstraeten, I.; Buyle, H.; Werbrouck, S.; Van Labeke, M.C.; Geelen, D. In Vitro shoot growth and adventitious rooting of *Wiekstromia gemmata* depends on light quality. *Isr. J. Plant Sci.* **2020**, *67*, 16–26. [CrossRef]
24. Zheng, L.; Van Labeke, M.C. Effects of different irradiation levels of light qualities on chrysanthemum. *Sci. Hortic.* **2018**, *233*, 124–131. [CrossRef]
25. Schroeter-Zakrzewska, A.; Kleiber, T.; Zakrzewski, P. The response of chrysanthemum (*Chrysanthemum* x *grandiflorum* Ramat./Kitam) cv Covington to a different range of fluorescent and led light. *J. Elem.* **2017**, *22*, 1015–1026. [CrossRef]
26. Dierck, R.; Dhooghe, E.; Huylenbroeck, J.V.; Straeten, D.V.D.; Keyser, E.D. Light quality regulates plant architecture in different genotypes of *Chrysanthemum morifolium* Ramat. *Sci. Hortic.* **2017**, *218*, 177–186. [CrossRef]
27. Kleiber, T.; Borowiak, K.; Schroeter-Zakrzewska, A.; Budka, A.; Osiecki, S. Effect of ozone treatment and light colour on photosynthesis and yield of lettuce. *Sci. Hortic.* **2017**, *217*, 130–136. [CrossRef]
28. Nissim-Levi, A.; Kitron, M.; Nishri, Y.; Ovaida, R.; Forer, I.; Shamir, M.O. Effects of blue and red LED lights on growth and flowering of *Chrysanthemum morifolium*. *Sci. Hortic.* **2019**, *254*, 77–83. [CrossRef]
29. Partch, C.L.; Sancar, A. Invited review-photochemistry and photobiology of chryptochrome blue light photopigments: The search for a photocycle. *Photochem. Photobiol.* **2005**, *81*, 1291–1304. [CrossRef]

30. Halliday, K.J.; Garcia, J.F.M.; Josse, E.M. Integration of light and auxin signaling. *Cold Spring Harb. Perspect. Biol.* **2009**, *1*, 1–11. [CrossRef]
31. Kaiser, E.; Ouzounis, T.; Giday, H.; Schipper, R.; Heuvelink, E.; Marcelis, L.F.M. Adding blue to red supplemental light increases biomass and yield of greenhouse-grown tomatoes, but only to an optimum. *Front. Plant Sci.* **2019**, *9*, 1–11. [CrossRef]
32. Schroeter-Zakrzewska, A. Influence of the light colour on the seedling quality of French marigold and Scarlet sage. *Bulg. J. Agric. Sci.* **2015**, *21*, 951–956.
33. Heo, J.; Lee, C.; Chakrabarty, D.; Paek, K. Growth responses of marigold and salvia bedding plants as affected by monochromic or mixture radiation provided by a Light Emitting Diode (LED). *Plant Growth Regul.* **2002**, *38*, 225–230. [CrossRef]

*agriculture*

MDPI

*Article*

# Comparative Assessment of Hydroponic Lettuce Production Either under Artificial Lighting, or in a Mediterranean Greenhouse during Wintertime

**Orfeas Voutsinos [1], Maria Mastoraki [1], Georgia Ntatsi [1], Georgios Liakopoulos [2] and Dimitrios Savvas [1,*]**

[1] Laboratory of Vegetable Crops, Department of Crop Science, Agricultural University of Athens, Iera Odos 75, 11855 Athens, Greece; vorfeas@aua.gr (O.V.); maria.mastoraki@wur.nl (M.M.); ntatsi@aua.gr (G.N.)
[2] Laboratory of Plant Physiology, Department of Crop Science, Agricultural University of Athens, Iera Odos 75, 11855 Athens, Greece; gliak@aua.gr
* Correspondence: dsavvas@aua.gr; Tel.: +30-210-529-4510

**Abstract:** Butterhead lettuce was grown hydroponically in a vertical farm under high (HLI) and low (LLI) light intensity (310, and 188 μmol $m^{-2}$ $s^{-1}$, respectively) and compared to hydroponically grown lettuce in a greenhouse (GT) during wintertime in Athens, Greece (144 μmol $m^{-2}$ $s^{-1}$). The highest plant biomass was recorded in the HLI treatment, whereas LLI and GT produced similar plant biomass. However, the LLI produced vortex-like plants, which were non-marketable, while the plants in the GT were normal-shaped and saleable. Net photosynthesis was highest in the HLI and higher in the LLI than in the GT, thereby indicating that light intensity was the dominant factor affecting photosynthetic performance. Nevertheless, the unsatisfactory performance of the LLI is ascribed, not only to reduced light intensity, but also to reduced light uniformity as the LED lamps were closer to the plants than in the HLI. Furthermore, the large solar irradiance variability in the GT resulted in substantially higher adaptation to the increased light intensity compared to LLI, as indicated by chlorophyll fluorescence measurements. Light intensity and photoperiod are believed to be the primary reasons for increased nitrate content in the GT than in the vertical farming treatments.

**Keywords:** artificial lighting; chlorophyll fluorescence; gas exchange; indoor farming; soilless culture

**Citation:** Voutsinos, O.; Mastoraki, M.; Ntatsi, G.; Liakopoulos, G.; Savvas, D. Comparative Assessment of Hydroponic Lettuce Production Either under Artificial Lighting, or in a Mediterranean Greenhouse during Wintertime. *Agriculture* **2021**, *11*, 503. https://doi.org/10.3390/agriculture11060503

Academic Editors: Athanasios Koukounaras and Filippos Bantis

Received: 18 May 2021
Accepted: 23 May 2021
Published: 29 May 2021



## 1. Introduction

Vertical farming is the procedure of growing vegetables in soilless culture, indoors with artificial lighting. It has been an existing concept for about 50 years, with the first commercial Plant Factory with Artificial Lighting (PFAL) to be established in Miura Nouen, in Shizuoka Prefecture, Japan in 1983 [1]. Due to the recent breakthroughs in the LED industry, vertical farming has become a feasible and scalable farming method. Meanwhile, the interest in developing efficient vertical farming systems is growing due to several factors, including the rise of the human population, extreme weather phenomena due to climate change, and ultimately the enormous pressure by consumers for high-quality fresh products. Freshness is especially important in cases of highly perishable goods, such as leafy vegetables. The ever-increasing demand for nutritious, fresh, safe for consumption and environmentally friendly food has been the driving force for the upscaling of the vertical farming industry, and the establishment of many vertical farming companies and startups.

Climatic conditions, especially temperature and light intensity, have a strong impact on growth, yield and nutritional quality of vegetables [2]. Light is not only the energy source for plants, but also an environmental signal modulating plant morphogenesis. Therefore, light can induce various physiological responses and affect growth and development, through its variations in intensity, photoperiod and spectrum [3–6]. Morphological and physiological changes occur when plants adapt to different light environments. Low photosynthetic photon flux density (PPFD) tend to induce shade-avoidance-like responses to plants, whereas high PPFD can enhance carbohydrate accumulation and net photosynthetic rate [7–9]. Moreover, the growth and morphology of plants are negatively influenced by reduced daily light integrals (DLI) [10], and thus, light intensity and photoperiod are limiting factors for glasshouse production during winter and early autumn. Extending the photoperiod can lead to increased fresh weight of lettuce [11].

Vertical farming can successfully address the problems arising from limited light integrals during wintertime as it provides unlimited opportunities to control the light intensity and duration. The environmental and nutritional control provide additional tools to manipulate crop growth and development [12], allowing stable produce, irrespective of the season or the outside environmental conditions. Lettuce (*Lactuca sativa* L.) is a model crop for studying the effect of lighting in vertical farms due to its fast growth and short production cycle [13,14]. Moreover, lettuce is one of the most important leafy vegetables worldwide, as it is considered a rich source of vitamins (A, C, E, K), polyphenols, and antioxidant compounds [15]. Due to its short size and production cycle, lettuce is a model plant for vertical farming studies and attracts a high interest for commercial production in vertical farming systems [16]. Nevertheless, lettuce is often accused of accumulating nitrate at levels which can be harmful for humans if consumed at excessive levels [17]. Nitrate accumulation in plants is affected by the environmental conditions and is also influenced by genetic factors. Of the environmental factors, light intensity seems to exert the strongest influence on nitrate accumulation in plant tissues [18]. Previous studies have shown that increasing the light intensity up to a certain level enhances the growth and quality of lettuce, and it has been suggested that the most efficient light intensity for lettuce in a plant factory is between 200–400 μmol $m^{-2}$ $s^{-1}$ [19–24].

The measurement of chlorophyll fluorescence is a proven, quantitative, non-invasive, powerful tool of assessing the properties of the photosynthetic apparatus [25], especially when it is used in combination with other non-invasive measurements such as gas exchange analysis [26]. Therefore, chlorophyll fluorescence coupled with gas exchange measurements have been used to assess the impact of artificial lighting at different intensities on the performance of the photosynthetic apparatus in plants grown in vertical farms. However, comparisons in growth, yield and photosynthetic performance between plants grown either in vertical farming systems or in conventional hydroponic Mediterranean greenhouses are lacking.

Despite the above-mentioned advantages of vertical over conventional farming, this cropping system raises skepticism, especially in countries with abundant sunlight, such as Greece, Spain, or Italy. However, the high light intensity might be utilized to provide an additional advantage if part of the energy needs for artificial lighting are covered by collecting solar energy through photovoltaic panels. Considering the above background, in the current study we compare a vertical farming system partly powered by solar energy with a conventional, hydroponic system when both are used for winter lettuce production under Mediterranean climatic conditions. In the current paper, leaf nitrate concentrations leaf anatomical characteristics and leaf photosynthetic parameters of plants grown in the vertical farming system under two different light intensities, a high light intensity (HLI) and a low light intensity (LLI), as well as in plants cultivated in a standard glasshouse (GT) during wintertime are reported. Details concerning the electricity consumption of the vertical farming system, used for this study, and the possibility to reduce electricity costs by using a hybrid-solar lighting system with photovoltaic panels, were reported in a previous paper [27].

## 2. Materials and Methods

### 2.1. Plant Material and Experimental Setup

Lettuce seedlings (*Lactuca sativa* L. cv. Glory) were provided from a commercial nursery (Plantas S.A, Chalandri, Greece) at the three-leaf stage. Prior to transplanting the growing medium was removed from the root surface area. The seedlings were transplanted into plastic cups using a slit sponge capable of holding the plants at the height of the hypocotyl. The plastic cups were then placed inside Nutrient Film Technique (NFT) gullies of a vertical farming system, henceforth termed Photon Rack (PR), as well as in similar gullies placed in a heated glasshouse at the Agricultural University of Athens (37°58′57.8″ N, 23°42′14.3″ E). The Photon Rack (PR) system was constructed by K. Dekoulis Lab, Kallithea, Athens, Greece.

The PR was a 1.9 m high, 3-layer rack with four 2-m long NFT gullies per layer. Each gully accommodated 12 plants. Each shelve was 200 cm long and 78 cm wide. Hence the density was 30 plants per $m^2$. The LED tubes were clipped above the hydroponic gullies on an aluminum fixture which was designed to ascend and/or descend depending on the wanted light intensity. The nutrient solution was delivered from a 90-L tank to the gullies of the top layer by a Hailea pump (HX8850, 4900 L $h^{-1}$, 100 W, Guangdong Hailea Group Co., Ltd., Chaozhou, Guangdong, China). The pump operated daily on a 24-h basis. Airflow was accomplished by using fans on the two ends of the rack. The PR was placed indoors in an acclimated room. Pictures of the PR can be found in Appendix A.

The vertical farming production of lettuce took place in the PR from 20 November 2017 to 19 December 2017 by applying two different treatments. The first treatment was a high light intensity treatment (HLI), consisting of either 16 hybrid-solar or conventional LED, which provided an average of 310 μmol $m^{-2}$ $s^{-1}$ irradiance. The LED tubes of the HLI treatment maintained their distance from the gully level throughout the cultivation period as seen in Figure 1. Details regarding the differences in energy consumption of the hybrid-solar and the conventional LED tubes have been provided in a previous paper [27]. The second treatment was a low light intensity treatment (LLI) with 8 LED tubes providing 188 μmol $m^{-2}$ $s^{-1}$ irradiance. The LED tubes were placed 20 cm above the NFT gully level and ascended manually during the cultivation period. At the end of the cultivation cycle the LED tubes had the same distance from the gully level, 40 cm, as the High Light Intensity (HLI) treatment (Figure 1). The light spectrum chosen was white broad spectrum (Figure 2) as various studies have shown that it is more beneficial for plant growth compared to narrow spectrum LEDs, like red, blue or their combinations [13,28–34]. In both treatments of the vertical farming system, a 12-h photoperiod, an average temperature (T) of 22 ± 1.5 °C, an average relative humidity (RH) of 90 ± 10% and an average $CO_2$ concentration of 400 ppm were maintained. To evaluate the outcome of the vertical farming method, a Glasshouse Treatment (GT) was also carried out during the 22 December 2017–20 January 2018 period as comparison. The Glasshouse treatment (GT) was completely depended on sunlight. Given that the conditions inside the climate chamber were fully controlled, and thus, independent of the outside environment, the outcome of the experiment would be the same regardless of the time it was carried out. The climatic conditions of the Greenhouse treatment were as follows; average light intensity was 144 μmol $m^{-2}$ $s^{-1}$, the photoperiod was 10 h $d^{-1}$ while the other climatic conditions; T, RH, $CO_2$ where 20 ± 1.5 °C, 64 ± 10% and 400 ppm respectively. The afore mentioned climatic conditions are summarized in Table 1. The chemical composition of the nutrient solution (NS) supplied to replenish plant uptake (replenishment NS) was as follows: K: 8 mmol $L^{-1}$, Ca: 4.8 mmol $L^{-1}$, Mg: 1.3 mmol $L^{-1}$, $NO_3^-$: 16.4 mmol $L^{-1}$, $NH_4^+$: 1.3 mmol $L^{-1}$, $H_2PO_4^-$: 1.8 mmol $L^{-1}$, Fe: 20 μmol $L^{-1}$, Mn: 6 μmol $L^{-1}$, Zn: 5 μmol $L^{-1}$, Cu: 0.75 μmol $L^{-1}$, B: 30 μmol $L^{-1}$, Mo: 0.5 μmol $L^{-1}$. The electrical conductivity and the pH of the recirculating nutrient solution were monitored every day and maintained to 2.4 dS $m^{-1}$, and 5,5–6,5, respectively, by adding appropriate amounts of replenishment NS, and nitric acid, respectively. In addition, the recirculating NS was renewed every week.

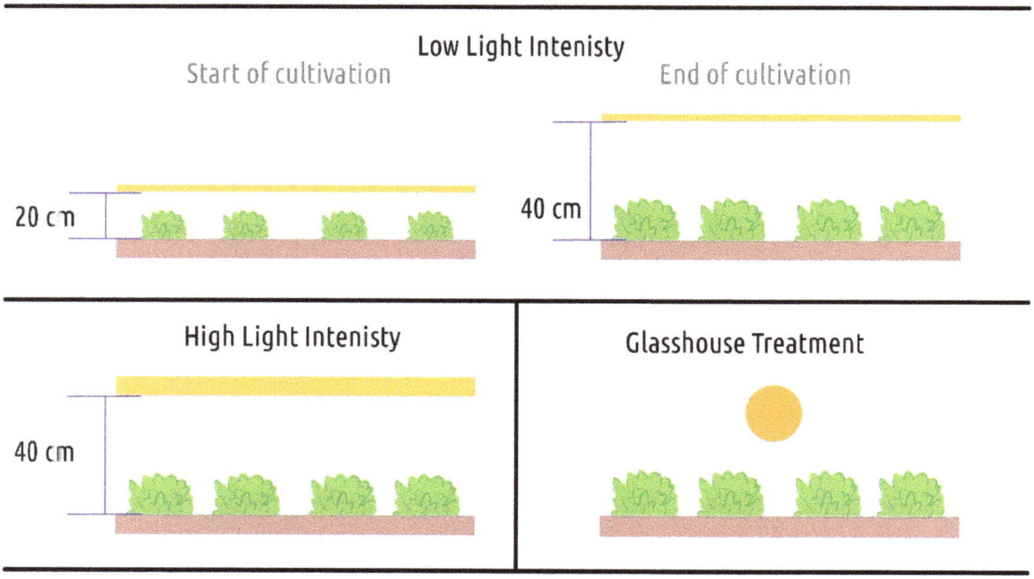


**Figure 1.** Schematic representation of the treatments. Low Light Intensity (LLI), High Light Intensity (HLI), Glasshouse Treatment (GT).

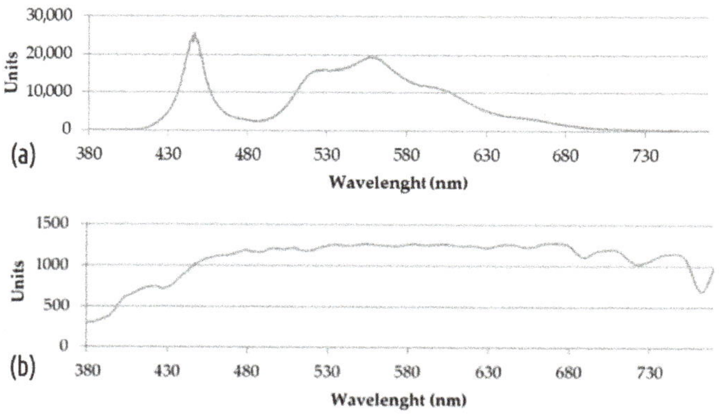


**Figure 2.** (**a**) Spectrum of LED lights used for this study, (**b**) Sun's spectrum, during noon.

Light intensity was measured using a photometer Li-Cor (LI-188B Integrating quantum/Radiometer/Photometer, LI-COR INC, Lincoln, NE, USA) at plant height. The spectrum was measured using a spectroradiometer (USB2000+, Ocean Optics Inc., Dunedin, FL, USA) at a close distance from the LED chips. The environmental parameters were measured using the Sigrow Pro sensor (Sigrow B.V., Wageningen Campus, Wageningen, The Netherlands). Growth characteristics, shoot fresh weight, leaf number and leaf area were measured during harvest.

**Table 1.** Overview of the treatments' characteristics and climate conditions studied.

| Treatment | Environmental Conditions | Average Light Intensity | Location | Characteristics |
|---|---|---|---|---|
| HLI | T: 22 ± 1.5 °C, RH: 90 ± 10%, $[CO_2]$: 400 ppm, Photoperiod: 12 h | 310 μmol $m^{-2}$ $s^{-1}$ | PR upper layer | 16 Conventional LED tubes |
| LLI | T: 22 ± 1.5 °C, RH: 90 ± 10%, $[CO_2]$: 400 ppm, Photoperiod: 12 h | 188 μmol $m^{-2}$ $s^{-1}$ | PR lower layer | 8 Conventional LED tubes |
| GT | T: 20 ± 1.5 °C, RH: 64 ± 10%, $[CO_2]$: 400 ppm, Photoperiod: 10 h | 144 μmol $m^{-2}$ $s^{-1}$ | Glasshouse | Solar light |

HLI: High light intensity; LLI: Low light intensity; GT: Greenhouse treatment; T: Temperature; RH: Relative humidity; PR: Photon Rack.

### 2.2. Chlorophyll Fluorescence and Leaf Gas Exchange

The in vivo chlorophyll fluorescence parameters (operating efficiency of PSII photochemistry, $\Phi_{PSII}$; electron transport rate, ETR; photochemical quenching of PSII, qP; and non-photochemical quenching, qN) were measured once, as described by Liakopoulos et al. [35], at the end of the cultivation in fully developed lettuce leaves, during the light period (specifically between 8:00 a.m. and 12:30 p.m.), using a portable chlorophyll fluorometer (PAM-2100, Heinz Walz GmbH, Effeltrich, Germany). Each leaf was acclimated for 20 min before the measurements were taken, using dark leaf clips. The light response curve was measured using 7 light intensities in the range between 0 to 938 μmol $m^{-2}$ $s^{-1}$. The starting light intensity was 40 μmol $m^{-2}$ $s^{-1}$, followed by 74, 120, 192, 302, 412, 631 and 938 μmol $m^{-2}$ $s^{-1}$. Measurements of photosynthetic light curves and photosynthetic characteristics were carried out using white light from the PAMs' halogen light source. Fluorescence measurements were taken on the same morning with gas exchange measurements. Measurements of light-saturated net $CO_2$ assimilation rate and stomatal conductance were conducted on mature leaves exposed to each light treatment (HLI, LLI, GT), using a portable open-circuit gas-exchange instrument (LI-6400, Li-COR Inc., Lincoln, NE, USA), equipped with a broad leaf chamber enclosing 6 $cm^2$ of leaf area. Temperature and relative air humidity inside the chamber were 30 ± 3 °C, and 30 ± 2%, respectively. Gas exchange parameters (net rate of $CO_2$ assimilation, A, transpiration rate, E; intercellular $CO_2$, ci; and stomatal conductance to $H_2O$, gs) were measured at ambient $CO_2$ atmospheric concentration under 7 different photosynthetic photon flux densities, supplied by the LED light of the instrument's chamber, ranging from 0 μmol $m^{-2}$ $s^{-1}$ to 1840 μmol $m^{-2}$ $s^{-1}$ after acclimation for 180 s. The starting light intensity was 0 μmol $m^{-2}$ $s^{-1}$, followed by 46, 92, 184, 460, 920, and 1840 μmol $m^{-2}$ $s^{-1}$. Six replicates for each treatment were measured (three readings at steady-state conditions were recorded per replicate and per light level). Water Use Efficiency (WUE) was calculated as "instantaneous WUE" between A and E (A/E) as described by Medrano et al. [36].

### 2.3. Fresh Weight, Leaf Number, Leaf Area, and Leaf Nitrate Concentration

Shoot fresh weight (sFW, g) and root fresh weight (rFW, g) were measured using the Mettler PE-3600 (Mettler Toledo LLC, Columbus, Ohio, USA) scale, after wiping the plant parts with paper to remove the surface water. The number of leaves were determined by destructive sampling during the harvest stage. The leaf nitrate concentration was determined by measuring nitrite after reduction of nitrate to nitrite by copperised cadmium (Cu-CD) columns and subsequent colorimetric determination of nitrite by a Griess diazo-coupling reaction as described by Novozamsky et al., [37]. The percentage of the daily nitrate intake in each treatment per average lettuce head was calculated by multiplying the average fresh weight of each lettuce head (kg) with the mean nitrate concentration (mg $kg^{-1}$) and then dividing by the human threshold of acceptable daily intake (ADI), which is 220 mg $d^{-1}$ [38]. The percentage of the daily intake of nitrate of each treatment per 100 g of fresh produce was estimated by dividing the mean nitrate concentration for 100 g of lettuce by the human threshold of ADI.

### 2.4. Statistical Analysis

The experiment was set up as a completely randomized design, with three treatments. Each treatment consisted of 4 NFT gullies. Due to lack of space, each plant constituted one replication. For the statistical analysis, 10 replication samples per treatment were collected randomly to minimize the position effect. The data were statistically evaluated by applying one-way ANOVA using the STATISTICA software package, version 9.0 (TIBCO Software Inc, Palo Alto, CA, USA) for Windows (Microsoft, Redmon, WA, USA). When ANOVA was significant for one measured parameter, the treatment means were separated using the Duncan's Multiple Range Test ($p \leq 0.05$). Data were presented in figures and tables as means $\pm$ SE of ten replicates.

## 3. Results

### 3.1. Quality and Biomass of Lettuce Plants Grown under Different Lighting Designs

The fresh biomass of the epigeous plant part in the HLI treatment was almost double as high as in the other two treatments, while the difference was not significant between the LLI and GT (Table 2). The fresh to dry weight ratio in the epigeous biomass did not differ significantly between the treatments (data not shown). The root biomass was higher in the HLI followed by the LLI treatment while GT had by far the lowest root biomass. The leaf number was significantly higher in plants of the HLI treatment compared to the other two treatments, and significantly higher in the LLI than in the GT treatment. The leaf area was significantly higher in the HLI treatment compared to LLI and GT, while the latter two treatments did not differ significantly from each other. The appearance of the lettuce plants is sown in Figure 3. The lowest nitrate concentrations were measured in the HLI and LLI treatments, without any significant differences between them, while the leaf nitrate concentration measured in the GT treatment was significantly higher than those measured in the vertical farming system, irrespective of light intensity (Table 2).

**Figure 3.** Average appearance of lettuce plants from each growing treatment at the harvest stage. Judged by visual inspection, the low light intensity in the Photon Rack (LLI) produced "vortex-like" morphology. The plants in the HLI and GT treatments were morphologically "normal", while in the HLI they had a greener appearance. HLI, High light intensity; LLI, Low light intensity; GT, Glasshouse treatment.

**Table 2.** Effect of different light intensities and hydroponic cropping system on growth and development of lettuce 'Glory' grown in a vertical farming system and in a glasshouse during the winter. HLI, High light intensity; LLI, Low light intensity; GT, Glasshouse treatment.

| Treatment | Fresh Shoot Weight (g plant$^{-1}$) | Dry Shoot Weight (g plant$^{-1}$) | Fresh Root Weight (g plant$^{-1}$) | Leaf Number (plant$^{-1}$) | Leaf Area (cm$^2$ plant$^{-1}$) | Nitrate Concentration per Kg of Fresh Weight (mg kg$^{-1}$) |
|---|---|---|---|---|---|---|
| HLI | 123.3 a | 8.77 a | 17.8 a | 21.0 a | 2005.7 a | 1250 b |
| LLI | 64.9 b | 4.82 b | 9.4 b | 17.8 b | 1457.6 b | 1748 b |
| GT | 58.1 b | 4.12 b | 5.4 c | 15.2 c | 1358.9 b | 3578 a |
| Statistical significance | *** | *** | *** | *** | *** | *** |

Mean ($n = 10$) followed by different letters indicate significant differences (for each comparison criteria stated) according to the Duncan's multiple range test ($p < 0.05$), *** significant at $p < 0.001$.

### 3.2. Physiological Characteristics: Chlorophyll Fluorescence and Gas Exchange

The chlorophyll fluorescence analysis showed that, based on the $\Phi_{PSII}$ and ETR parameters, plants of the HLI treatment were more capable of utilizing the light for photosynthesis than those of the other treatments, whereas LLI and GT plants did not differ significantly (Figure 4).

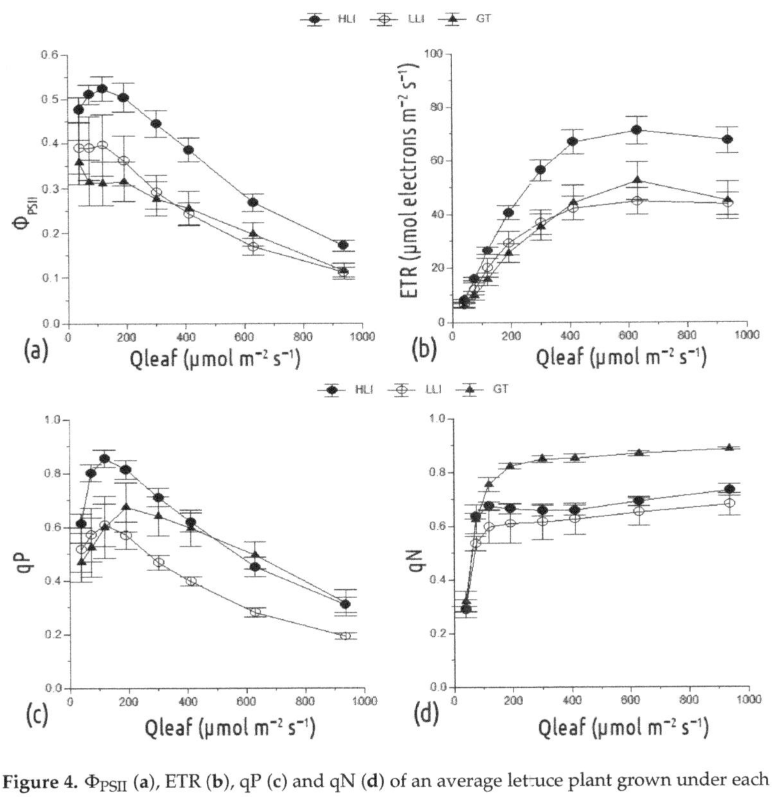


**Figure 4.** $\Phi_{PSII}$ (**a**), ETR (**b**), qP (**c**) and qN (**d**) of an average lettuce plant grown under each different treatment. HLI, High light intensity; LLI, Low light intensity; GT, Glasshouse treatment. Vertical bars indicate ± standard errors of means.

Looking further into the qP values, it was shown that, for PPFD lower than 300 μmol $m^{-2}$ $s^{-1}$, the HLI treatment was significantly higher than the other two treatments (Figure 4c), while at PPFD higher than 300 μmol $m^{-2}$ $s^{-1}$, the qP values of the HLI treatment were significantly higher only in comparison with those recorded in the LLI. Furthermore, the qP in the GT treatment, representing the energy ratio distributed to photosynthetic electron transport, was significantly lower than in the LLI treatment under PPFD below 150 μmol $m^{-2}$ $s^{-1}$, similar to that measured in the LLI for PPFD equal to 150 μmol $m^{-2}$ $s^{-1}$, and higher than in the LLI for PPFD higher than 150 μmol $m^{-2}$ $s^{-1}$. The measurements of the qN, which represents energy dissipation at the PSII antenna level due to the xanthophyll cycle and other photoprotective or regulatory processes, revealed significantly higher values in the GT compared to the other treatments (Figure 4d).

As shown in Figure 5, the highest net photosynthetic rates and the lowest transpiration rates were measured in the HLI treatment compared to the other two treatments. The highest levels of transpiration and stomatal conductance were measured in plants of the LLI, and the differences were significant compared to both the HLI and the GT. On the other hand, as seen in Figure 4 the intercellular $CO_2$ was significantly lower in the GT compared to both the HLI and LLI, while the latter treatments did not differ significantly from each other. Finally, the HLI exhibited the highest water use efficiency (Figure 6) compared to the LLI and the GT, while the latter two treatments did not differ significantly from each other.

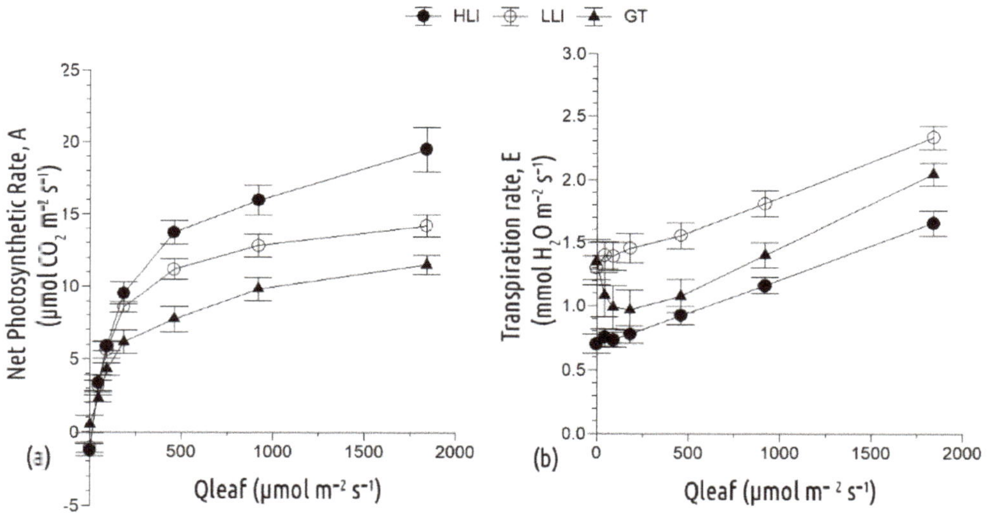


**Figure 5.** Net Photosynthetic Rate, A, (**a**), Transpiration rate, E, (**b**) of an average lettuce grown under each treatment's conditions. HLI, High light intensity; LLI, Low light intensity; GT, Glasshouse treatment. Vertical bars indicate ± standard errors of means.

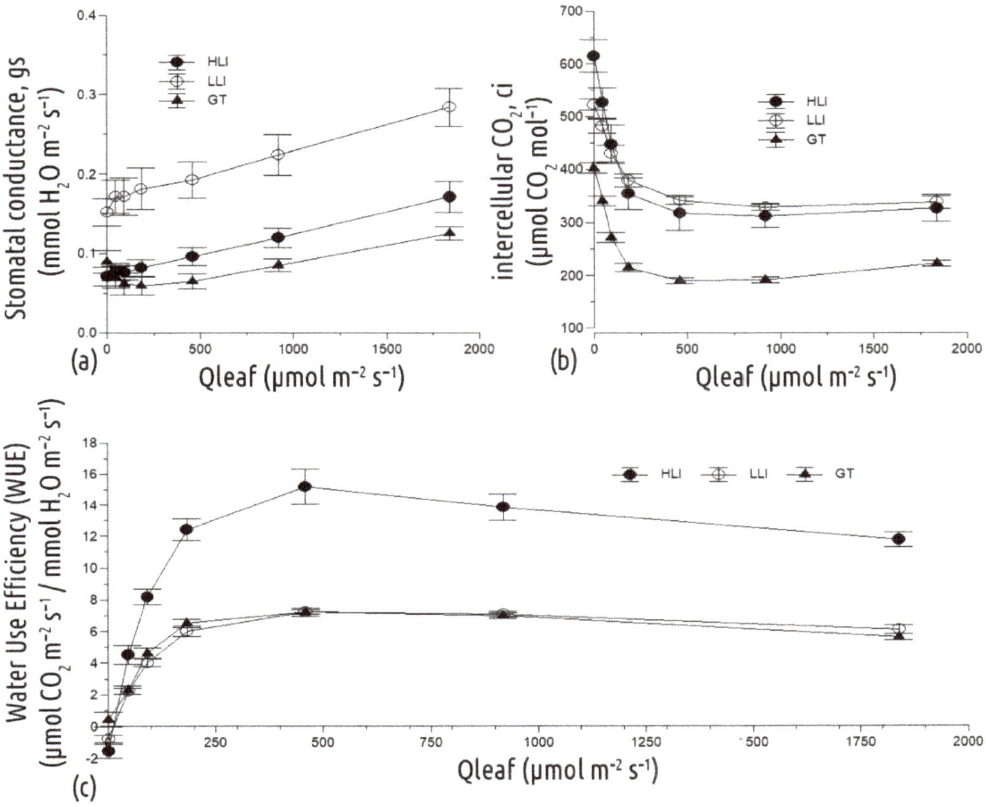


**Figure 6.** Stomatal conductance, gs, (**a**) intercellular $CO_2$, ci, (**b**,**c**) water use efficiency (WUE) of an average lettuce grown under each treatment's conditions. HLI, High light intensity; LLI, Low light intensity; GT, Glasshouse treatment. Vertical bars indicate ± standard errors of means.

## 4. Discussion

Light intensity strongly affects growth and quality of lettuce as reported by Kang et al. [4], and Fu, et al. [19]. The present study showed that an irradiance level of 188 μmol $m^{-2}$ $s^{-1}$ applied constantly for 18 h in a vertical hydroponic system was insufficient for lettuce, while a PPFD of 310 μmol $m^{-2}$ $s^{-1}$ could produce large lettuce heads containing less nitrates compared to those produced hydroponically in a Mediterranean greenhouse during wintertime. However, apart from the light intensity, the light uniformity had also a strong impact on lettuce growth in the current study. Indeed, the GT treatment, which took place in a Mediterranean hydroponic greenhouse during wintertime with an average PPFD of 144 μmol $m^{-2}$ $s^{-1}$ and a 10 h photoperiod, produced morphologically salable lettuces, albeit with a lower biomass than in the HLI. Morphological differences at the harvest stage between plants originating from the three treatments are shown in Figure 3. Apparently, the shape of the plants in the LLI was peculiar, while, based on the shape, HLI plants could not be distinguished from those of the GT treatment. In contrast, the LLI treatment produced morphologically non-marketable lettuce heads characterized by a vortex-like morphology although the mean light intensity was higher than that prevailing in the GT. The vortex-like morphology was presumably a result of poor light uniformity in the LLI treatment, which has a similar effect with that imposed by competition among neighboring plants on the Red:Far red ratio as reported by Ballaré [39–41], and Nagashima and Hikosaka [42].

The placement of the LED tubes closer to the plants in the LLI treatment led to areas exposed to high light intensity alternated by areas of lower light intensity on the upper surface of the plants and inside the plant canopy. As reported by Marchior et al. [43], self-shading can lead to exposure of a large part of plant leaves to low light levels, and concomitantly to severe restrictions in the rate of net assimilation. Nevertheless, it is worth to note that the vortex-like morphology would not be a problem if the lettuces were produced for fresh cut salads.

In addition to the light uniformity, differences in the light quality had also a strong impact on growth of lettuce. As reported by Dougher and Bugbee [44], Lin et al., [13], Li and Kubota [14], and other researchers [32,45–47], the spectrum can greatly affect the growth of a crop. In the vertical farming system white LEDs were used in both the LLI and the HLI treatment, while in the GT treatment the plants were receiving natural sunlight. On this basis, it is suggested that the biomass differences between the vertical farming treatments on the one and the GT on the other were partly imposed by differences in the light spectrum.

The higher leaf nitrate concentrations in the HLI treatment compared to the LLI and the GT are reasonable, given that the assimilation rate of nitrates into the plant cells is primarily dictated by the activity of nitrate reductase, which is depending on the light conditions [48,49]. As reported by Viršile et al. [50] an increased light intensity provides more energy available to photochemistry, provided that $CO_2$ supply from the atmosphere is ample, leads to enhanced carbohydrate production and accelerated nitrate assimilation to amino acids. However, the light intensity alone cannot explain the huge difference in leaf nitrate concentration between the LLI and the GT. Indeed, the difference in light intensity between the LLI and the GT is relatively small, compared to that between the HLI and the LLI. However, the difference in leaf $NO_3^-$ concentration is much larger between the LLI and the GT than between HLI and LLI. This lack in proportionality between light levels and leaf nitrate concentrations indicates that the leaf nitrate concentration was influenced by both light intensity and light quality. Indeed, as reported by Chen et al. [30], blue light boosts the synthesis of nitrate reductase in directly or indirectly ways. Therefore, the lower leaf nitrate concentrations in the vertical farming treatments of the current study may be associated with a high proportion of blue light in the light spectrum applied in these treatments.

As shown in Figure 7, when the nitrate values were studied per plant, the nitrate values of each treatment were all below the human threshold of acceptable daily intake (ADI), which is 220 mg day$^{-1}$ for a person weighing 60 kg [38]. The consumption of an average lettuce originating from the HLI treatment covered the 70% of ADI, while LLI covered just 52% and GT the 94%. However, taking a mean daily consumption of 100 g as a calculation basis, the GT greatly surpassed the ADI limit, providing 163%, while LLI and HLI remained safe for consumption with 79%, and 57% of the ADI, respectively. This indicates that the consumption of greenhouse-produced lettuce in winter is associated with a higher risk of surpassing the safety threshold. Nevertheless, this may be not associated with a higher health risk, as recent investigations dispute the harmful effects of nitrate and its derivate nitrite on human health [51].

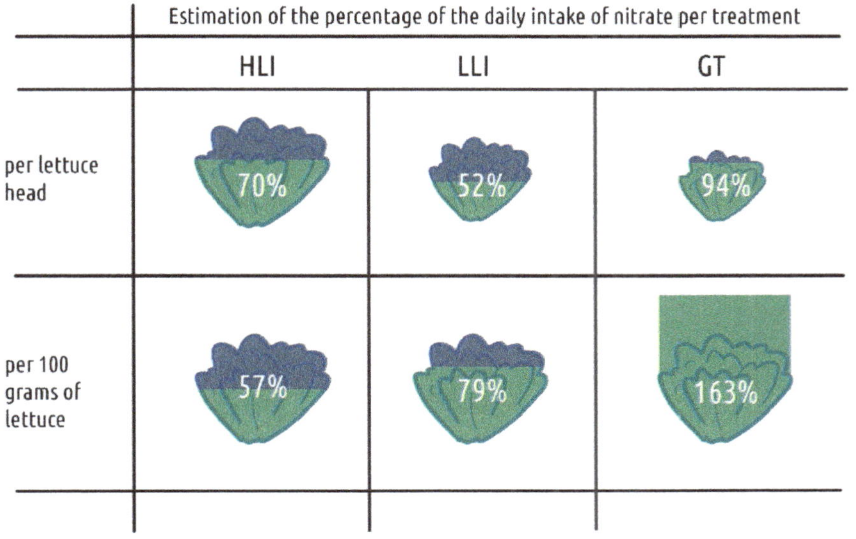


**Figure 7.** Estimation of percentage of the daily intake of nitrate of each treatment per average lettuce head and per 100 g of fresh produce. HLI, High light intensity; LLI, Low light intensity; GT, Glasshouse treatment.

The measurements of chlorophyll fluorescence and gas exchange parameters showed that HLI was photosynthetically the most efficient treatment. Furthermore, the results concerning chlorophyll fluorescence parameters at different light levels demonstrate the acclimation effect in the lettuce plants grown under each treatment. The plants of the HLI treatment were growing under a stable PPFD of around 300 μmol $m^{-2}$ $s^{-1}$, while the LLI plants were exposed to a stable but lower PPFD of around 190 μmol $m^{-2}$ $s^{-1}$. However, the GT plants were growing in an environment with unstable, natural lighting corresponding to an average PPFD of around 140 μmol $m^{-2}$ $s^{-1}$, but maximum PPFD levels in the greenhouse reached 500 μmol $m^{-2}$ $s^{-1}$ or more at times during some days. These conditions may have been partially responsible for the increasing efficiency of the photosynthetic activities in the GT under increased light intensity in comparison to LLI, the efficiency of which dropped with increased PPFD (point of 300 μmol $m^{-2}$ $s^{-1}$ for $\Phi_{PSII}$ and ETR, and point of 100–150 μmol $m^{-2}$ $s^{-1}$ for qP and qN) as shown in Figure 4.

As stated by Walters R.G [52], plants have evolved several mechanisms enabling them to adapt to changes in growth conditions. The adaptation mechanisms include both morphological changes improving light interception on the long term (Ballare' [41]; Weston et al. [53]), and adjustments in the functioning of individual proteins maximizing the efficiency of the photosynthetic apparatus, which operate on timescales ranging from seconds to hours (Demmig-Adams and Adams [54]). The HLI and LLI treatments of the vertical farm were adapted to stable environmental conditions, and thus, were unable to adapt quickly to any deviation from those conditions as seen from the chlorophyll fluorescence measurements (Figure 4). The unstable lighting conditions of the GT treatment in the greenhouse helped the plants adapt in more diverse lighting conditions. This can be seen from the qP measurements Figure 4c, where the GT treatment appears to be more capable of utilizing high light intensities than the LLI and HLI.

The high intercellular $CO_2$ concentrations in the LLI are ascribed to the significantly high levels of stomatal conductance. However, the low net photosynthesis in the LLI despite the high levels of intercellular concentration indicate that the restriction of plant biomass in this treatment originated from limitations in the photosynthetic apparatus. In contrast, the lower rates of net photosynthesis in the GT compared to the other two treatments were associated with reductions in both the stomatal conductance and the intercellular $CO_2$ concentrations, which indicates that the plant biomass in this treatment was partly restricted by stomatal limitations. The reduced stomatal conductance in the GT is ascribed to the lower relative humidity in the greenhouse air compared to than maintained in the vertical farming treatments (Table 1). As has been reported in a previous paper [27]), stomatal length did not differ significantly among treatments whereas the stomatal density of the GT treatment was significantly lower in comparison to that measured in the other treatments. These results highlight another advantage of lettuce cultivation in vertical farming systems, namely their inherent ability to allow for maintenance of the air humidity to optimal levels for plant growth.

The highest WUE was observed in the HLI treatment, which had the largest leaf area as well. For the LLI treatment it was expected to have low WUE since it had decreased chlorophyll fluorescence and increased stomatal conductance. The lettuces of the HLI treatment grew under high relative humidity conditions (RH = 90%), which led to decreased transpiration in comparison to the GT were the relative humidity was lower (RH = 64%) and therefore the transpiration was greater. This also had an impact on the calcium ascension since in both the HLI and LLI treatments, there were lettuce plants that suffered from tip burn. An unexpected point regarding the LLI was that the stomatal conductance and the transpiration rate were higher than in the other two treatments despite the high relative humidity (RH = 90%). It is speculated that the decreased distance between the LEDs and the plant canopy led to uncontrolled high illuminance at certain parts of the leaves. This high light intensity perhaps acted as a signal, tricking the plant into activating a high light intensity response at plant level.

## 5. Conclusions

The aim of the present work was to evaluate the commercial benefits of growing lettuce indoors with artificial lighting compared to glasshouse production under Mediterranean climatic conditions. During this experiment, morphological and physiological characteristics were also measured. A significant increase of the biomass and quality characteristics as well as the photosynthetic capacity was observed in lettuces grown in the high PPFD treatment (HLI) of the vertical farm. Nitrate content analysis showed that when studying lettuce heads, the percentage of the daily nitrate intake were below the human threshold for all treatments. Whereas, in consuming 100 g of lettuce, only the vertical farming treatments (HLI and LLI) were below that threshold and therefore safe for consumption. These results further support the superiority of lettuces grown in vertical farms. Even though the outcome of this study supported that the use of the HLI treatment was able to produce lettuce plants of higher quality, in comparison to LLI and GT treatments, during winter in Athens, a large-scale experiment in a year-round cultivation period should be further investigated to draw accurate conclusions for the feasibility and appropriateness of vertical farming in Greece.

**Author Contributions:** Conceptualization, O.V., M.M. and D.S.; methodology, G.L. and D.S.; software, O.V., M.M. and G.N.; validation, O.V., M.M., G.N. and G.L.; formal analysis, O.V., M.M., G.N. and G.L.; investigation, O.V., M.M., G.N. and G.L.; resources, D.S. and G.L.; data curation, O.V., M.M., G.N. and G.L; writing—original draft preparation, O.V.; writing—review and editing, O.V., M.M., G.N., G.L. and D.S.; visualization, O.V.; supervision, D.S.; project administration, D.S.; funding acquisition, D.S. All authors have read and agreed to the published version of the manuscript.

**Funding:** This research received no external funding.

**Institutional Review Board Statement:** Not applicable.

**Informed Consent Statement:** Not applicable.

**Conflicts of Interest:** The authors declare no conflict of interest.

## Appendix A. Photon Rack

**Figure A1.** Constriction of the Photon Rack (PR) system at K.Dekoulis Lab in Kallithea, Athens, Greece.

**Figure A2.** The photon flux density measurements were carried out using LI-188B Integrating quantum/Radiometer/Photometer, LI-COR INC, Lincoln, NE, USA. The LED tubes were attached on an aluminum base that allowed them to ascend and descend manually. The measuring tape indicated the distance between the LED tubes and the layer's surface. The photometer was placed on a white A4 paper that was separated into 9 squares. Readings were recorded for each of those squares to completely map out the PPFD of each layer in relation to the LED tube's distance which ranged from 10 cm to 40 cm.

**Figure A3.** The LED tubes of the High Light Intensity treatment (HLI) were stationary. The distance between the NFT gullies and the LED tubes was 40 cm. The distance between the plant canopy and LED tubes was decreased as the plants grew due to their increase in height, unlike in the LLI treatment were the LED tubes ascended from 20 cm to 40 cm distance from the NFT gullies as the plants grew, until the point the reached the ceiling of the layer.

## References


1. Kozai, T.; Fang, W.; Chun, C.; Yang, Q.; Tong, Y.; Cheng, R.; Kubota, C.; Lu, C. Chapter 3—PFAL Business and R&D in the World: Current Status and Perspectives. In *Plant Factory*; Kozai, T., Niu, G., Takagaki, M., Eds.; Academic Press: San Diego, CA, USA, 2016; pp. 35–68.
2. Savvas, D.; Passam, H.C. *Hydroponic Production of Vegetable and Ornamentals*; Embryo Publications: Athens, Greece, 2002.
3. Abidi, F.; Girault, T.; Douillet, O.; Guillemain, G.; Sintes, G.; Laffaire, M.; Ahmed, H.B.; Smiti, S.; Huché-Thélier, L.; Leduc, N. Blue light effects on rose photosynthesis and photomorphogenesis. *Plant Biol.* **2013**, *15*, 67–74. [CrossRef]
4. Kang, J.H.; KrishnaKumar, S.; Atulba, S.L.S.; Jeong, B.R.; Hwang, S.J. Light intensity and photoperiod influence the growth and development of hydroponically grown leaf lettuce in a closed-type plant factory system. *Hortic. Environ. Biotechnol.* **2013**, *54*, 501–509. [CrossRef]
5. Zhang, X.; He, D.; Niu, G.; Yan, Z.; Song, J. Effects of environment lighting on the growth, photosynthesis, and quality of hydroponic lettuce in a plant factory. *Int. J. Agric. Biol. Eng.* **2018**, *11*, 33–40. [CrossRef]
6. Inada, K.; Yasumoto, Y. Effects of light quality, daylength and periodic temperature variation on the growth of lettuce and radish plants. *Jpn. J. Crop Sci.* **1989**, *58*, 689–694. [CrossRef]
7. Fu, Y.; Li, H.; Yu, J.; Liu, H.; Cao, Z.; Manukovsky, N.S.; Liu, H. Interaction effects of light intensity and nitrogen concentration on growth, photosynthetic characteristics and quality of lettuce (Lactuca sativa L. Var. youmaicai). *Sci. Hortic.* **2017**, *214*, 51–57. [CrossRef]
8. Dou, H. Niu, G.; Gu, M.; Masabni, J.G. Responses of sweet basil to different daily light integrals in photosynthesis, morphology, yield, and nutritional quality. *HortScience* **2018**, *53*, 496–503. [CrossRef]
9. Lu, N.; Bernardo, E.L.; Tippayadarapanich, C.; Takagaki, M.; Kagawa, N.; Yamori, W. Growth and accumulation of secondary metabolites in perilla as affected by photosynthetic photon flux density and electrical conductivity of the nutrient solution. *Front. Plant Sci.* **2017**, *8*, 1–12. [CrossRef] [PubMed]
10. Hernández, R.; Kubota, C. Growth and morphological response of cucumber seedlings to supplemental red and blue photon flux ratios under varied solar daily light integrals. *Sci. Hortic.* **2014**, *173*, 92–99. [CrossRef]
11. Yan, Z.; He, D.; Niu, G.; Zhai, H. Evaluation of growth and quality of hydroponic lettuce at harvest as affected by the light intensity, photoperiod and light quality at seedling stage. *Sci. Hortic.* **2019**, *248*, 138–144. [CrossRef]
12. Kozai, T.; Niu, G.; Takagaki, M. *Plant Factory: An Indoor Vertical Farming System for Efficient Quality Food Production*; Academic Press: Cambridge, MA, USA, 2015.
13. Lin, K.-H.; Huang, M.-Y.; Huang, W.-D.; Hsu, M.-H.; Yang, Z.-W.; Yang, C.-M. The effects of red, blue, and white light-emitting diodes on the growth, development, and edible quality of hydroponically grown lettuce (*Lactuca sativa* L. var. *capitata*). *Sci. Hortic.* **2013**, *150*, 86–91. [CrossRef]
14. Li, Q.; Kubota, C. Effects of supplemental light quality on growth and phytochemicals of baby leaf lettuce. *Environ. Exp. Bot.* **2009**, *67*, 59–64. [CrossRef]
15. Senizza, B.; Zhang, L.; Miras-Moreno, B.; Righetti, L.; Zengin, G.; Ak, G.; Bruni, R.; Lucini, L.; Sifola, M.I.; El-Nakhel, C.; et al. The Strength of the Nutrient Solution Modulates the Functional Profile of Hydroponically Grown Lettuce in a Genotype-Dependent Manner. *Foods* **2020**, *9*, 1156. [CrossRef] [PubMed]
16. Touliatos, D.; Dodd, I.C.; Mcainsh, M. Vertical farming increases lettuce yield per unit area compared to conventional horizontal hydroponics. *Food Energy Secur.* **2016**, *5*, 184–191. [CrossRef] [PubMed]
17. Cometti, N.N.; Martins, M.Q.; Bremenkamp, C.A.; Nunes, J.A. Nitrate concentration in lettuce leaves depending on photosynthetic photon flux and nitrate concentration in the nutrient solution. *Hortic. Bras.* **2011**, *29*, 548–553. [CrossRef]
18. Chen, Z. Shah Jahan, M.; Mao, P.; Wang, M.; Liu, X.; Guo, S. Functional growth, photosynthesis and nutritional property analyses of lettuce grown under different temperature and light intensity. *J. Hortic. Sci. Biotechnol.* **2021**, *96*, 53–61. [CrossRef]
19. Fu, W.; Li, P.; Wu, Y.; Tang, J. Effects of different light intensities on anti-oxidative enzyme activity, quality and biomass in lettuce. *Hortic. Sci.* **2012**, *39*, 129–134.
20. Ferrón-Carrillo, F.; Guil-Guerrero, J.L.; González-Fernández, M.J.; Lyashenko, S.; Battafarano, F.; da Cunha-Chiamolera, T.P.L.; Urrestarazu, M. LED Enhances Plant Performance and Both Carotenoids and Nitrates Profiles in Lettuce. *Plant Foods Hum. Nutr.* **2021**, *76*. [CrossRef]
21. Min, Q.; Marcelis, L.F.M.; Nicole, C.C.S.; Woltering, E.J. High Light Intensity Applied Shortly Before Harvest Improves Lettuce Nutritional Quality and Extends the Shelf Life. *Front. Plant Sci.* **2021**, *12*, 76. [CrossRef]
22. Nicole, C.C.S.; Charalambous, F.; Martinakos, S.; Van De Voort, S.; Li, Z.; Verhoog, M.; Krijn, M. Lettuce growth and quality optimization in a plant factory. *Acta Hortic.* **2016**, *1134*, 231–238. [CrossRef]
23. Bantis, F.; Ouzounis, T.; Radoglou, K. Artificial LED lighting enhances growth characteristics and total phenolic content of Ocimum basilicum, but variably affects transplant success. *Sci. Hortic.* **2016**. [CrossRef]
24. Song, J.; Huang, H.; Hao, Y.; Song, S.; Zhang, Y.; Su, W.; Liu, H. Nutritional quality, mineral and antioxidant content in lettuce affected by interaction of light intensity and nutrient solution concentration. *Sci. Rep.* **2020**, *10*, 1–9. [CrossRef]
25. Fu, W.; Li, P.; Wu, Y. Effects of different light intensities on chlorophyll fluorescence characteristics and yield in lettuce. *Sci. Hortic.* **2012**, *135*, 45–51. [CrossRef]

26. Baker, N.R. Chlorophyll Fluorescence: A Probe of Photosynthesis In Vivo. *Annu. Rev. Plant Biol.* **2008**, *59*, 80–113. [CrossRef] [PubMed]
27. Savvas, D.; Voutsinos, O.; Mastoraki, M.; Liakopoulos, G.; Dekoulis, K.; Ntatsi, G. Exploring the possibility to use energy from solar panels to provide artificial light through LEDs in a vertical hydroponic crop of lettuce. *Acta Hortic.* **2020**, *1296*, 943–950. [CrossRef]
28. Chen, X.-L.; Guo, W.-G.; Xue, X.-Z.; Wang, L.-C.; Qiao, X.-J. Growth and quality responses of "Green Oak Leaf" lettuce as affected by monochromic or mixed radiation provided by fluorescent lamp (FL) and light-emitting diode (LED). *Sci. Hortic.* **2014**, *172*, 168–175. [CrossRef]
29. Chen, X.-L.; Xue, X.-Z.; Guo, W.-Z.; Wang, L.-C.; Qiao, X.-J. Growth and nutritional properties of lettuce affected by mixed irradiation of white and supplemental light provided by light-emitting diode. *Sci. Hortic.* **2016**, *200*, 111–118. [CrossRef]
30. Snowden, M.C.; Cope, K.R.; Bugbee, B. Sensitivity of Seven Diverse Species to Blue and Green Light: Interactions with Photon Flux. *PLoS ONE* **2016**, *11*, e0163121. [CrossRef]
31. Park, Y.; Runkle, E.S. Spectral effects of light-emitting diodes on plant growth, visual color quality, and photosynthetic photon efficacy: White versus blue plus red radiation. *PLoS ONE* **2018**, *13*, 1–14. [CrossRef]
32. Mickens, M.A.; Skoog, E.J.; Reese, L.E.; Barnwell, P.L.; Spencer, L.E.; Massa, G.D.; Wheeler, R.M. A strategic approach for investigating light recipes for 'Outredgeous' red romaine lettuce using white and monochromatic LEDs. *Life Sci. Sp. Res.* **2018**, *19*, 53–62. [CrossRef]
33. Kim, H.-H.; Goins, G.D.; Wheeler, R.M.; Sager, J.C. Green-light Supplementation for Enhanced Lettuce Growth under Red- and Blue-light-emitting Diodes. *HortScience* **2004**, *39*, 1617–1622. [CrossRef]
34. Swan, B.V.; Bugbee, B. *Increasing Blue Light from LED 's Reduces Growth of Lettuce;* Research Capitol Hill, Utah State University: Logan, UT, USA, 2017; p. 56.
35. Liakopoulos, G.; Spanorigas, I. Foliar anthocyanins in Pelargonium × hortorum are unable to alleviate light stress under photoinhibitory conditions. *Photosynthetica* **2012**, *50*, 254–262. [CrossRef]
36. Medrano, H.; Tomás, M.; Martorell, S.; Flexas, J.; Hernández, E.; Rosselló, J.; Pou, A.; Escalona, J.M.; Bota, J. From leaf to whole-plant water use efficiency (WUE) in complex canopies: Limitations of leaf WUE as a selection target. *Crop J.* **2015**, *3*, 220–228. [CrossRef]
37. Novozamsky, I.; Houba, V.J.G.; van der Eijk, D.; van Eck, R. Notes on determinations of nitrate in plant material. *Neth. J. Agric. Sci.* **1983**, *31*, 239–248. [CrossRef]
38. Brkić, D.; Bošnir, J.; Bevardi, M.; Gross Bošković, A.; Miloš, S.; Lasić, D.; Krivohlavek, A.; Racz, A.; Mojsović-Ćuić, A.; Trstenjak Uršulin, N. Nitrate in Leafy Green Vegetables and Estimated Intake. *Afr. J. Tradit. Complement. Altern. Med. AJTCAM* **2017**, *14*, 31–41.
39. Ballaré, C.L.; Scopel, A.L.; Jordan, E.T.; Vierstra, R.D. Signaling among neighboring plants and the development of size inequalities in plant populations. *Proc. Natl. Acad. Sci. USA* **1994**, *91*, 10094–10098. [CrossRef] [PubMed]
40. Ballaré, C.L. Illuminated behaviour: Phytochrome as a key regulator of light foraging and plant anti-herbivore defence. *Plant Cell Environ.* **2009**, *32*, 713–725. [CrossRef] [PubMed]
41. Ballaré, C.L. Keeping up with the neighbours: Phytochrome sensing and other signalling mechanisms. *Trends Plant Sci.* **1999**, *4*, 97–102. [CrossRef]
42. Nagashima, H.; Hikosaka, K. Plants in a crowded stand regulate their height growth so as to maintain similar heights to neighbours even when they have potential advantages in height growth. *Ann. Bot.* **2011**, *108*, 207–214. [CrossRef]
43. Marchiori, P.E.R.; Machado, E.C.; Ribeiro, R.V. Photosynthetic limitations imposed by self-shading in field-grown sugarcane varieties. *Field Crops Res.* **2014**, *155*, 30–37. [CrossRef]
44. Dougher, T.A.O.; Bugbee, B. Long-term blue light effects on the histology of lettuce and soybean leaves and stems. *J. Am. Soc. Hortic. Sci.* **2004**, *129*, 467–472. [CrossRef]
45. Hytönen, T.; Pinho, P.; Rantanen, M.; Kariluoto, S.; Lampi, A.; Edelmann, M.; Joensuu, K.; Kauste, K.; Mouhu, K.; Piironen, V.; et al. Effects of LED light spectra on lettuce growth and nutritional composition. *Light. Res. Technol.* **2018**, *50*, 880–893. [CrossRef]
46. Pinho, P.; Jokinen, K.; Halonen, L. The influence of the LED light spectrum on the growth and nutrient uptake of hydroponically grown lettuce. *Light. Res. Technol.* **2017**, *49*, 866–881. [CrossRef]
47. Brazaitytė, A.; Ulinskaitė, R.; Duchovskis, P.; Samuolienė, G.; Siksnianienė, J.B.; Jankauskienė, J.; Sabqievienė, G.; Baranouskis, K.; Stanienė, G.; Tamulaitis, G.; et al. Optimization Of Lighting Spectrum For Photosynthetic System And Productivity Of Lettuce By Using Light-Emitting Diodes. *Acta Hortic.* **2006**, *711*, 183–188. [CrossRef]
48. Lillo, C. Light regulation of nitrate reductase in green leaves of higher plants. *Physiol. Plant.* **1994**, *90*, 616–620. [CrossRef]
49. Rajasekhar, V.K.; Gowri, G.; Campbell, W.H. Phytochrome-Mediated Light Regulation of Nitrate Reductase Expression in Squash Cotyledons. *Plant. Physiol.* **1988**, *88*, 242–244. [CrossRef] [PubMed]
50. Viršilė, A.; Brazaitytė, A.; Vaštakaitė-Kairienė, V.; Miliauskienė, J.; Jankauskienė, J.; Novičkovas, A.; Samuolienė, G. Lighting intensity and photoperiod serves tailoring nitrate assimilation indices in red and green baby leaf lettuce. *J. Sci. Food Agric.* **2019**, *99*, 6608–6619. [CrossRef]
51. Habermeyer, M.; Roth, A.; Guth, S.; Diel, P.; Engel, K.H.; Epe, B.; Fürst, P.; Heinz, V.; Humpf, H.U.; Joost, H.G.; et al. Nitrate and nitrite in the diet: How to assess their benefit and risk for human health. *Mol. Nutr. Food Res.* **2015**, *59*, 106–128. [CrossRef] [PubMed]

52. Walters, R.G. Towards an understanding of photosynthetic acclimation. *J. Exp. Bot.* **2004**, *56*, 435–447. [CrossRef]
53. Weston, E.; Thorogood, K.; Vinti, G.; López-Juez, E. Light quantity controls leaf-cell and chloroplast development in Arabidopsis thaliana wild type and blue-light-perception mutants. *Planta* **2000**, *211*, 807–815. [CrossRef] [PubMed]
54. Demmig-Adams, B.; Adams, W.W. Photoprotection and Other Responses of Plants to High Light Stress. *Annu. Rev. Plant Physiol. Plant Mol. Biol.* **1992**, *43*, 599–626. [CrossRef]

MDPI
St. Alban-Anlage 66
4052 Basel
Switzerland
Tel. +41 61 683 77 34
Fax +41 61 302 89 18
www.mdpi.com

*Agriculture* Editorial Office
E-mail: agriculture@mdpi.com
www.mdpi.com/journal/agriculture

www.ingramcontent.com/pod-product-compliance
Lightning Source LLC
Chambersburg PA
CBHW051612210926
51850CB00009B/2297